The Biology of Traditions
Models and Evidence

Socially maintained behavioral traditions in nonhuman species hold great interest for biologists, anthropologists, and psychologists. This book treats traditions in nonhuman species as biological phenomena that are amenable to the comparative methods of inquiry used in contemporary biology. Chapters in the first section define behavioral traditions and indicate how they can arise in nonhuman species, how widespread they may be, how they may be recognized, and how we can study them. The second part summarizes the latest research programs seeking to identify traditions in diverse taxa, with contributions from leading researchers in this area. The book ends with a comparison and evaluation of the alternative theoretical formulations and their applications presented in the book and makes recommendations for future research building on the most promising evidence and lines of thinking. *The Biology of Traditions* will be essential reading for students and researchers in the fields of anthropology, biology, and psychology.

DOROTHY M. FRAGASZY is Professor of Psychology and the Chair of the Neuroscience and Behavior Program at the University of Georgia, where she teaches development, developmental psychobiology, and comparative cognition. She has also co-authored *The Complete Capuchin: The Biology of the Genus* Cebus (2003; ISBN 0 521 66116 1 (hb) and 0 521 66768 2).

SUSAN PERRY is Assistant Professor in the Department of Anthropology at the University of California, Los Angeles, and Head of the Junior Independent Research Group in Cultural Phylogeny at the Max Planck Institute for Evolutionary Anthropology in Leipzig, Germany. Her research focuses currently on the role of social influence on the behavioral development of wild capuchin monkeys.

The Biology of Traditions

Models and Evidence

Edited by
DOROTHY M. FRAGASZY
University of Georgia, Athens, USA
and
SUSAN PERRY
University of California, Los Angeles, USA and Max-Planck Institute for Evolutionary Anthropology, Leipzig, Germany

PUBLISHED BY THE PRESS SYNDICATE OF THE UNIVERSITY OF CAMBRIDGE
The Pitt Building, Trumpington Street, Cambridge, United Kingdom

CAMBRIDGE UNIVERSITY PRESS
The Edinburgh Building, Cambridge CB2 2RU, UK
40 West 20th Street, New York, NY 10011-4211, USA
477 Williamstown Road, Port Melbourne, VIC 3207, Australia
Ruiz de Alarcón 13, 28014 Madrid, Spain
Dock House, The Waterfront, Cape Town 8001, South Africa

http://www.cambridge.org

First published 2003

Printed in the United Kingdom at the University Press, Cambridge

Typefaces Lexicon No.2 9/13 pt and Lexicon No. 1 *System* LaTeX 2ε [TB]

A catalog record for this book is available from the British Library

Library of Congress Cataloging in Publication data

The biology of traditions : models and evidence / Dorothy M. Fragaszy and
Susan Perry, editors.
p. cm.
Includes bibliographical references.
ISBN 0 521 81597 5
1. Social behavior in animals. 2. Learning in animals. I. Fragaszy, Dorothy M.
(Dorothy Munkenbeck), 1950– II. Perry, Susan, 1965–
QL775 .B56 2003
591.5 – dc21 2002036833

ISBN 0 521 81597 5 hardback

In loving memory of my mother and father, Joan and Russ Munkenbeck, and my sister Lynne and brother-in-law Jim Wallace, whose traditions, as well as *joie de vivre*, I try to maintain.

Dorothy M. Fragaszy

In memory of Guapo, Lomas Barbudal's great innovator, and the other capuchin monkeys whose striking social conventions opened my eyes to the importance of traditions as an exciting research topic. I thank my husband Joe and daughter Kate for their patience as I worked on this project.

Susan Perry

Contents

Contributors

Elsa Addessi
Istituto di Scienze e Tecnologie della Cognizione, Consiglio Nazionale delle Richerche, Via Ulisse Aldrovandi 16/B, 00197 Rome, Italy
and
Dipartimento di Biologia Animale e dell'Uomo, Università di Roma "La Sapienza', P. le A. Moro, 5-00185 Rome, Italy

Mary Baker
Department of Anthropology, Whittier College, 13406 Philadelphia St., Whittier, CA 90608, USA

Sue Boinski
Department of Anthropology, 1112 Turlington Hall, PO Box 117305, University of Florida, Gainesville, FL 32611-7305, USA

Julie Bouchard
Department of Biology, McGill University, 1205 Avenue Docteur Penfield, Montreal, Québec H3A 1B1, Canada

Gwen Dewar
Department of Anthropology, University of Michigan, Ann Arbor, MI 48109, USA

Linda Fedigan
Department of Anthropology, University of Calgary, Calgary, Alberta T2N 1N4, Canada

Dorothy Fragaszy
Department of Psychology, University of Georgia, Athens, GA 30602-3013, USA

Bennett G. Galef, Jr.
Department of Psychology, McMaster University, Hamilton, Ontario L8S 4K1, Canada

Julie Gros-Louis
Department of Psychology, University of Indiana, 1101 E. 10th St., Bloomington, IN 47405, USA

MaLinda Henry
Behavior, Ecology, Evolution and Systematics Program, University of Maryland, College Park, MD 20742, USA

Satoshi Hirata
Great Ape Research Institute, Hayashibara Biochemical Laboratories Inc., 952-2 Nu, Tamano Okayama 706-0316, Japan

Michael A. Huffman
Primate Research Institute, Kyoto University, 41-2 Kanrin, Inuyama Aichi 484-8508, Japan

Katherine Jack
Department of Anthropology, Appalachian State University, Boone, NC 28608, USA

Vincent M. Janik
Centre for Social Learning and Cognitive Evolution, School of Biology, University of St. Andrews, St. Andrews, Fife KY16 9TS, UK

Jeremy R. Kendal
Sub-Department of Animal Behaviour, University of Cambridge, Madingley, Cambridge, CB3 8AA, UK

Kevin N. Laland
Centre for Social Learning and Cognitive Evolution, School of Biology, University of St. Andrews, St. Andrews, Fife KY16 9TS, UK

Louis Lefebvre
Department of Biology, McGill University, 1205 avenue Docteur Penfield, Montréal, Québec H3A 1B1, Canada

Katherine C. MacKinnon
Department of Sociology and Criminal Justice, Saint Louis University, 3500 Lindell Boulevard, St. Louis, MO 63103, USA

Janet Mann
Department of Biology, Georgetown University, Reiss Science Building, Washington, DC 20057, USA

Joseph Manson
Max-Planck-Institut für evolutionäre Anthropologie, Inselstraße 22, 04103 Leipzig, Germany

Melissa Panger
Department of Anthropology, George Washington University, 2110 G St. NW, Washington, DC 20052, USA

Susan Perry
Department of Anthropology, University of California at Los Angeles, Los Angeles, CA 90095, USA
and
Max-Planck-Institut für evolutionäre Anthropologie, Inselstraße 22, 04103 Leipzig, Germany (July–December 2002)

Kendra Pyle
Department of Biology, University of Pennsylvania, Philadelphia, PA 19104, USA

Robert P. Quatrone
Trevor Day School, 1 West 88th st., New York, NY 10024, USA

Simon M. Reader
Behavioural Biology, Utrecht University, Padualaan 14, PO Box 80086, 3508 TB, Utrecht, The Netherlands

Lisa M. Rose
Department of Anthropology and Sociology, University of British Columbia. 6303 NW Marine Drive, Vancouver, British Columbia, V6T 1Z1, Canada

Anne E. Russon
Psychology Department, Glendon College of York University, 2275 Bayview Avenue, Toronto, Ontario M4N 3M6, Canada

Lara Selvaggi
42 Lexington Avenue, Apt 3, Greenwich, CT 06830, USA

Brooke Sergeant
Department of Biology, Georgetown University, Reiss Science Building, Washington, DC 20057, USA

Peter J. B. Slater
School of Biology, University of St. Andrews, St. Andrews, Fife, KY16 9TS, UK

Claudia M. Stickler
College of Natural Resources and Environment, Box 116455, 105 Black Hall, University of Florida, Gainesville, FL 32611-6455, USA

Karen Sughrue
205 Forest Resources Laboratory, Pennsylvania State University, University Park, PA 16802, USA

Carel P. van Schaik
Biological Anthropology and Anatomy, Duke University, Box 90383, Durham, NC 27708-0383, USA

Elisabetta Visalberghi
Istituto di Scienze e Tecnologie della Cognizione, Consiglio Nazionale delle Ricerche, Via Ulisse Aldrovandi 16/B, 00197 Rome, Italy

Preface

For many decades, the scientific discussion about social learning in nonhuman animals has been dominated by two concerns: (1) whether any nonhuman species, but ape species in particular, possess "culture", and (2) which nonhuman species exhibit imitation, assumed by many to be a prerequisite or at the least an important support for culture. However, from a biological point of view, these questions only narrowly address fundamental issues about social learning in nonhuman animals. Their link to functional, developmental, and evolutionary questions is not obvious, for example. We wanted to know about these latter topics, as well as more broadly about mechanisms supporting social learning, so we set about asking our colleagues what they thought. We got many answers that we felt were worthy of better dissemination than they were receiving in the literature or in the classroom. This book is the result.

This book is intended for individuals interested in understanding social learning (the common short-hand phrase for what is more precisely called socially aided learning) in animals from a biological perspective. We focus on one outcome of social learning, traditions, as an element in behavioral ecology. By tradition, we mean a distinctive behavior pattern shared by two or more individuals in a social unit, which persists over time and that new practitioners acquire in part through socially aided learning. The process of social learning does not lead inevitably to enduring traditions, however. Ultimately, we would like to understand how particular environments, social attributes, and life ways contribute to the appearance and persistence of traditions in particular taxa. Such an understanding will help us to appreciate the contribution of social learning to biology. It will also help us to appreciate the roots of human traditions in the intersection of particular social propensities and ecological circumstances in the past and present of our species.

Traditions have long been considered as one element of culture, and the relationships among social learning, traditions, and culture in primates have been hotly debated (e.g., Boesch and Tomasello, 1998; Matsuzawa *et al.*, 2001; McGrew, 1998; van Schaik, Deaner, and Merritt, 1999). Efforts to analyze traditions in nonhuman primates began with studies of Japanese macaques but recently have focused particularly on the great apes, and, more particularly, on a single species in one genus of great apes (the common chimpanzee, *Pan troglodytes*). Collations of findings across the several sites where chimpanzees have been studied for decades have documented many instances of putative traditional behaviors (Boesch and Tomasello, 1998; Whiten *et al.*, 1999, 2001). Unfortunately, the intense focus on a single species, and on a single issue (the degree to which chimpanzees possess "culture"), has restricted discourse about social learning in nonhuman animals in an unhealthy manner. A truly biological understanding of social learning requires broader treatment, both taxonomically and theoretically (cf. Marler, 1996; Kamil, 1998).

The contributors to this volume broaden the playing field for discussions of "culture" in nonhuman animals by considering the evidence for traditions in nonhuman primates alongside the evidence for traditions in two other orders of mammals (rodents and cetaceans) and one other class of vertebrates (birds). The contributions in this volume do not focus exclusively on transmission patterns within one group (the usual focus of experimental social learning studies) nor exclusively on intraspecific variation across groups (the usual focus of observational studies in natural settings), but rather the intersection of the two topics.

In the early chapters (Chs. 1–5) of the book, we highlight theoretical and conceptual issues in the study of traditions, and of social learning in general, in nonhuman species. We begin by presenting an explicitly biological approach to the phenomenon of traditions. We lay out what kinds of empirical evidence are necessary and sufficient to conclude that behavioral variants within or between groups reflect social transmission (i.e., are traditions), and we review the options for obtaining these sorts of evidence from nonhuman animals in common research settings (in nature and in the laboratory). Two contributions review general theoretical models for investigating the circumstances under which individuals are expected to rely on social learning. The authors devote particular attention to considering how these models can be operationalized to make specific predictions that can be tested in real-world settings: that is, setting the research agenda to make use of the power of general models. Lastly, two contributions examine the relations among relative brain size and the distribution of reports in the peer-reviewed scientific literature about social learning and unusual (innovative) behaviors in nonhuman primates and in birds. In Chapters 6–14, we present empirical evidence for within- and between-group variability that may qualify as traditions in rodents, cetaceans, birds, and primates (Japanese macaques, orangutans, chimpanzees, and capuchin monkeys). The contributions span laboratory and field studies and include a wide spectrum of interests and methodological approaches. Three chapters concern capuchin monkeys, a genus of special interest to the editors of this volume, and one which we believe will be very rewarding to study in this regard. In the concluding chapter, we highlight the shared viewpoints and findings presented by the contributors to build a picture of the state of the science in this area. Then we consider how most productively to test theoretical models and point out some areas where we think critical thinking is needed to make headway in this area of science. We intend that our readers will come away from this book with a richly synthetic appreciation of social learning and of traditions as potential outcomes of social learning. We also want our readers to appreciate that traditions in nonhuman animals have important implications for biology, including evolution and ecology. We will have succeeded if our efforts inspire vigorous and rigorous examination and refinement of this view of traditions.

References

Boesch, C. and Tomasello, M. 1998. Chimpanzee and human cultures. *Current Anthropology*, **39**, 91–604.

Kamil, A. 1998. On the proper definition of cognitive ethology. In *Animal Cognition in Nature*, ed. R. Balda, I. Pepperberg, and A. Kamil, pp. 1–28. New York: Academic Press.

Marler, P. 1996. Social cognition. Are primates smarter than birds? In *Current Ornithology*, Vol 13, ed. V. Nolan Jr. and E. D. Ketterson, pp. 1–32. New York: Plenum Press.

Matsuzawa, T., Biro, D., Humle, T., Inoue-Nakamura, N., Tonooka, R., and Yamakoshi, G. 2001. Emergence of culture in wild chimpanzees: education by master–apprenticeship. In *Primate Origins of Human Cognition and Behavior*, ed. T. Matsuzawa, pp. 557–574. Tokyo: Springer Verlag.

McGrew, W. C. 1998. Culture in nonhuman primates? *Annual Review of Anthropology*, **27**, 301–328.

van Schaik, C., Deaner, R., and Merrill, M. 1999. The conditions for tool use in primates: implications for the evolution of material culture. *Journal of Human Evolution*, **36**, 719–741.

Whiten, A., Goodall, J., McGrew, W., Nishida, T., Reynolds, V., Sugiyama, Y., Tutin, C., Wrangham, R., and Boesch, C. 1999. Cultures in chimpanzees. *Nature*, **399**, 682–685.

Whiten, A., Goodall, J., McGrew, W., Nishida, T., Reynolds, V., Sugiyama, Y., Tutin, C., Wrangham, R., and Boesch, C. 2001. Charting cultural variation in chimpanzees. *Behaviour*, **138**, 1481–1516.

Acknowledgments

We thank the Office of the Vice President for Academic Affairs and Provost of the University of Georgia for providing financial support, through a grant from the State-of-the-Art Conference Program, for the conference that gave rise to this book. We thank the contributors for their stimulating discussion at the conference, and for providing thoughtful peer review of one another's chapters. Some contributors to this volume were asked to contribute after the conference had taken place; we thank them for their gracious acceptance of our request and their expeditious preparation of their chapters. Gwen Dewar deserves special thanks for preparing the index. Sarah Cummins-Sebree and Louise Seagraves helped extensively with copy editing and formatting. We thank Tracey Sanderson of Cambridge University Press for guiding us through the process of producing an edited volume.

DOROTHY M. FRAGASZY AND SUSAN PERRY

1

Towards a biology of traditions

1.1 Introduction

One who sees things from the beginning will have the finest view of them

ATTRIBUTED TO ARISTOTLE

In late 1997, a series of exchanges occurred on the internet bulletin board established by Linda Fedigan a year earlier to facilitate communication among the select circle of individuals studying capuchin monkeys (genus *Cebus*, in the family Cebidae of the New World monkeys). Someone posted a description of a strikingly odd behavior she had noticed in her main study group of about two dozen white-faced capuchin monkeys (*C. capucinus*). The behavior, a pattern of two individuals interacting in an apparently affiliative manner, had not been described in the literature for any other animal species. Several members of the group performed this behavior with each other routinely over a period of seven years, and it appeared a perfectly familiar aspect of their social behavior that field season, as if they always did this odd thing (see Ch. 14, for more details about the mystery behavior). Nevertheless, they had not done this during the first year of the study, nor had she observed the behavior in the neighboring group. The researcher was understandably curious whether anyone else had ever seen anything like it, or had any ideas on how it might have originated or its function. A flurry of messages ensued over the next few weeks, with several researchers confirming the first person's suspicion that this behavior was not a universal behavior in white-faced capuchins, and not known at all in other species of capuchins. These respondents, moreover, provided their own examples of odd social behaviors common in their groups, which they had assumed were present in other groups but were now wondering if that assumption were premature. At the conclusion of the on-line discussion, the correspondents were left with a tantalizing list of potentially group-unique behaviors in the genus, and the distinct impression that some of these might be traditions. As those who work with capuchins, including the two authors of this chapter, are firmly convinced that these monkeys are socially responsive as well as brash and intrepid individuals, we were all intrigued by the possibility that these monkeys might have behavioral traditions. To make such a claim publicly, and to place the phenomenon into the biological framework we were convinced was necessary, was obviously going to be a substantial project requiring the ideas and efforts of many people.

1.2 More than a question of culture

Behavioral scientists have often considered social learning in nonhuman animals as a precursor of culture as we know it in humans (e.g., Bonner, 1980). Culture has many meanings in anthropology, including belief systems, codes of conduct, and so forth, that we do not expect to exist in nonhuman species. The only essential element of human culture potentially shared with nonhuman species is the continuation of behavioral practices across generations through social learning. Although anthropologists generally agree that sharing this single domain with humans is not a sufficient basis to attribute culture to nonhuman animals (cf. Boesch and Tomasello, 1998; McGrew, 1992, 1998), the convergence still fascinates behavioral biologists. Early contributions suggesting a parallel between traditions in nonhuman animals and human culture were provided by Japanese zoologists conducting many of the first longitudinal observational studies of monkeys in natural conditions (Itani and Nishimura, 1973; Kawai, 1965; Kawamura, 1965; also see de Waal, 2001 for an overview). These researchers were very interested in the appearance of novel behaviors in groups of monkeys and the fact that other individuals eventually displayed behaviors that initially had been the province of a single "inventor". Their term for the phenomenon was translated from Japanese into English as "protocultural", "precultural", and "subcultural"; and the debate was on. A vigorous controversy has brewed ever since over what is necessary for a behavior pattern shared among members of a group to be identified as "cultural", which species might be said to "have culture", and which learning mechanisms are necessary to claim that a particular practice qualifies as "cultural". Discussions of social learning in nonhuman primates, and particularly chimpanzees, have been at the forefront of these controversies. The rate of discussion has now reached a feverish pitch. A sampling of titles of publications in the last four years alone at the time we are writing include, for example, "Cultural primatology comes of age" (de Waal, 1999), "Cultures in chimpanzees" (Whiten *et al.*, 1999), "Charting cultural variation in chimpanzees" (Whiten *et al.*, 2001), "Chimpanzee and human cultures" (Boesch and Tomasello, 1998), "Chimps in the wild show stirrings of culture" (Vogel, 1999), "Culture in nonhuman primates?" (McGrew, 1998), "Emergence of culture in wild chimpanzees: education by master–apprenticeship" (Matsuzawa *et al.*, 2001), "Primate culture and social learning" (Whiten, 2000), and "Orangutan cultures and the evolution of material culture" (van Schaik *et al.*, 2003). The *New York Times Magazine* issue on 9 December 2001, in an article entitled "The year in ideas", included an essay "Apes have culture too". In part, this torrent of interest is motivated by the concern that apes are losing the battle for survival in nature; the call is out to prevent "culturecide" as populations are decimated by human activities in their home areas. In part it is because we are just coming to realize things about apes that bring them ever closer, behaviorally, to the threshold that many have set dividing humans from nonhuman relatives.

This debate, regardless of its origins or purpose, is driven largely by anthropocentric, not biological, concerns about the meanings of culture. These anthropocentric concerns are outside the scope of our efforts here. Rather, we are interested in traditions as features of behavior in nonhuman animals without regard to whether these traditions meet any particular set of criteria for nomination as "cultural". We define traditions as enduring behavior patterns shared among members of a group that depend to a measurable degree on social contributions to individual learning, resulting in shared practices among members of a group. If there were another, less value-laden, term than traditions to describe such behavioral phenomena we would use that term. However, we do not have an alternative term at our disposal without creating a new word that would not be understood outside of our own small readership. So long as the term "tradition" captures best those aspects of shared practice that we are interested in here, we shall continue to use this term.

Arguments in favor of according a special status to primates in regard to social learning, and the probability that shared behaviors reflect social influences on learning (i.e., that primates have traditions), are often rooted in a simple notion of phylogenetic association. This notion is that species that share a more recent link with human ancestors in evolutionary history are likely to share with humans more elaborated social learning. Alternatively, social learning might be more important in the lives of members of these species. However, phylogenetic association with humans is not predictive of social learning propensities (Box and Gibson, 1999; Fragaszy and Visalberghi, 1996). No distinctive form of social learning is unique to humans, or to humans and closely related primates (Russon *et al.*, 1998; see also Fritz and Kotrschal, 1999; Voelkl and Huber, 2000; Zentall, Sutton, and Sherburne, 1996). This strong statement applies even to "true imitation", according to Russon *et al.* (1998). Social learning in many forms is apparently widespread in the animal kingdom, although we have not looked for it intensively in many species. Box and Gibson (1999) urge us to look widely for possible cases of social learning in natural settings; many of the chapters in their book suggest why we should look for social learning in a variety of mammalian taxa where previously few had thought to look for such evidence. Social learning must be examined as an element in the behavioral biology of animals, rather than as a lead-up to, or incomplete version of, a (possibly) uniquely human characteristic (Box and Gibson, 1999; de Waal, 2001; Giraldeau, 1997; Avatal and Jablonski, 2000; Laland *et al.*, 2000).

Phylogenetic trends in the size and organization of the nervous system are useful supports for theories about behavioral evolution. For example, birds that store and retrieve thousands of nuts have an enlarged hippocampus, a part of the brain involved in memory formation, compared with closely related nonstoring species (Basil *et al.*, 1996; Krebs *et al.*, 1989). Relative forebrain size and absolute forebrain size both correlate positively with the number of reported instances of social learning and of behavioral innovations across taxa in nonhuman primates (see Reader, this volume; Reader and Laland, 2001). Similarly, the corresponding variable in birds (the relative size of the neostriatum and hyperstriatum ventrale) correlates positively with the frequency of reported feeding innovations across taxa (Lefebvre *et al.*, 1997), although evidently not to social learning of foraging habits (Ch. 4). Covariance between brain size and propensity to innovate and (in primates) to develop traditions would suggest that social learning is part of a functionally seamless whole reflecting overall neural power in a general sense, rather than specialized capacities for social learning or for innovation. This conclusion makes good sense if social learning is understood as modulation of learning through social context, as we argue below, rather than a set of specific learning abilities. Big brains afford more modulated learning.

The extraordinarily conservative patterns of neurogenesis across broad taxonomic mammalian groups (Finlay, Darlington, and Nicastro, 2001) lead to the powerful conclusion that brains and behavior co-evolved in a most general way, rather than in accord with selective pressures for specific behavioral attributes (such as enhanced social learning propensities or propensities to innovate). In this view, we should expect behavioral flexibility and social sophistication in many forms in any species with relatively large brains, regardless of their membership in any particular taxonomic order. If Finlay *et al.* (2001) are correct that the size of all parts of the brain reflect conservative growth patterns, virtually always independent of specific selective pressures, then we should expect behavioral flexibility (afforded by a large isocortex) to be enhanced even in taxa where we cannot identify any particular selective pressure for a certain form of flexibility. In other words, capacities supporting social learning, like all forms of learning, may simply come along with brain size. What use specific taxa make of these abilities is likely to vary in accord with a constellation of ecological and social variables. This is our concern in this volume. What contributions to behavioral biology and to evolution might traditions confer on those taxa where they occur, and where might traditions occur?

1.3 The biological significance of traditions

Our particular concern in this volume is with traditions as one outcome of social learning. The claim is often made that humans, through culture, are the only species whose behavior has effectively modified natural selection (for example, through agriculture or medicine). However, a human-centered perspective on the relation between culture and biological evolution is misleadingly narrow. Species modify their environments through their behavior, a process labeled "niche construction" by Laland *et al.* (2000; see also Lewontin, 1978; Odling-Smee, Laland, and Feldman, 1996). One consequence of niche construction is that behavior is conceptualized as more than the target of natural selection. It also modifies the environment for subsequent generations, so that now behavior is conceptualized as participating in the process of selection. As Laland *et al.* (2000 p. 135) put it, the evolutionary significance of niche construction rests on the feedback that it generates: "In the presence of niche construction, adaptation ceases to be a one-way process, exclusively a response to environmentally imposed problems: instead, it becomes a two-way process, with populations of organisms setting as well as solving problems."

One outcome of niche construction can be a shift in the genetic make-up of a population. A clear example of niche construction affecting regional genetic characteristics has been described by Durham (1991): human pastoralist groups are able to digest lactose and can eat dairy products and drink milk; human groups with other subsistence methods (e.g., hunter-gatherers, agriculturalists) lack the appropriate digestive enzyme and are lactose intolerant.

Niche construction in a very wide sense is potentially possible in all orders of living creatures, reflecting biological processes as varied as overt behavior (e.g., beavers constructing dams) to metabolic activity in microorganisms impacting the properties of the soil in which they live (Pulliam, 2000). Pulliam has modeled the consequences for microorganisms of altering their chemical surroundings, assuming two character types for the organism (constructors and nonconstructors). These models show that where niche construction occurs, niche constructors will come to dominate the population over a range of cost scenarios (where costs are incurred by the presence of nonconstructors). In other words, self-constructed ecosystems can over time come to be dominated by self-maintaining, mutualistic constructors. In this way, niche-construction processes can provide a benefit for all members of a community and can support multilevel selection as Sober and Wilson (1998) envision it occurs. Pulliam (2000) suggested that niche construction is an important feature driving the evolution of species assemblages (communities) dominated by mutualistic constructors, as observed in mutualistic communities of microorganisms living in the soil, for example.

Niche construction is more likely, in evolutionary terms, where its effects remain local, so that the benefits of niche construction are available to the individuals paying the costs of producing the effects. Niche construction is, therefore, most likely to evolve in species with certain types of social system and settlement pattern or in certain environments where movement is slow (Pulliam, 2000). In mobile animals, niche construction processes are more likely in species where individuals remain near one another or otherwise encounter the products of each other's activity on a regular basis. Social learning (which occurs within groups, so that its impacts on the environment remain local for that group) is clearly one mechanism supporting niche construction and enhancing its feedback potential in natural selection. Niche construction may produce "key innovations" that enable a species to make use of a resource which it previously could not use.

Behavioral traditions are one element of constructed niches; they are biologically significant for this fundamental reason. Traditions may support the maintenance of mundane but adaptive practices (such as using certain travel routes) among members of a living group. They may also result in the spread of a specific innovation, for example a new method of processing food, inclusion of a new item in the diet, or a new means of regulating temperature or constructing shelter. Both the continuation of familiar practices and the dissemination of new practices are biologically important, but the key role of behavioral innovation in speciation has generated more interest recently on the part of (quantitative) evolutionary modelers. Most contributors to this volume are concerned to a greater or lesser degree with the role of social learning in generating new traditions founded on a behavioral innovation that appears rarely in the population; two chapters in this volume (Chs. 3 and 4) address this issue primarily.

Several other contributions in this volume concern the evidence for traditions in various mammalian taxa, and what the behaviors in question contribute to the ecology of the groups where they are found. To most biologists, the controversies over whether or not an individual, population, or species exhibits "culture" are of no concern, but the possibility that traditions impact behavioral ecology, fitness, and evolution is of riveting interest.

We consider social learning and traditions from the perspective of ethology. Ethology is that part of biology most directly concerned with behavior. Ethology was established as a distinct branch of biology in the early years of the twentieth century and has matured into a vigorous field in the intervening century. As laid out by Tinbergen (1963), ethology is concerned with questions about behavior cast broadly in terms of causation (mechanism), ontogeny, evolution (phylogenetic history), and survival value (adaptive function). Since Tinbergen's (1963) seminal statement framing the scope of ethology, scientists studying the behavior of animals have recognized multiple levels of explanation as necessary for a comprehensive biological understanding of any behavior. Moreover, explanations at one level must be compatible with explanations at other levels: the organism is an integrated whole, with an unbroken connection to its individual and phylogenetic past and to its current circumstances. The power of this integrative perspective is evident in the contemporary vigor of ethology and its ability to interface substantively with other areas of biology (Kamil, 1998). We believe that explicitly treating social learning from this perspective will aid us in producing coordinated, complementary data across field and laboratory projects that will speak powerfully to contemporary questions about social learning in all animals, including humans.

1.4 Definitions of social learning

Behavioral scientists define social learning, in its broadest meaning, as changes in the behavior of one individual that result, in part, from paying attention to the behavior of another (Box, 1984). A broad definition of social learning encompasses one individual learning about the world from simply accompanying another. For example, when a naïve individual accompanies its social group on travels through the home range, it can learn the locations of resources, and habitual paths among them, as guppies (*Poecilia reticulata*) do (Laland and Williams, 1997, 1998). In this example, the behavior of the others allows the "learner" to generate experiences and encounter resources it would not otherwise; the others have by their behavior enabled the learner to learn.

A broad definition of social learning also covers the acquisition of social skills that involve direct interaction with partners. Individuals can learn specific, and sometimes idiosyncratic, modes of interacting with others (such as the affiliative behaviors of the kind described in Ch. 14). When the behaviors acquired through direct interaction are typical of the species, we describe this learning process as socialization (Box, 1984). When the behaviors are idiosyncratic to a dyad or a group, we describe the process as conventionalization (Tomasello, 1990). Some authors prefer to incorporate additional strictures to this very general definition, specifically to exclude behavioral changes that accompany, for example, direct social interactions (such as displaying submission to a more dominant individual, or coordinated sequences of social interaction during courtship) as social learning (Galef, 1988). Perhaps we will eventually develop phrases to distinguish these various settings for social learning: one to refer to social learning that is directly dependent on another's actions, but not interactive (i.e., learning from demonstrations); another for social learning that is dependent on direct interaction between participants; and yet another for social learning arising through passive exposure merely from accompanying others. For our purposes in this book, we accept the broadest definition, in accord with our interest in all the ways that animals can develop shared behaviors that depend in some way upon the social context for their repeated generation.

Some theorists challenge the notion, sometimes implicit but more often explicit in most contemporary treatments of social learning, that social learning occurs through the "transfer" of "information" from one individual to another. Information, after all, is not a thing. Learning does not entail the transfer of particles of information, unchanged during transfer across the space between heads (Ingold, 1998). An alternative view, well represented in contemporary anthropology and psychology, considers cognition as the process of organizing and maintaining streams of activity rather than the process of managing particles of knowledge (e.g., Gibson, 1966, 1986; Johnson, 1987; Reed, 1996; Thelen and Smith, 1994; van Gelder, 1998). In this view, activities of organisms are always grounded in ongoing engagement with the environment. All experience occurs in a background of meaning, and that meaning is a composite of social as well as asocial elements, and encompasses the current emotional and motivational state of the individual (D'Amasio, 1994). Knowledge and practice (behavior) are inseparable. Consequently, knowledge *per se* cannot be "transferred". Rather, an individual is continuously seeking meaning in others' perceived activities as well as all aspects of its own engagement with the current environment, and it alters its own behavior in accord with ongoing experience. In this framework, there is no possibility to separate "social" from "asocial" learning, or to consider learning processes as distinctive to one or the other (Fragaszy and Visalberghi, 2001; Ingold, 1998). What is distinctive about individuals acting in social settings is that they can generate behaviors that are similar to one another. The social learning process of concern to us is one of generation, not transmission. Adopting this perspective, what distinguishes social learning and traditions across species derives from the depth of meaning afforded by the social component of the environment, and the likelihood of generating similar practices (see Matsuzawa *et al.* (2001) for a convergent view).

Russon (1997) has suggested a similar interpretation of social learning in terms that are perhaps more familiar to biologists. In Russon's wording, a social partner alters the experience of the learner compared with experience without the social partner. The trajectory of action and perception through time is different in social versus nonsocial conditions. This could arise through increased salience of experiences that occur in presence of others, for example. Social partners generate particular experiences: they are animate, active agents, and they produce behaviors that are particularly salient to conspecifics. Learners may attend preferentially to conspecifics and may be predisposed to respond in particular ways to particular "signals" the conspecifics generate or behaviors in which certain individuals engage. This notion seems relevant to many proposed mechanisms of social learning, including those grounded in information-processing language and those grounded in Pavlovian conditioning (Byrne, 1999; Domjan *et al.*, 2000, Fragaszy, 2000; Fragaszy and Visalberghi, 2001; Russon, 1999). Here we note that social context is a rich and ever-changing background for individual activity. The added experiential aspect arising from social context can channel and scaffold individual efforts to acquire expertise. Social context constitutes a means of focusing behavior more effectively or differently than would have occurred in an asocial context.

The contribution of social context to skill development and decision making is likely to vary as a function of the social relationships of participants in the setting (Coussi-Korbel and Fragaszy, 1995). This aspect of theory in social learning is addressed by several contributions in this volume. For example, van Schaik (Ch. 11; see also van Schaik *et al.*, 1999) discusses how social tolerance contributes to the appearance of technological traditions in apes. Perry *et al.* (Ch. 14) present exciting new data on the relation between extent of proximity and likelihood of sharing specific social interactional patterns and foraging behaviors in capuchin monkeys. Mann and Sargeant (Ch. 9) present information on similarities in foraging methods in mother and offspring dyads in dolphins. The significance of social tolerance to effective social learning is a central theme of many contributions in our volume.

It cannot be stated too often that social learning is not distinguished as a different kind of learning process than other learning. As far as we now know, there is no distinctive learning mechanism associated with social learning: there is no separate neural tissue devoted to social learning and there is no evidence for a "social learning module", as has sometimes been proposed by those adopting a modular perspective on cognition (e.g., Cosmides and Tooby, 1992). Nor is there any competition, so to speak, within the individual between reliance on social learning and reliance on individual learning. Sometimes quantitative modelers make an assumption that socially biased learning is distinctive in function or process from individual learning, but this is merely a convenient assumption used to explore the evolutionary consequences of different organizations of learning (e.g., Richerson and Boyd, 2000; Laland, Richerson, and Boyd, 1996). Our categorization of "social learning" as distinctive from "asocial learning" arises from the contextual elements only. A more accurate characterization of these processes is the term socially biased learning (Fragaszy and Visalberghi, 2001).

The reader might at this point wonder about the issue of imitation, wherein an individual reproduces sequences of actions after observing another perform these sequences. Understanding how attention to observed action is coupled with the production of matching actions (as occurs during imitation), whether the actions are novel or familiar, is an important goal for cognitive scientists and neuroscientists (e.g., Byrne, 1999; Heyes and Ray, 2000; Myowa-Yamakoshi and Matsuzawa, 1999; Rizzolatti *et al.*, 1999; Whiten, 1998, 2000). Understanding the developmental trajectories, functional outcomes, and evolutionary pathways leading to imitation are also of value, particularly because imitation is a rare phenomenon. However, we can dismiss the notion that imitation is the *sine qua non* for traditions (shared behavioral patterns maintained in part by socially supported learning). A complete understanding of imitation will not lead to understanding how socially maintained practices arise in humans or any other taxon (Heyes, 1993; Heyes and Ray, 2000; Ingold, 1998). "Copying" behavior of others (as in imitation) is not a sufficient basis to produce skill; rather, skill requires repeated individual practice (Bernstein, 1996). Traditional practices are generated by each individual; they cannot be handed down as "units" from one individual to another, any more than the corporal bodies that perform them can be handed down (Ingold, 1998). Understanding how traditional knowledge and practice can be maintained requires a dynamic conception of the individual as engaged with its world, both social and asocial elements, in ongoing commerce.

In short, to understand the genesis of traditions we should strive to understand the nature of social bias in learning (where learning is broadly construed to include skill development). Nevertheless, in accord with the literature in this field, we use the term social learning to refer to the process in which social context contributes to skill development and decision making. When we understand how the social aspect of experience enables individuals to generate skills and adopt practices similar to those of their social partners, we may decide that some other label captures the process better. Until then, let us retain the categorical concept of social learning for comparative analyses of this phenomenon, realizing that it represents a construct about the context of learning, and not about the mechanisms of learning or distinctive neural structure. To conduct comparative analyses of social learning, we need to identify behaviors across species that share a common benefit from exposure to, or interaction with, social partners for their generation.

1.5 Definition of tradition, and a model of "tradition space"

We focus on traditions in this work because they are an obvious link between social learning and evolutionary processes. A tradition is a behavioral practice that is relatively enduring (i.e., is performed repeatedly over a period of time), that is shared among two or more members of a group, and that depends in part on socially aided learning for its generation in new practitioners. Prototypically, a tradition is shared among most or all members of a group, although it could be maintained by just one dyad or just one class of individuals (e.g., members of one matriline, only juvenile females, etc.). A particular behavior cannot be identified as a tradition without inferring that socially aided learning supports its shared presence across individuals. The extent to which social influence affects the generation of shared practice can vary, however, and this definition does not specify what extent of shared practice reflects social influence. Similarly, how long a behavioral practice must persist to qualify as "enduring" is a matter of debate. Some theorists acknowledge ephemeral traditions (shared behavior practices lasting a few days to a few months), in humans as well as other species (Bikhchandani, Hirshlifer, and Welch, 1998; Boesch and Tomasello, 1998; Laland *et al.*, 2000); others restrict the term to behaviors that persist across generations (Heyes, 1993; McGrew, 1998; Sugiyama, 1993; Whiten *et al.*, 1999). In short, the temporal dimension of persistence of a shared practice can range from brief to the remainder of an individual's life and beyond (in other practitioners); the shared behavior can be evident in as few as two individuals or extend to an entire group, and the extent to which social influences affect the generation of the practice in new individuals can vary from minimally helpful to absolutely necessary. For our purposes, a measurable social contribution to the generation of the practice in new practitioners is necessary for a behavior to qualify as a tradition.

In this view, traditions can vary along three orthogonal dimensions (duration, distribution, and extent of contribution of social influences to the expression of the behavior across individuals within a group). Traditions can thus be conceived as occurring within a "tradition space", as illustrated in Figure 1.1 under the heading of the group process model, to emphasize that traditions are identified according to properties of

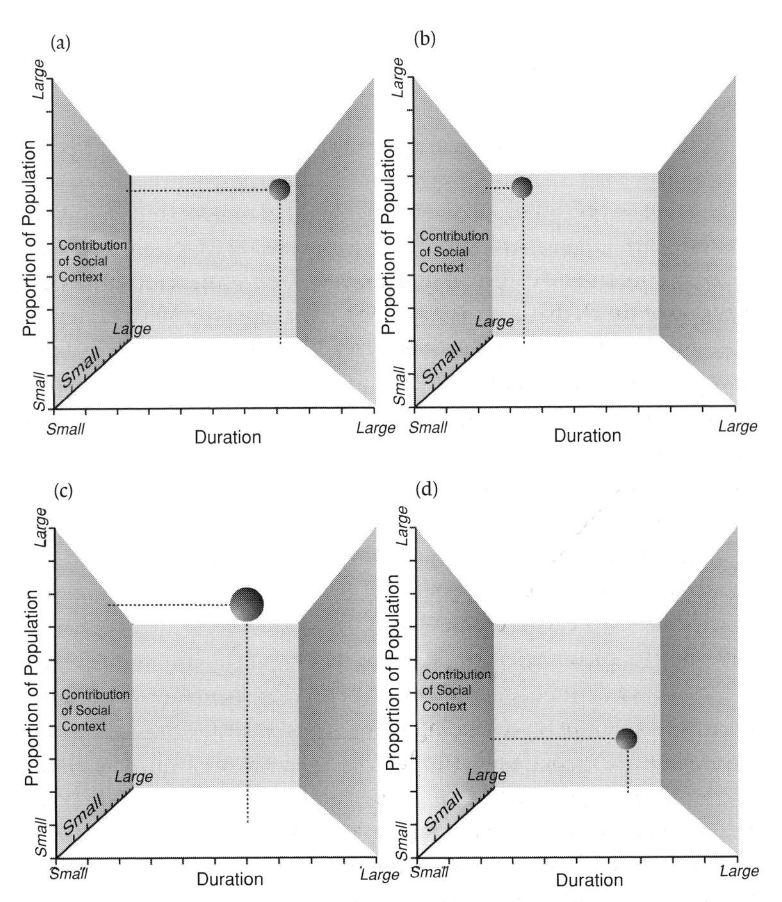

Fig. 1.1. The process model of traditions, conceived as a three-dimensional space. The defining axes are the duration of the behavior within the group (x axis), the proportion of the population displaying the behavior at any one time (y axis), and the contribution of social context to the acquisition of the behavior by new practitioners (z axis). Any distinctive behavior can, in principle, be placed into a unique location in this space. (a) A prototypical tradition: a behavior that is long-enduring, evident in most members of the group, and largely dependent on social context for its acquisition. (b–d) more problematic cases, where the behavior is evident only for a short time (b), social context provides a measurable but small contribution to the generation of the practice (c), or only a small proportion of the population exhibits the behavior (d).

behavior observed within a group. We use the term model here to mean a conceptual representation. Here the three orthogonal dimensions are represented as x, y, and z axes. Now traditions can be seen as falling along a scale in each dimension. Behaviors that are long lasting, are present in most or all members of a group, and are strongly dependent on social influences for their generation in new practitioners occupy one quadrant of this space (as in Fig. 1.1a). Behaviors meeting these criteria fall clearly within the common meaning of the term tradition. How far down or out from this quadrant can we go in tradition space and still identify a behavioral practice as a tradition? To give three examples, what about behaviors that are relatively ephemeral but widespread and highly dependent on social influences (depicted in Fig. 1.1b)? Or behaviors that are long lasting and widespread within a group but are not strongly dependent on social influences (in other words, that are often independently generated; as depicted in Fig. 1.1c)? Or behaviors that are clearly dependent on social influences for their generation but appear only in a few individuals within a group (depicted in Fig. 1.1d)? Of these last three examples, can we call all three traditions? Do we need to subdivide this concept to do justice to these three dimensions? Different contributors to this volume express different points of view on this related set of problems. The debate is useful grist for our efforts to develop theoretical models of traditions as biologically important phenomena.

This perspective on traditions is at variance with the usual way comparative biologists have approached the problem of identifying candidate traditions. Most discussions in the contemporary literature on traditions or culture in nonhuman animals, particularly primates, are grounded in a comparison of a completely different set of attributes, namely, (a) the degree of similarity of the behaviors seen in different social groups, (b) the (usually hypothetical) degree of genetic and behavioral exchange among members of different groups, and (c) the extent of environmental similarity across sites inhabited by different groups. We shall refer to this paradigm as the group contrast model of traditions, also called regional contrast by Dewar (Ch. 5) and method of elimination by van Schaik (Ch. 11). The argument goes like this.

1. Group X and group Y are currently or until very recently members of a single breeding population (i.e., genetically similar).
2. Group X performs an action in one form and group Y either does not perform it or performs it in a distinctively different form.
3. No obvious environmental difference limits the two groups from exhibiting the same form of the behavior.

This model relies on characteristics unrelated to an essential feature of traditions: their dependence on social context for acquisition by new practitioners of the practice in question. However, this is the model that underlies, for example, the listing of behavioral variations in chimpanzees studied at different field sites published by a consortium of field observers (Whiten *et al.*, 1999), or the compendium of behavioral variations seen in cetaceans published by Rendell and Whitehead (2001). It is evident in several of the chapters in this volume as well, as a starting point to identify candidate traditions (e.g., Chs. 11 and 14). McGrew (1998) suggests that field primatologists in particular adopt this approach because their subjects of study are too long lived to adopt an ontogenetic, or process, approach, as exemplified most elegantly in the work of Terkel and Aisner with rats (Terkel, 1996; see Ch. 6).

While the group contrast model may be a useful starting place to identify candidate traditions, it cannot be the ending point. Comparisons of extant behaviors, no matter how different the behaviors appear across groups, no matter how similar the environments or how similar in genetic makeup the populations, are *never sufficient* to resolve the question, "Is behavior X traditional in population Y?" A tradition is not confirmed *until* one can show that social learning contributes to the generation of a practice in new practitioners. The group-comparison data only set the stage by indicating some behaviors that are likely to be acquired in part through social learning. As Dewar (Ch. 5) points out, however, traditions are not limited to behaviors that vary across groups, and we may be seriously limiting our search by looking only at such behaviors. Huffman and Hirata (Ch. 10) discuss this issue in relation to the phenomenon of stone rubbing observed in many free-living groups of Japanese macaques.

The standard model of identifying traditions is illustrated in Fig. 1.2 under the heading group comparison model as a three-dimensional space, where the axes are degree of phylogenetic relatedness (genetic similarity), degree of behavioral similarity, and degree of environmental/ecological similarity. Here, the similarity between two or more groups is measured at one point in time. The small ball shows the ideal situation for identifying a candidate tradition according to this conception: two groups are highly related phylogenetically (indeed, are members of a single breeding population), they inhabit similar microhabitats, but they vary distinctly in the form of behavior X. Often the behavior pattern is widely evident in each population, and there is usually an attempt to verify longevity of the pattern. However, most often there is no evidence bearing on the ontogeny of the behavior in new practitioners. This model, we reiterate, can suggest candidate traditions but it does not get at the essence of what a tradition is: a behavior pattern shared among members of a group that

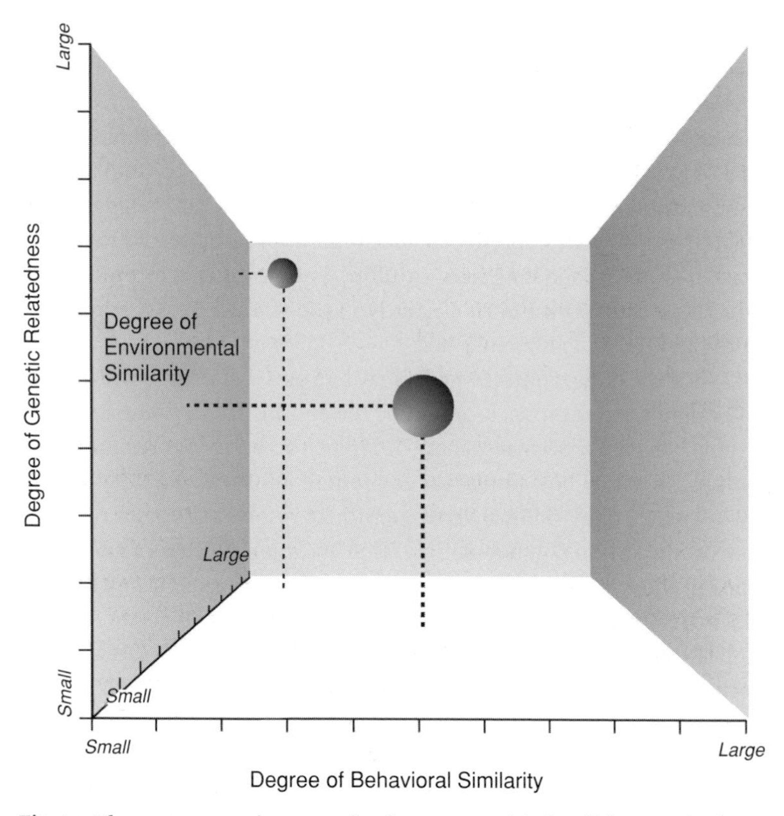

Fig. 1.2. The group comparison or regional contrast model of tradition conceived as a three-dimensional space. The location in space here defines the relation between two or more groups: the degree of similarity for the behavior of interest (*x* axis), the degree of genetic similarity for the groups under comparison (*y* axis), and the degree of environmental similarity for the groups under comparison (*z* axis). The small ball indicates a case that would be identified as a strong candidate for the label "traditional": a behavior showing strong differences across genetically similar groups living in similar environments. The larger ball illustrates a more problematic case: a behavior that is moderately different in groups with moderately different gene pools and that live in moderately different environments.

depends to a measurable degree on social contributions to the generation of the behavior in new practitioners. The model identifies one possible outcome of the process: behavioral differences between groups. Unfortunately, other processes besides social learning can lead to the same outcome, and this model cannot discriminate false positives (behavioral differences that are dependent on asocial factors and independent of socially aided learning). It is also prone to false negatives because it cannot identify behaviors that are dependent on socially aided learning but are similar across groups.

The large ball in Fig. 1.2 illustrates a common and visibly problematic situation. In this case, groups are judged to be somewhat differentiated genetically, to live in somewhat varying habitats, and to exhibit some degree of behavioral variation. What can this model now predict about the likelihood that the variations between the groups in behavior pattern X are supported by social learning? It cannot speak to this issue at all. It is important to note that drawing a conclusion from this model in this situation is no more problematic, on logical grounds, than drawing conclusions in what is considered the ideal situation, indicated by the small ball.

All too often this model has been accepted as the best method available for identifying traditions in nonhuman animals. We argue instead that the group comparison model is logically inadequate to allow strong conclusions about the status of any behavior as a tradition. To accept that a behavior is a tradition with this model one must confirm two null hypotheses: no genetic differences and no environmental differences (sufficient to account for the observed difference). These can never be "proven" to the skeptics' satisfaction. On top of that problem of logic, the notion that explanations of differences at the genetic or environmental level can support or rule out explanations at the *ontogenetic* level is clearly mistaken. This notion was thoroughly discredited years ago by the compelling arguments of Lehrman (1970) and others, who argued for an epigenetic understanding of individual development. Ontogenetic phenomena require an explanation in terms of individual ontogeny, not static notions of environmental conditions or genetic endowments. Social learning occurs during an individual's life; traditions are an outcome of several individuals' development. Social learning phenomena must (ultimately) be explained in terms of their development.

Why do behavioral scientists still feel compelled to exclude a genetic explanation for a behavioral character before they can consider how a behavior is acquired? Probably because they are confused as to what level of explanation a "genetic difference" affords. As Lerhman (1970) points out, the terms innate, inherited; and their relatives (e.g., hereditary, heritable) have two quite different meanings that are often confused. One meaning, used by geneticists, is that a character is inherited if variations of this character across individuals can be shown to arise from differences in the genetic constitution of the different individuals. The term is reserved for observed variability of the character in individuals with different genomes observed in the same environment; it implies "achievable by natural selection or by artificial selection". It does *not* address the question of whether variations in the environment during development would have an effect on the adult phenotype. The same genome can be associated with entirely different phenotypes in differing environments, and an environmental difference that greatly influences phenotypic development of one genome may have no effect on another, as demonstrated by Haldane more than half a century ago (Haldane, 1946). A straightforward example of the first principle is the variation across individuals and populations in parthenogenetic species, for example in whiptail lizards as documented by Taylor, Walker and Cordes (1997).

The second meaning often attributed to the terms innate, hereditary, or inherited is that of developmental fixity, the notion that a behavioral character is so developmentally canalized that it appears reliably even in the face of highly variable individual circumstances. This is an entirely different concept to the one discussed above; it has no bearing on "achievable by natural selection". It also has no bearing on what processes play a role in the behavior's development. Not keeping the two meanings distinct can lead to confusion. For example, to use the observation that the behavior of hybrid offspring matches that of both of its parents (in accord with first meaning accorded by geneticists) as evidence that a behavior is "innate" and, in the same sentence, that learning does not appear to influence the development of the behavior (in accord with the second meaning) reveals confusion about what innate means. (See Gottlieb (1992) for further discussion of how these concepts have been confused in the history of genetics and psychology.) Further, the pernicious and false notion that every element of behavior ought, on logical grounds, to be classifiable as "innate" or "learned" obscures serious consideration of how behaviors of interest develop. As Lerhman (1970, p. 33) said, "The distinction between 'innate' and 'acquired' is an inadequate set of concepts for analyzing development."

We believe that the logical inadequacies of the group comparison model (or as Dewar, (Ch. 5) labels it, the regional contrast model) are partially responsible for the frustrations that many have expressed with the task of trying to confirm that behaviors of particular interest are or are not traditions, and the equal frustration of those who see claims of tradition as over-rated. The model implies that a "genetic differences" explanation can supercede an "acquired" explanation as the source of a behavioral difference between groups. In actuality, these explanations are independent of one another. The model also requires the logically impossible procedure of confirming null hypotheses. As the model is logically inadequate, alternative interpretations can never be excluded, and the claim for tradition is necessarily weak. But is this unsatisfactory state of affairs necessary? We don't believe so. The process model of traditions does not suffer from these flaws, and we can indeed collect evidence from both field and laboratory that can be addressed with that model. In the next section and in our concluding chapter, we consider what kinds of evidence we should be collecting that can bear more deeply on the question of whether traditions exist in nonhuman species.

1.6 The comparative method in ethology

As MacLarnon (1999) reminds us, John Mill (1872/1967) explicated the principles of logical induction that govern the scientific enterprise today. Mill laid out four methods of inductive reasoning using comparative evidence: agreement, disagreement, residues, and concomitant variation. The first two methods rest on the principles that we can conclude that a causal relationship, or an enabling relationship, exists between a certain condition and the phenomenon under study by comparing (a) two instances in which a phenomenon occurs and the comparison groups have only one element in *common* (agreement) or (b) two instances in which a phenomenon occurs in one group but does not occur in another, where only one element is *different* between the comparison groups (disagreement). The method of disagreement is the familiar logic of experimental design, where one independent variable is manipulated to determine its effect on one or more dependent variables, holding other independent variables constant. Combining these two methods produces the joint method of agreement and difference wherein if both a set of dissimilar circumstances save one element X (agreement) and a set of similar circumstances save the same element X (disagreement) show the expected relation of presence and absence of phenomenon P, we can draw a strong conclusion about the necessity of element X to the occurrence of phenomenon P.

Phenomena in the natural world, where experimental manipulations are less frequently possible, rarely lend themselves to the strict standards of evidence required by the methods of agreement or disagreement, or their union (joint agreement and disagreement). In the natural world, multiple factors influence the occurrence of virtually all phenomena. Hence, the second two principles take on great importance for studies of naturally occurring phenomena. In these methods, we measure the magnitude of a phenomenon, rather than its presence or absence. In the method of residues (Mill's third method), one subtracts the magnitude of a phenomenon known to be associated with one set of conditions from its magnitude observed in a different but closely related set of conditions (ideally, similar conditions with one categorical difference). We attribute the difference, or *residual*, in the magnitude of the phenomenon to the differing conditions. For example, we may be interested in the frequency of grooming between groups that vary (ideally, only) with respect to the presence or absence of a particular kind of parasite. The logic of this method parallels that of the recently developed CAIC method (comparative analysis by independent contrasts: Harvey and Purvis, 1991; Purvis and Rambaut, 1995) used in phylogenetic contrasts, which takes into account the degree of relatedness of the various taxonomic groups used in the analysis (see Ch. 3).

The method of concomitant variations (Mill's fourth method) similarly relies upon a comparison of the size of a phenomenon between two or more circumstances. In this method, one scales the magnitude of a particular relevant variable that is always present but varies in scalar fashion (say, risk of predation) with the magnitude of the phenomenon of interest (say, group size). In the case of the relationship between risk of predation and group size, the group is the unit of analysis. Van Schaik (Ch. 11) uses this logic to evaluate the relationship between party size and the presence of putative traditions in chimpanzees. This method can also be used to evaluate the concordance between behavioral similarity in pairs of animals within a group, such as the use of a particular foraging technique, and some other aspect of their behavior with each other, such as the proportion of time they spend in proximity to one another, as illustrated in Ch. 14. In this case, the pair is the unit of analysis.

Neither of these methods provides the clear evidence of causal or conditional relationship that the first two methods do. Rather, they allow us to make the best use of available information; they provide correlational evidence concerning categorical or scalar variations of relevant variables across conditions. They allow us to identify that a relationship exists between the degree of some condition between groups, or between dyads within a group, and the probability that the dyad shares a behavioral characteristic.

The comparisons envisioned by Mill to identify the contributions of some condition to the occurrence of a phenomenon are widely used in ethology and other sciences. They do not exhaust analytical strategies, however. We have an arsenal of other methods that support analysis of development. Developmental analyses are concerned with *how* a characteristic comes about: how something changes through time in an individual. In the case of behavior, longitudinal observations of an individual, or a set of individuals, provide the most powerful analyses. Data of this sort relevant to understanding the origins of traditions in nonhuman animals come from studies of vocal learning in many taxa, but especially in birds (see Ch. 8). The now-classic developmental studies of Terkel (Terkel, 1996) demonstrating the development of pine-cone stripping in young black rats whose mothers use this method of feeding have already been mentioned above. Mann and Sargeant (Ch. 9) provide data of this type for bottlenose dolphins. In nonhuman primates, the best examples of developmental studies relevant to understanding the origins of shared practices are those of stone handling in Japanese macaques (Ch. 10), the development of nut cracking in young chimpanzees (Inoue-Nakamura and Matsuzawa, 1997), and the development of various feeding techniques in young orangutans (Russon, 2003 and Ch. 12).

1.7 Standards of evidence: experimental and observational

What evidence do we require to determine that social learning has occurred? In the laboratory, social learning can be documented by its outcome in accord with the methods of agreement, disagreement, or joint agreement and disagreement. In a common design, we compare two groups of subjects. In the first group (the "social learning group"), individuals differ measurably at the outset in the manner of achieving something (e.g., finding food). Subsequently two or more individuals jointly behave in the same environment, either simultaneously or sequentially. In the second group (the "individual learning group"), individuals do not behave jointly in the same environment; they encounter the same circumstances on their own. Thus the individuals' exposure to the circumstances is the same across groups, but the social context of their experience is different. We conclude that social learning has occurred if members of the social learning group alter their behavior to be more similar to their social partner's behavior following joint exposure, compared with subjects that encounter the same problems individually. Usually in this design, one or more individual(s) in the social learning group is more proficient at the task (and serves as a "demonstrator" to the others). The hypothesis to be tested is that the less-proficient individuals in the social learning group will become more proficient following exposure to the demonstrators than will members of the individual learning group following equivalent exposure to the problem, but without a demonstrator. In other words, exposure to the situation with a demonstrator allows the learner to behave more like the demonstrator more quickly than a solo learner alters its behavior to be more proficient (and more like the social group's demonstrator). Galef's studies of social learning of food preferences in rats (Ch. 6) and Visalberghi and Addessi's studies of food choices in capuchin monkeys (Ch. 7) illustrate the subtleties of experimental design that can follow from this logic.

A second common experimental design in social learning experiments is the "two action design", in which two or more groups of subjects encounter the same problem with a demonstrator–partner. The solutions practiced by the proficient partner vary in key ways between the groups; for example, in one group the demonstrator may pull up a lid to open a container, and in another group the demonstrator may push the lid down to open the container. In this design, one seeks differential shifts towards the more expert partner's behavior on the part of the less-proficient partner in all groups, and the form of change is predicted to vary between the groups. Zentall (1996, p. 232) provides several examples of studies using this design (see also Fritz and Kotrschal, 1999; Voelkl and Huber, 2000).

Regardless of design or circumstance, as an individual acts in the environment, the consequences of its actions will impact whether or how often the behavior is performed again. Socially learned behaviors produce a history of consequences, as do all behaviors (Galef, 1992). In this sense, the methods of agreement and disagreement are not a perfect fit to the problem of demonstrating social learning, as behavior has a historical component that these logical principles do not encompass. For example, over time, a behavior may become modified or may become performed more selectively as a consequence of continuing practice, or it may be abandoned by some individuals. Unfortunately for the scientist interested in assessing the likelihood that a behavior is a tradition, all these processes have the net effect of masking the differences in behavior between groups that experienced different learning contexts at an earlier time. Comparisons of groups according to the consequences of experience at a single, earlier period may thus become muddied, especially as the temporal distance between the different learning context and the evaluation of performance increases.

What evidence for social learning can we expect to collect from naturalistic observations? It is not possible to obtain the same evidence that we can obtain in experimental situations. Field observers cannot train an individual to serve as a demonstrator to others, nor can they group animals by skill levels on a given task. Observers of animals in natural settings cannot determine with certainty that the changes in behavior they observe across time in an individual's proficiency or form at some particular task reflect social influence on learning, because they cannot rule out asocial influences by comparison with a control group. Changes in performance may also reflect some concurrently varying feature of the situation (such as seasonal changes in the availability of resources, physical changes in the individual, and so on). This could be ruled out with a control group in the same context but shielded from social influence, but this is not possible in natural circumstances. Moreover, unlike in experimental studies, it is usually impossible to know any individual's level of experience with a task prior to the start of observations. Nevertheless, field observers can document social contexts in which behaviors occur and changes over time in individual performance; they can document intragroup variation in behavior at a particular time, and they can seek comparable evidence about specific practices in other groups of the same species or of related species. The contributions by field researchers in this book illustrate the application of these forms of logic to the analysis of behavior of many species of animal in natural settings.

Identification of locale-specific behaviors is not sufficient to conclude that social learning is a necessary element in the generation of those behaviors. Multiple pathways lead to similar behaviors in many instances. For example, Galef (1980) examined how rats could develop the habit of swimming under water to collect shellfish from the riverbank. This unusual manner of foraging is (or was) common in rats living at a certain location along the River Po in Italy (Gandolfi and Parisi, 1973). Diving for food seemed a strong candidate for a behavior fully dependent upon social learning for its persistence in the population. Nevertheless, Galef (1980; see also Ch. 6) showed experimentally that juvenile rats could acquire this habit readily without any social scaffolding in conditions similar in relevant ways to their riverbank habitat. Social learning might aid individuals to develop the behavior but the behavior is not necessarily dependent on social learning for its generation. We still do not know the extent to which social learning does in fact contribute to the continuance of this practice in rats along the River Po (the residual in the method of residues), or indeed whether it contributes at all. It seems plausible, but it is not necessary.

Sometimes those conducting naturalistic observations argue that demonstrating the necessity of social learning in the generation of similar behaviors in different individuals requires excluding all plausible alternative explanations (usually, environmental sources, such as resource availability, and presumed genetic differences) (Boesch, 1996; McGrew, 1998). Unfortunately, it is a logical impossibility to exclude all other mechanisms besides social learning that might produce similar behaviors in two or more individuals on the basis of observations of spontaneous behavior in natural settings. Field observations simply cannot provide the data necessary for such strong inferences. This is a misguided attempt to use the logic of the method of agreement when the elements needed to use this logic are not available (see also Ch. 5). It is logically possible, however, to adopt the method of residues or of concomitant variations and to show that social learning *aids* the generation of similar behaviors. This can be done, for example, by documenting the development of skill as a function of the extent of social support during learning (correlating rate of skill development in several individuals with extent of social support). To confirm that social learning aids in the generation of similar behaviors, we need to document the spread of a specific behavior to multiple new practitioners in a variety of circumstances. Considering each new practitioner as a new link, and a series of links as a transmission chain, we can evaluate (a) how rapidly new practitioners develop the behavior with differing forms of social support, (b) how close the behavioral resemblance remains across links, and (c) how different the patterns are in different social units. This task is easier if the behavior is present in some groups and not others, and logically even easier if a behavioral innovation is observed at the outset, and its spread followed within a group. It is still possible, however, even if the behavior is present in all groups.

Some authors emphasize the persistence of a behavioral pattern across biological generations as necessary to accord it the status of a tradition (e.g., McGrew, 1998; Whiten *et al.*, 1999). As may be surmised from the traditions space model provided in Fig. 1.1 and discussed earlier, we find this requirement too restrictive. Temporal stability is surely important for the evolutionary significance of a particular pattern. Traditions allow one generation to impact the conditions of natural selection of the next generation; the selective environment is scaffolded for the next generation by the behavior of the previous one. Traditions can contribute to constructed niches (Laland *et al.*, 2000) and thus have effects on fitness. However, in theory, even ephemeral traditions (lasting only a portion of the individual's lifespan) can have fitness consequences. Vocal traditions in many taxa drift in less than a lifespan; degree of adherence to the traditional song of the moment can still influence individual fitness. As Perry *et al.* argue (see Ch. 14), other forms of social conventions may also have this consequence.

Documentation of socially aided learning by animals in natural settings is likely to remain challenging, whatever method is adopted for this purpose. Shifting social context and ongoing behavior of several individuals are not easy to record in real time. Even documenting intergroup variation in the presence or absence of a specific behavior can be difficult, because of the difficulty in interpreting negative evidence. Although statistical methods can be used to examine the probability of noting a behavior in one population given its rate of occurrence in another population, to evaluate whether the two populations produce the behavior at equivalent rates (e.g., see Ch. 11), interpreting behaviors seen at extremely low frequencies remains problematic.

However, the situation is far from hopeless. As contributors to this volume show, there are many different forms of evidence that can be brought to bear on the question of the third dimension in traditions, that of social contributions to the generation of the behavior in new practitioners. We anticipate that the sample efforts presented in this volume will generate new ideas for those studying many different taxa and forms of behavior about how to evaluate the contribution of social influences in their own cases of interest.

1.8 Conclusions

Our principal aim is to understand traditions as biological phenomena in order to improve our understanding of their contribution to the evolution and current life ways of various taxa. We have adopted an ethological stance to this problem, noting that we should recognize explanations at different levels (evolution, function, mechanism, and development) as having complementary value, and that explanations at these different levels should be compatible with one another. Ideally, we would like to create a model that effectively predicts when and in what domains traditions will appear in a particular species, and how social influences will support the generation of shared behaviors, taking into account the species' constellation of ecological, social, and behavioral characteristics. We would like to model evolutionary trends in the occurrence of traditions, as well as ontogenetic patterns governing the acquisition of shared behaviors. We are far from reaching all these goals at present. Nevertheless, we are encouraged by the energy and creativity of the research community around this issue as represented by the contributions to this volume.

One of our central concerns in this chapter has been to lay out a definition for traditions that permits empirical rigor. To this end, we have suggested conceptualizing traditions as behaviors located within a specific region of the three-dimensional space defined by the axes of temporal duration, proportion of population displaying the practice, and contribution of social influences on the generation of new practitioners (the process model; see Fig. 1.1). This heuristic model makes it clear why documenting group specificity and long (even intergenerational) duration, currently the most frequently used data to argue for or against the status of a behavior as traditional in a particular group, will never be sufficient to make a strong claim for that status. The third dimension (contribution of social influence) must be examined in its own right; it is neither derivative of nor predicted by the other two dimensions. We do not yet have a principled basis to specify numerical values defining the area of traditions; that awaits further theoretical developments. However, the process model nudges us to look for ways to measure the effects of social influence on acquisition, to achieve adequate definitional rigor for the phenomenon. This task is important no matter what level of explanation is under consideration.

Behavioral scientists work in settings ranging from the laboratory (where virtually every aspect of social context, individual history, and environmental circumstance can be monitored and controlled) to field conditions, where the observer must make do with incomplete information. Therefore, we must be prepared to make the best use of very different kinds of information. We must acknowledge the different forms of comparison enabled by the different circumstances we face in these different conditions of scientific inquiry, and we must adapt our analytical goals to the data supported by each condition. For those who study social learning, this means adopting the method of residues or the method of concomitant variation, to use Mill's terminology, to examine the critical dimension of social contribution to shared practice when we cannot manipulate the relevant variables of social context and solo practice. Those who have the luxury (and the burden) of designed experiments can adopt the methods of agreement and disagreement (that is, traditional experimental methods). All of us have the responsibility to adopt longitudinal methods where possible, as developmental analyses are necessary to understand how shared practices arise.

Understanding traditions as biological phenomena requires the collaborative efforts of scientists working from diverse theoretical and methodological realms (modeling, experiments, and observations of behavior in natural settings). Although field observations will virtually never support the use of the stronger methods of agreement and disagreement, they can be a very rich source of information supporting other methods. In particular, two forms of information from naturalistic observations are relevant to studies of traditions: (a) behavioral variation within groups, in conjunction with patterns of social affiliations or (a less powerful method) across sites, and (b) longitudinal data on the generation of skilled practice by new practitioners. Longitudinal data relevant to acquisition will enable us to identify traditions more rigorously than has been the case previously.

This chapter is followed by contributions by Laland and Kendall, Reader, Lefebvre and Bouchard, and Dewar (Chs. 2–5) addressing evolutionary, comparative, and process models of traditions. Chapters 6–14 cover a variety of taxa and of empirical approaches to the study of traditions. Several contributions illustrate the logic and power of analyzing naturally occurring patterns of variation with moderately longitudinal data (Chs. 9–12, 14). In Ch. 13, Boinski *et al.* describe the starting point for studies of traditions, a behavioral phenomenon that seems likely in their estimation to rely on social context for some aspects of its development. Contributions from experimental scientists (Ch. 6 by Galef, and Ch. 7 by Visalberghi and Addessi) illustrate how complementary use of the different comparative methods aids a full understanding of complex biological phenomena, and both provide cautionary examples of behaviors that seem likely to be dependent on social learning but that can arise rather easily in other ways. Janik and Slater (Ch. 8) review traditional aspects of vocal communication in birds and mammals to round out the topical and taxonomic coverage. In the final chapter, we draw out shared themes evident in these contributions to suggest directions for future work, and to highlight opportunities for fruitful collaboration.

At the end of the day, we must recognize that social learning, leading to traditions, is a central participatory feature in behavioral biology; it deserves our concentrated attention even though it is no more amenable to easy comprehension than any other aspect of behavioral biology. Developing clear conceptual and methodological approaches is a necessary first step in creating a rigorous field of study devoted to this subject. We intend that this book will stimulate progress in this endeavour.

1.9 Acknowledgements

We thank the University of Georgia State-of-the-Art Conference Program for funding the conference that gave rise to this volume. The various contributors to the conference and the book provided many useful comments and stimulating discussion; we particularly thank V. Janik and E. Visalberghi for their comments. We also thank I. Bernstein, P. Gowaty, J. Mann, L. McCall, D. Promislow, and M. West for commenting on earlier versions of this chapter.

References

Avital, E. and Jablonka, E. 2000. *Animal Traditions: Behavioural Inheritance in Evolution*. New York: Cambridge University Press.

Basil, J. A., Kamil, A. C., Balda, R. P., and Fite, K. V. 1996. Differences in hippocampal volume among food storing corvids. *Brain, Behavior, and Evolution*, **47**, 156–164.

Bernstein, N. 1996. *Dexterity and its development*. Hillsdale, NJ: Erlbaum.

Bikhchandani, S., Hirshlifer, D., and Welch, I. 1998. Learning from the behavior of others: conformity, fads, and informational cascades. *Journal of Economic Perspectives*, **12**, 151–170.

Boesch, C. 1996. Three approaches for assessing chimpanzee culture. In *Reaching into Thought*, ed. A. Russon, K. Bard, and S. Parker, pp. 404–429. Cambridge: Cambridge University Press.

Boesch, C. and Tomasello, M. 1998. Chimpanzee and human cultures. *Current Anthropology*, **39**, 91–604.

Bonner, J. 1980. *The Evolution of Culture in Animals*. Princeton, NJ: Princeton University Press.

Box, H. 1984. *Primate Behaviour and Social Ecology*. London: Chapman & Hall.

Box, H. and Gibson, K. 1999. *Mammalian Social Learning: Comparative and Ecological Perspectives*. Cambridge: Cambridge University Press.

Byrne R. W. 1999. Imitation without intentionality: using string parsing to copy the organization of behavior. *Animal Cognition*, **2**, 63–72.

Cosmides, L. and Tooby, J. 1992. Cognitive adaptations for social exchange. In *The Adapted Mind: Evolutionary Psychology and the Generation of Culture*, ed. J. H. Barkow, L. Cosmides, and J. Tooby, pp. 163–228. New York: Oxford University Press.

Coussi-Korbel, S. and Fragaszy, D. 1995. On the relation between social dynamics and social learning. *Animal Behaviour*, **50**, 1441–1453.

D'Amasio, A. 1994. *Descartes' Error. Emotion, Reason, and the Human Brain*. New York: Putnam.

de Waal, F. 1999. Cultural primatology comes of age. *Nature*, **399**, 635–636.

de Waal, F. 2001. *The Ape and the Sushi Master*. New York: Basic Books.

Domjan, M., Cusato, B., and Villarreal, R. 2000. Pavlovian feed-forward mechanisms in the control of social behavior. *Behavioral and Brain Sciences*, **23**, 235–249.

Durham, W. 1991. *Coevolution: Genes, Culture, and Human Diversity*. Stanford, CA: Stanford University Press.

Finlay, B., Darlington, R., and Nicastro, N. 2001. Developmental structure in brain evolution. *Behavioral and Brain Sciences*, **24**, 263–308.

Fragaszy, D.M. 2000. Extending the model: Pavlovian social learning. *Behavioral and Brain Sciences*, **23**, 255–256.

Fragaszy, D. and Visalberghi, E. 1996. Primate primacy reconsidered. In *Social Learning in Animals: The Roots of Culture*, ed. C. Heyes and B. Galef, pp. 65–84. New York: Academic Press.

Fragaszy, D. and Visalberghi, E. 2001. Recognizing a swan: socially-biased learning. *Psychologia*, **44**, 82–98.

Fritz, J. and Kotrschal, K. 1999. Social learning in common ravens, *Corvus corax*. *Animal Behaviour*, **57**, 785–793.

Galef, B.G., Jr. 1980. Diving for food: analysis of a possible case of social learning in wild rats (*Rattus norvegicus*). *Journal of Comparative and Physiological Psychology*, **94**, 416–425.

Galef, B.G., Jr. 1988. Imitation in animals: History, definition, and interpretation of data from the pyschological laboratory. In *Social Learning: Psychological and Biological Perspectives*, ed. T. Zentall and B. G. Galef Jr., pp. 3–28. Hillsdale, NJ: Erlbaum.

Galef, B.G., Jr. 1992. The question of animal culture. *Human Nature*, **3**, 157–178.

Gandolfi, G. and Parisi, V. 1973. Ethological aspects of predation by rats, *Rattus norvegicus* (Berkenhout) on bivalves, *Unio pictorum*, L. and *Cerastoderma lamarcki* (Reeve). *Bollettino di Zoologia*, **40**, 69–74.

Gibson, J.J. 1966. *The Senses Considered as Perceptual Systems*. Boston, MA: Houghton Mifflin.

Gibson, J.J. 1986. *The Ecological Approach to Visual Perception*. Hillsdale, NJ: Erlbaum.

Giraldeau, L.-A. 1997. The ecology of information use. In *Handbook of Behavioral Ecology*, ed. N. Davies and J. Krebs, pp. 42–68. Oxford: Blackwell Scientific.

Gottlieb, G. 1992. *Individual Development and Evolution: The Genesis of Novel Behavior.* New York: Oxford University Press.

Haldane, J.S. 1946. The interaction of nature and nurture. *Annals of Eugenics*, **13**, 197–205.

Harvey, P.H. and Purvis, A. 1991. Comparative methods for explaining adaptations. *Nature*, **351**, 619–624.

Heyes, C. 1993 Imitation, culture and cognition. *Animal Behavior*, **46**, 999–1010.

Heyes, C.M. and Ray, E.D. 2000. What is the significance of imitation in animals? *Advances in the Study of Behavior*, **29**, 215–245.

Ingold, T. 1998. Commentary on Boesch and Tomasello, Chimpanzee and human cultures. *Current Anthropology*, **39**, 605–606.

Inoue-Nakamura, N. and Matsuzawa, T. 1997. Development of stone tool use by wild chimpanzees. *Journal of Comparative Psychology*, **111**, 159–173.

Itani, J. and Nishimura, A. 1973. The study of infrahuman culture in Japan. In *Symposia of the Fourth International Congress of Primatology*, Vol. 1, ed. E. W. Menzel Jr., pp. 20–26. Basel: Karger.

Johnson, M. 1987. *The Body in the Mind*. Chicago, IL: University of Chicago Press.

Kamil, A. 1998. On the proper definition of cognitive ethology. In *Animal Cognition in Nature*, ed. R. Balda, I. Pepperberg, and A. Kamil, pp. 1–28. New York: Academic Press.

Kawai, M. 1965. Newly-acquired pre-cultural behavior of a natural troop of Japanese monkeys on Koshima Island. *Primates*, **6**, 1–30.

Kawamura, S. 1965. Sub-culture among Japanese macaques. In *Monkeys and Apes: Sociological Studies*, ed. S. Kawamura and J. Itani, pp. 237–289. Tokyo: Chuokoronsha.

Kinzey, W. 1997. *New World Primates: Ecology, Evolution, and Behavior*. Hawthorne, NY: Aldine de Gruyter.

Krebs, J. R., Sherry, D. F., Healy, S. D., Perry, H., and Vaccerino, A. L. 1989. Hippocampal specialization of food-storing birds. *Proceedings of the National Academy of Sciences, USA*, **86**, 1388–1392.

Laland, K. and Williams, K. 1997. Shoaling generates social learning of foraging information in guppies. *Animal Behaviour*, **53**, 1161–1169.

Laland, K. and Williams, K. 1998. Social transmission of maladaptive information in the guppy. *Behavioral Ecology*, **9**, 493–499.

Laland, K., Richerson, P., and Boyd, R. 1996. Developing a theory of animal social learning. In *Social Learning in Animals: The Roots of Culture*, ed. C. H. Heyes and B.G. Galef Jr., pp. 129–154. New York: Academic Press.

Laland, K., Odling-Smee, J., and Feldman, M. 2000. Niche construction, biological evolution, and cultural change. *Behavioral and Brain Sciences*, **23**, 131–146.

Lefebvre, L., Whittle, P., Lascaris, E., and Finkelstein, A. 1997. Feeding innovations and forebrain size in birds. *Animal Behaviour*, **53**, 549–560.

Lehrman, D. 1970. Semantic and conceptual issues in the nature–nurture problem. In *Development and Evolution of Behavior*, ed. L. Aronson, E. Tobach, D. Lehrman, and J. Rosenblatt, pp. 17–52. San Francisco, CA: Freeman Press.

Lewontin, R. 1978. Adaptation. *Scientific American*, **239**, 156–169.

MacLarnon, A. 1999. The comparative method: principles and illustrations from primate socioecology. In *Comparative Primate Socioecology*, ed. P. C. Lee, pp. 5–22. Cambridge: Cambridge University Press.

Matsuzawa, T., Biro, D., Humle, T., Inoue-Nakamura, N., Tonooka, R., and Yamakoshi, G. 2001. Emergence of culture in wild chimpanzees: education by master–apprenticeship. In *Primate Origins of Human Cognition and Behavior*, ed. T. Matsuzawa, pp. 557–574. Tokyo: Springer Verlag.

McGrew, W. C. 1998. Culture in nonhuman primates? *Annual Review of Anthropology*, **27**, 301–328.

McGrew, W. C. 1992. *Chimpanzee Material Culture: Implications for Human Evolution*. Cambridge: Cambridge University Press.

Mill, J. S. 1872/1967. *A System of Logic: Ratiocinative and Inductive. Being a Connected View of the Principles of Evidence and the Methods of Scientific Investigation*. London: Longman.

Myowa-Yamakoshi, M. and Matsuzawa, T. 1999. Factors influencing imitation of manipulatory actions in chimpanzees (*Pan troglodytes*). *Journal of Comparative Psychology*, **113**, 128–136.

Odling-Smee, F. J., Laland, K. N., and Feldman, M. W. 1996. Niche construction. *American Naturalist*, **147**, 641–648.

Pulliam, R. 2000 (November). Gaia, evolution, and the theory of niche construction. Colloquium presented to the Institute of Ecology, University of Georgia, Athens, GA.

Purvis, A. and Rambaut, A. 1995. Comparative analysis by independent contrasts (CAIC): an Apple Macintosh application for analysing comparative data. *Computer Applications in Bioscience*, **11**, 247–251.

Reader, S. M. and Laland, K. N. 2001. Primate innovation: sex, age and social rank differences. *International Journal of Primatology*, **22**, 787–805.

Rizzolatti, G., Fadiga, L., Fogassi, L., and Gallese, V. 1999. Resonance behaviors and mirror neurons. *Archives Italiennes de Biologie*, **137**, 85–100.

Reed, E. S. 1996. *Encountering the World*. New York: Oxford University Press.

Rendell, L. and Whitehead, H. 2001. Culture in whales and dolphins. *Behavioral Brain Sciences*, **24**, 309–382.

Richerson, P. J. and Boyd, R. 2000. Climate, culture and the evolution of cognition. In *The Evolution of Cognition*, ed. C. Heyes and L. Huber, pp. 329–346. Cambridge, MA: MIT Press.

Russon, A. 1997. Exploiting the expertise of others. In *Machiavellian Intelligence II: Extensions and Evaluations*, ed. R. Byrne and A. Whiten, pp. 174–206. Cambridge: Cambridge University Press.

Russon, A. E. 1999. Naturalistic approaches to orangutan intelligence and the question of enculturation. *International Journal of Comparative Psychology*, **12**, 181–202.

Russon, A. E. 2003. Comparative developmental perspectives on culture: the great apes. In *Between Biology and Culture: Perspectives on Ontogenetic Development*, ed. H. Keller, Y. H. Poortinga, and A. Schoelmerich. Cambridge: Cambridge University Press.

Russon, A., Mitchell, R., Lefebvre, L., and Abravanel, E. 1998. The comparative evolution of imitation. In *Piaget, Evolution, and Development*, ed. J. Langer and M. Killen, pp. 103–143. Mahwah, NJ: Erlbaum.

Sober, E. and Wilson, D. S. 1998. *Unto Others: The Evolution and Psychology of Unselfish Behavior.* Cambridge, MA: Harvard University Press.

Sugiyama, Y. 1993. Local variation of tools and tool use among wild chimpanzee populations. In *The Use of Tools by Human and Nonhuman Primates*, ed. A. Berthelet and J. Chavaillon, pp. 175–187. New York: Oxford University Press.

Taylor, H., Walker, J. M., and Cordes, J. 1997. Reproductive characteristics and body size in the parthenogenetic teiid lizard *Cnedmidophorous tesselatus*: comparison of sympatric color pattern classes C and E in De Baca County, New Mexico. *Copeia*, **1997**, 863–868.

Terkel, J. 1996. Cultural transmission of feeding behavior in the black rat (*Rattus rattus*). In *Social Learning in Animals: The Roots of Culture*, ed. C. M. Heyes and B. G. Galef Jr., pp. 267–286. San Diego, CA: Academic Press.

Thelen, E. and Smith, L. 1994. *A Dynamical Systems Approach to the Development of Cognition and Action*. Cambridge, MA: MIT Press.

Tinbergen, N. 1963. On aims and methods of ethology. *Zeitschrift für Tierpsychologie*, **20**, 410–433.

Tomasello, M. 1990. Cultural transmission in the tool use and communicatory signalling of chimpanzees? In *"Language" and Intelligence in Monkeys and Apes*, ed. S. Parker and K. Gibson, pp. 274–311. Cambridge: Cambridge University Press.

van Gelder, T. 1998. The dynamical hypothesis in cognitive science. *Behavioral and Brain Sciences*, **21**, 615–665.

van Schaik, C., Deaner, R., and Merrill, M. 1999. The conditions for tool use in primates: implications for the evolution of material culture. *Journal of Human Evolution*, **36**, 719–741.

van Schaik, C. P., Ancrenaz, M., Borgen, G., Galdikas, B., Knott, C. D., Singleton, I., Suzuki, A., Utami, S. S., and Merrill, M. 2003. Orangutan cultures and the evolution of material culture. *Science*, **299**, 102–105.

Voelkl, B. and Huber, L. 2000. True imitation in marmosets. *Animal Behaviour*, **60**, 195–202.

Vogel, G. 1999. Chimps in the wild show stirrings of culture. *Science*, **284**, 2070–2073.

Whiten, A. 1998. Imitation of the sequential structure of actions by chimpanzees (*Pan troglodyates*). *Journal of Comparative Psychology*, **112**, 270–281.

Whiten, A. 2000. Primate culture and social learning. *Cognitive Science*, **24**, 477–508.

Whiten, A., Goodall, J., McGrew, W. M., Nishida, T., Reynolds, V., Sugiyama, Y., Tutin, C. G., Wrangham, R. W. W., and Boesch, C. 1999. Cultures in chimpanzees. *Nature*, **399**, 682–685.

Whiten, A., Goodall, J., McGrew, W. C., Nishida, T., Reynolds, V., Sugiyama, Y., Tutin, C. E. G., Wrangham, R. W. W., and Boesch, C. 2001. Charting cultural variation in chimpanzees. *Behaviour*, **138**, 1481–1516.

Zentall, T. 1996. An analysis of imitative learning in animals. In *Social Learning in Animals: The Roots of Culture*, ed. C. Heyes and B. Galef, pp. 2221–2243. New York: Academic Press.

Zentall, T. R., Sutton, J. E., and Sherburne, L. M. 1996. True imitative learning in pigeons. *Psychological Science*, **7**, 343–346.

KEVIN N. LALAND AND JEREMY R. KENDAL

2

What the models say about social learning

2.1 Introduction

All too often theoretically minded scientists soar off into an abstract mathematical world that seemingly makes little contact with empirical reality. The field of animal social learning and tradition has its very own assortment of theory, although in truth it is a somewhat paltry portion, and the mathematics rarely get that sophisticated. Nonetheless, the modelers and the empirical scientists, while perhaps converging, have for the most part yet to meet in any consensus of shared goals and understanding. As the most effective mathematical models in science are undoubtedly those making clear, empirically testable predictions, it would obviously be of value if the mathematics had some utility to other researchers in the field of animal social learning. Moreover, as the best models are those with assumptions well informed by empirical findings, it would also clearly help if social learning researchers collected the kind of information that was relevant to grounding the models.

The over-arching goal of this article is to contribute towards the further integration of empirical and theoretical work in animal social learning. While this is a worthy long-term objective, it is apparent that such an integration is unlikely to happen overnight. At the time of writing, most of the mathematical theory in our field has been developed without the benefit of a thorough understanding of animal social learning, in fact, largely without nonhuman animals in mind. Similarly, with one or two exceptions striking for their singularity (Laland and Williams, 1998; Wilkinson and Boughman, 1999), there has been virtually no experimental testing of the models' predictions.

As a modest step in the right direction, this chapter reviews, summarizes and explains in simple nonmathematical terms what the models have to say about social learning. We also focus on the areas that the models have thus far neglected, but which, in our view, may benefit from a theoretical perspective. In order to structure this information in an intuitive fashion, we present it in the form of pertinent questions of relevance to researchers on animal traditions. The next section describes questions for which the models have provided answers, albeit with varying degrees of utility. We attempt to draw out clear predictions from the theory. In the rare instances where the predictions have been subject to empirical test, we present the findings of these studies and discuss the model's performance. Where the theory has not been tested, we endeavor to illustrate how it might be. This is followed by a section focusing on as-yet neglected questions about social learning where theory and empirical work could usefully be integrated. We concentrate on what we believe to be the more tractable questions. In the final section, we discuss future directions for animal social learning theory.

2.2 Questions about social learning that the models have addressed

By social learning we refer to socially guided individual learning. Whilst most theoretical models have distinguished between asocial and social learning as if they were binary categories, in reality it may sometimes be more appropriate to regard cases of social learning as arrayed on a dimension with greater or lesser reliance on social cues (Laland *et al.*, 1993).

2.2.1 When should animals use social learning?

Several theoretical analyses have explored the circumstances under which natural selection should favor reliance on social learning, as opposed to asocial learning or evolved nonlearned behavior (Aoki and Feldman, 1987; Bergman and Feldman, 1995; Boyd and Richerson, 1985, 1988; Cavalli-Sforza and Feldman, 1983; Feldman, Aoki, and Kumm, 1996; Laland, Richerson, and Boyd, 1996; Rogers, 1988; Stephens, 1991). It is now well established that the issue hangs, in part, on patterns and rates of variability in the environment over evolutionary time. In an environment that is changing comparatively slowly, or that exhibits relatively little spatial heterogeneity in resources, populations are able to evolve appropriate behavior patterns through natural selection, and learning is of little adaptive value. At the other extreme, in rapidly changing or highly variable environments, asocial learning pays. Of course, there is a limit to how changeable the environment can get beyond which learning of any kind is worthless: in a randomly changing environment, learning is of no use at all and unlearned behavior will again be favored. However, provided the environment retains some semblance of predictability, asocial learning will generally be of value. In contrast, here unlearned behavioral traits cannot respond appropriately to environmental fluctuations, while social learning is unreliable because it may lead individuals to acquire outdated or locally inappropriate behaviors. Therefore, social learning is favored at intermediate rates of change as individuals can acquire relevant information without bearing the costs of direct interaction with the environment associated with asocial learning, but with greater phenotypic flexibility (Fig. 2.1a). Within this window of environmental variability, vertical transmission of information (social learning from parents) is generally thought to be an adaptation to slower rates of change than horizontal transmission (social learning among unrelated individuals of the same cohort), and this can be regarded as a rule of thumb (Fig. 2.1b). The models cover the entire spectrum from observer learning immediately after demonstrator to long periods (e.g., up to a generation) intervening between observer and demonstrator learning.

In fact, the relationship between the pathway of information flow and the rate of environmental change is actually more subtle than the above rule of thumb might imply. A more precise specification of the findings of the theory would be that the temporal distance between the demonstrator's initial acquisition of the trait and the observer's learning from this is inversely related to the rate of environmental change. Frequently there is a whole generation between parents and offspring acquiring learned traits; consequently learning such skills from parents is only useful if the relevant aspect of the environment changes slowly relative to the generational time of the species concerned, so that it remains valid from one generation to the next. It is also commonly the case that individuals of a similar age acquire information at roughly the same time. Thus, if the environment changes more frequently than the generational times of the species, learning from individuals of the same cohort may be of value. However, it is important to note that instances of vertical transmission in which offspring learn from their parents shortly after the parents have themselves learned

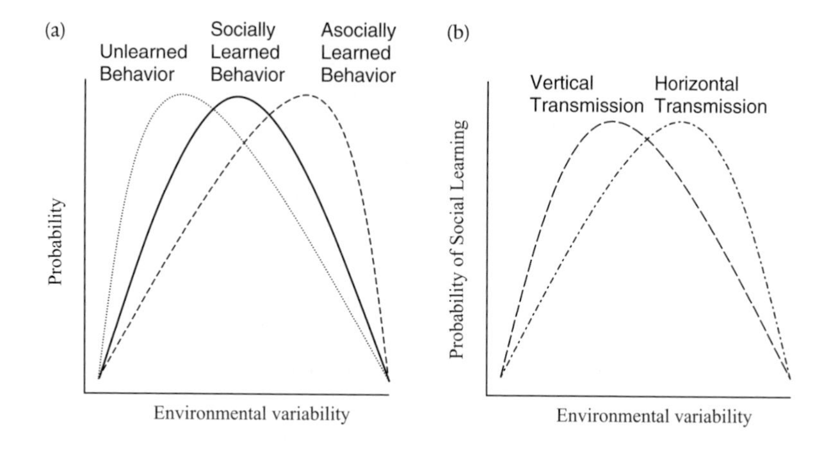

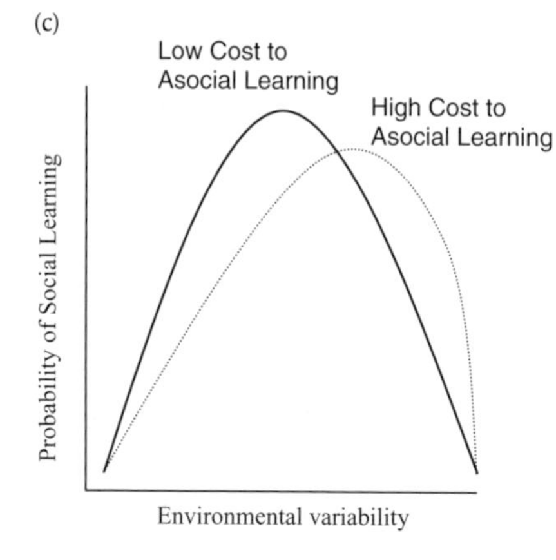

Fig. 2.1. A qualitative illustration of the kinds of circumstance under which social learning is expected (see text for details) as the probability of social learning with respect to spatial or temporal environmental variation. (a) Comparison with unlearned and asocially learned behavior; (b) comparison of vertical transmission (from parents) and horizontal transmission (from peers); and (c) when the cost of asocial learning is either low or high.

the trait might well be of utility in a rapidly changing environment. Likewise, a similarly aged observer may learn a behavior many years after its demonstrator.

In order to predict whether social or asocial learning is likely to flourish, we need to ask ourselves how similar would be the environment of the likely demonstrator and observer when each expresses the behavior? To the extent that it is similar, we should expect a greater reliance on social learning (or, perhaps, heightened sensitivity to others during learning), while where it is different we predict asocial learning. In general, the more environments change in space and time, the more likely it is that demonstrator and observer will experience different aspects of the environment. However, the utilization of social learning depends upon the type of trait under consideration. For instance, it may pay individuals to learn the location of a water source from parents if this water has been found at the same location for many years, but it may not pay to learn from parents the foraging range of a predator that annually moves its nest site. How to process an omnipresent food type can be learned even from an immigrating individual, but the processing of patchily distributed food types is perhaps best learned from a local.

Moreover, we also need to factor in the cost and probability of an individual acquiring the pertinent skill (Fig. 2.1c). For instance, we might expect more social learning of the identity of predators, which is likely to be extremely hazardous to learn asocially, than of the identity of prey, which commonly does not involve any direct hazard. This postulate receives empirical support among fishes (Brown and Laland, 2003). In contrast, we might expect more social learning of the location of a rare and patchily dispersed prey item, which would be difficult to find, than of the location of a common and homogeneously dispersed item, which would be easy to find. Although this is most likely if prey location were temporally invariant, in some cases we might anticipate a trade off between patterns of environmental variability and the cost of asocial learning. We might find individuals learn socially even in a very variable environment if the costs of learning asocially are high. In circumstances where socially transmitted information is likely to be reliable, we expect a broad range of conditions under which social learning will be utilized irrespective of the costs of asocial learning.

Although these findings are theoretically robust, they concern the efficacy of social learning over evolutionary time scales. Consequently, it is germane to ask whether contemporary populations of animals capable of social learning shift along the dimension from more to less reliance on social cues when learning, depending on the patterns of environmental variability that they experience in relevant resources. On the assumption that they do, a number of empirically testable predictions can be formulated from the above theory (see also Ch. 5).

Prediction 1. The probability that a trait is socially learned should increase with the probability that the demonstrator and observer experience similar aspects of the environment for which the trait is of utility.

Prediction 2. Skills and knowledge pertaining to resources that change their quality or location rapidly, or that are highly variable in their geographic distribution, should be less subject to social learning than skills and knowledge pertaining to resources that are comparatively constant in their quality, location, and distribution.

Prediction 3. Skills and knowledge that are relatively costly to acquire alone, where the cost could be measured in terms of time, energy, or hazard, should be more subject to social learning than skills and knowledge that can be acquired cheaply.

Prediction 4. Populations of animals capable of social learning might be predicted to increase their use of social learning when the pertinent environmental resources become less variable in their quality and location, and increase their use of asocial learning when those resources become more variable.

Note that whether the models predict that social learning will occur must be assessed on a trait-by-trait and not species-by-species basis. In other, words, these predictions should not be interpreted as implying that individuals of some species will be good social learners and others not. Rather the predictions suggest that individuals in some species will utilize social information to guide their learning in restricted circumstances and not in other circumstances, depending on the nature of the behavior and the variability in the relevant resources involved. Note too that if contemporary populations of animals are incapable of switching between social learning and asocial learning depending on the patterns of environmental variability that they experience in relevant resources, this does not invalidate the models in addressing events over evolutionary time scales.

To our knowledge, there have been only two empirical tests of these predictions. In the first, Wilkinson and Boughman (1999) used data from studies of evening bats, greater spear-nosed bats, and vampire bats to test the predictions of Laland *et al.*'s (1996) model. Wilkinson and Boughman used rates of following to food sites from a communal roost as an estimate of the degree to which social learning was employed by the bats to determine the location of food. They also used the number and variability of food items in the bats' diet to estimate the environmental variability and recorded the probability of successful feeding among the bats. This allowed them to assess whether the data conformed to the predictions of the model, with mixed results. The amount of following by the bats decreased with increasing environmental variability and increasing probability of successful feeding, consistent with predictions 1–3 above. However, Wilkinson and Boughman's (1999) estimates of the values of the parameters in the model put the bat populations in the region of the parameter space in which the model predicted no social learning (i.e., circumstances where bats should not have followed). This latter discrepancy may result either from a weakness in the model or from inaccuracy in the parameter estimates, and more innovative studies along the lines of Wilkinson and Boughman (1999) are required to determine which explanation is correction.

Given the comparatively embryonic nature of both the theory and the exercise of testing it, we might anticipate that qualitative predictions are more likely to be confirmed than precise quantitative model fitting. For instance, it is extremely difficult to quantify in any absolute sense the level of environmental variability in order to predict the occurrence of social learning, but any reliable measure of variability can be used to assess whether there is a qualitative relationship with the incidence of social learning. To determine whether a qualitative relationship holds, it matters little whether the measure of environmental variability is a complex multivariate formulae or a simple univariate character, for instance variability in number of food items or the density of prey or predators.

The prediction that social learning may be maladaptive (by which we mean that the social learning trait has lower relative fitness than an alternative) in an environment with rapid or sudden changes has been subject to controversy (Galef, 1995, 1996; Laland, 1996; Laland and Williams, 1998). Galef (1995, 1996) suggested that animals should not be expected to acquire outdated or inappropriate skills from knowledgeable individuals, even in a changeable environment, because the demonstrators will rapidly adjust their behavior according to patterns of reinforcement in the current environment. A debate has ensued on this issue, leading to an empirical test of the models (Laland and Williams, 1998), and a resulting clarification of the meaning of the term "maladaptive social learning" (see Table 2.1). In Laland and Williams' experiment, founder populations of fish were established that had been trained to take different routes to a food, and then the founders were gradually replaced with naïve animals in order to determine whether the route preferences remained. There were two routes to feed, with one route substantially longer than the other and associated with an energetic cost. The long route was designed to represent an environment in which the optimal route to feed had suddenly changed, with animals required to learn to track this change by switching preference. The experiment found that swimming with founders that had a prior preference for the long route slowed down the rate at which

Table 2.1. *Is social learning adaptive?*

Features of social learning	Adaptive nature
The capacity for social learning	This can be assumed to be adaptive in that it typically generates fitness-enhancing behavior; if it had not done so in the past it would not have evolved
Socially transmitted information	This may or may not be adaptive. While socially transmitted information is generally of utility, in environments that are spatially or temporally variable, individuals may acquire information from others that is outdated or inappropriate
The behavior influenced by social learning	Such behavior is typically adaptive in animals. While in humans socially learned behavior (e.g., use of contraception) may reduce absolute fitness (that is, the number of offspring an individual contributes to the breeding population), there are few, if any, known examples in nonhuman animals of socially learned behavior that reduce absolute fitness
Traditions	These may or may not be adaptive. In relatively slowly changing, and comparatively homogeneous environments, socially transmitted traditions will typically approach the local optimum. However, arbitrary or fitness-neutral behavior patterns may be maintained as traditions indefinitely, while, in environments that are spatially or temporally variable, traditions may not track environmental changes as effectively as individuals reliant on asocial learning. Animals may sometimes be locked into conventions in which deviations from the traditional behavior are penalized

subjects adjusted to the new patterns of reinforcement in their environment, relative to fish that swim alone. If this finding applies to natural populations of animals, where behavioral traditions lag behind environmental fluctuations, the lag may be greater for animals that aggregate and rely on social information than it would be for isolates. While the findings of this experiment support the theoretical predictions, they should be interpreted with caution. The experiment provide evidence for the social transmission of maladaptive information (i.e., take the long route), and for a suboptimal behavioral tradition (for taking the long route), but neither the behavior of the fish (where it pays to shoal for protection from predators) nor the general capacity for social learning (which is typically advantageous) should be described as maladaptive. Table 2.1 provides further clarification of these distinctions.

Giraldeau, Caraco and Valone (1994) developed a mathematical model to explore how the costs and benefits of social learning are affected by scrounging (acquiring food made accessible by others), motivated by observations of social foraging in pigeons. They concentrated on within-generation social learning of a trait that enhances resource production, assuming both frequency-dependent asocial learning (which decreases as a result of scrounging, as an animal that scrounges reduces its opportunity for learning through its own experiences) and frequency-dependent social learning (which increases with the number of demonstrators). The acquired trait results in an increased ability to find resource clumps, relative to a baseline rate. Giraldeau *et al.* (1994) found that social learning increased the expected number of individuals foraging at the elevated rate relative to asocial learning, and with no social learning there was a significant fitness cost to group foraging. They hypothesized that the adaptive function of social learning may be to allow individuals to circumvent some of the inhibitory effects that scrounging has on individual learning of a foraging skill. This is an example of an elegant piece of work in which theory and empirical findings have been neatly combined to produce insightful and sometimes counterintuitive results.

It is easy to conceive of other factors that might affect reliance on social learning and that might warrant both experimental and observational studies with animals. For instance, is social learning more likely when individuals are confronted with a complex rather than a simple problem? Is the likelihood of social learning increased if the problem or context is unfamiliar? Is social learning more likely in a threatening than in a benign environment? Is social learning about resources more or less likely in populations that compete for access for those resources? When should animals actively transmit information to favor learning in others? Is recruitment, or advertising of resource finds, likely to favor social learning? We would like to encourage researchers to collect such data so that they could then be incorporated into more sophisticated theoretical treatments.

One further kind of empirical test of the aforementioned theory would be particularly valuable. It would be of great interest to know to what extent the transmission or acquisition of socially learned information really does affect fitness. Any case study that could provide a direct measure of the reproductive advantage accrued by acquisition of a skill through social learning, or even an indirect measure of fitness, such as foraging success, would be enormously valuable. For example, Terkel's (1996) study of black rats showed that social learning enables them to survive in pine forests, while Beck and Galef (1989) provided evidence that social learning enables young Norway rats to survive in environments where protein is hard to find. This type of finding would not only allow the most direct testing of theoretical predictions to-date but also may facilitate the development of a new branch of theory employing a life-history approach, which may be particularly amenable to field testing (Sibly, 1999).

2.2.2 How do novel learned traits spread through populations?

A second class of models predicts the pattern of spread of novel traits as a result of social learning processes. Researchers have speculated as to whether the shape of the diffusion curve may reveal something about the learning processes involved. Most models predict that the diffusion of cultural traits will exhibit a sigmoidal pattern over time, with the trait initially increasing in frequency slowly, then going through a period of rapid spread, and finally tailing off (Boyd and Richerson, 1985; Cavalli-Sforza and Feldman, 1981). The reason this pattern is anticipated is that as the trait spreads the number of demonstrators increases (enhancing the opportunity for social learning in the remaining observing individuals), but the number of individuals left to learn decreases. Early and late on in the process, the opportunities for social learning are limited because there are too few demonstrators and then too few observers, respectively; however, growth is rapid during the intermediate stages.

Many researchers have been interested in whether it might be possible to "reverse-engineer" from the dynamics of the diffusion processes to infer about the processes responsible for them (see Ch. 10). Much discussion has centered on whether the shape of the diffusion curve may allow social and asocial learning processes to be distinguished. It would certainly be extremely useful if social learning of a trait within a particular ecological context carried with it a signature pattern of diffusion that could be easily distinguished, and this would throw a new light on field data for the spread of innovations, such as potato washing in macaques. Unfortunately, this discussion has been carried out from a position of almost total ignorance of what patterns might be predicted if particular learning processes are operating. Most strikingly, there has been little consideration given as to what kind of diffusion curve might be expected when exclusively asocial learning processes are in operating in a population. It would seem that many researchers have assumed that asocial learning would result in a linear, nonacceleratory, or at least nonsigmoidal, increase in frequency (Galef, 1991; Lefebvre, 1995a,b; Roper, 1986), which would allow it readily to be distinguished from a diffusion dependent on social learning. For instance, Roper (1986) concluded that a sigmoid curve "rules out" the possibility that animals are learning independently, while Galef (1991) used the nonacceleratory characteristic of the spread of sweet potato washing in macaques to argue that social learning is not involved. Unfortunately, it is easy to conceive of how asocial learning could generate a sigmoidal pattern: if there is a normal distribution of learning ability, then the trait would increase slowly initially as there are relatively few smart individuals, then rapidly as the majority of learners of average ability acquire the skill, and then the increase will fall off again when only the few really slow learners are left (Lefebvre, 1995a). The assumption that asocial learning will result in a linear increase over time ignores any normal residual variation in learning rates that might exist between individuals. Ironically, individual variability in learning has also been neglected in the models of social learning, which raises the possibility that social learning processes may not generate a sigmoidal diffusion after all. Better models are badly needed. In our judgment, reasoning as to the nature of the learning processes underlying the diffusion of an innovation on the basis of the shape of its diffusion curve is premature in the absence of a truly satisfactory body of theory that makes detailed predictions based on an extensive modeling of the relevant processes. The suggestions that asocial learning is likely to lead to linear increase over time and that only social learning can generate acceleratory or sigmoidal diffusion should now be regarded as discredited.

There have been various attempts to fit models to diffusion data, ranging from formal curve fitting to casual argument, based on both experimental and observational data collected in primates, birds, and fish. In general, the sigmoid prediction of the models is not supported, which could be interpreted to imply that the models are fine but that the animals are not learning socially; however, in our judgment, this is much more likely to reflect a weakness in the models, or the poverty of the data. The diffusion curves for learned traits in natural populations exhibit a diversity of patterns, including linear, exponential logarithmic, quadratic, and hyperbolic sine functions, as well as some sigmoid patterns (Laland *et al.*, 1996; Lefebvre, 1995a,b; Reader, 2000; Reader and Laland, 2000).

Reasons for these discrepancies probably include (a) the models are too simple, neglecting the effects of population's social structure and directed social learning (Coussi-Korbel and Fragaszy, 1995); (b) the models do not incorporate both social and asocial learning processes (Galef, 1995); and (c) the data are largely unsatisfactory. Models are required that incorporate factors such as kin subgroups, the effect of age, social rank and gender differences in information transmission, and competition for resources (Laland *et al.*, 1996; Lefebvre and Giraldeau, 1994; Reader, 2000). Reader (2000) showed that a hyperbolic sine function may have been the result of a sex difference in learning performance among fish. Here each within-sex population appeared to exhibit a sigmoidal pattern, but because the females on average were substantially faster at learning than the males, the curves were not aligned and combining the data resulted in a non-sigmoidal pattern. Consequently, it is conceivable that the data and the models can be reconciled when population structure and directed social learning are considered.

There is also a problem with the impoverished nature of the data. In virtually all cases, the diffusion curves are based on observations of a single population, over long periods of time, and where there are no clues as to whether social learning is operating. Reliable curves require replicate populations exposed to the same novel trait. Experimental studies may generate more useful data than isolated observations from the field, since established methods can be employed to determine if there is evidence for social learning, and because the same task can be presented to replicate populations, increasing the reliability of the findings. We would particularly welcome diffusion studies in which trained demonstrators are introduced into replicate populations with two alternative means of solving a problem. In addition, models require behavioral data that reflect or measure the population structure, such as affiliative or aggressive interactions and proximity between individuals. These approaches would generate data from which it is possible to establish whether social learning is taking place, and perhaps who is learning from whom, while at the same time generating reliable patterns of diffusion based on processes occurring in many populations, not just one.[1]

In spite of these problems, there are still grounds to be upbeat about the possibility of using diffusion data to interpret the underlying learning processes. Predictions based exclusively on the cumulative number of

[1] We are currently engaged in developing more sophisticated models that we hope can be used by researchers to elucidate the nature of the processes underlying the diffusion dynamics. We would welcome collaboration with persons with suitable datasets so that the models can be developed as a useful and practical deductive tool.

individuals that express a trait over time utilize only a subset of the available information, and methods that take account of the distribution of the trait in space and the relatedness and patterns of association of trait users are likely to be more powerful. We are investigating two approaches along these lines that may ultimately prove useful in distinguishing between behaviors that are learned with the aid of more or less social contribution. The first is the use of agent-based models to assess whether the spatial pattern of diffusion can provide information about the underlying learning processes. This approach allocates rules as to how individual agents behave and allows a population of agents to interact within a virtual environment. We anticipate that socially transmitted behavior will exhibit a more aggregated distribution and a weaker level of covariation with ecological distributions than asocially learned behavior. The second exploits the concept of directed social learning, examining whether the route of transmission can reveal information about the transmission processes. Although these analyses are ongoing, preliminary findings suggest that simple statistical methods may prove useful in many instances (see Section 2.3.1).

2.2.3 How does social learning affect the evolutionary process?

A great deal of theory has considered how social learning and tradition might affect evolution by generating a second system of inheritance, namely cultural inheritance, which can modify the selection pressures acting on genes. In most cases, these models assume the stable cultural transmission of information from one generation to the next over long periods of time. For this reason, most of this theory is probably unlikely to apply outside of hominids (see Feldman and Laland (1996) for a review). Possible exceptions are theoretical models of mate-choice copying (Kirkpatrick and Dugatkin, 1994; Laland, 1994), birdsong (Lachlan and Slater, 1999), and sexual imprinting (Laland, 1994). In all these cases, learning processes are predicted to generate stable selection pressures that favor natural or sexual selection, and key assumptions of the models and theoretical predictions are ripe for the testing. For instance, White and Galef (2000) have investigated mate-choice copying in quail and found that females that socially learn a preference for a particular male will generalize their preference to other males with the same plumage characteristics. This experiment confirmed the plausibility of a key premise of the theoretical models that assume that mate-choice copying could drive sexual selection.

Lachlan and Slater (1999) have developed a "cultural trap" hypothesis to explain why birdsong is learned, which they explore using a theoretical model. Their hypothesis is based on the idea that alleles that widen the "band width" over which songs are acquired by males and preferred by females are more likely to invade an avian population than alleles that narrow the band width. This is because when widening alleles are rare, mutant "wide" males will copy the songs of "narrow" males and, therefore, will be at no selective disadvantage relative to such narrow males, while wide females will mate with narrow males, and again be at no selective disadvantage relative to narrow females. However, when narrow alleles are rare, mutant narrow females will not recognize the songs of some wide males, and the narrow allele will be selected against. Although Lachlan and Slater's model was developed with birds in mind, its findings may generalize to aspects of the communication systems of other taxonomic groups to the extent that these assumptions are justified. The model assumes that male song preferences and female mating preferences are based on the same alleles, that males choose the most frequently heard song, that wide birds are no more likely than narrow birds to produce inappropriate songs or mate with heterospecifics, and that females do not prefer some recognized songs over others.

Several contributors to this volume (notably Fragaszy and Perry (Ch. 1) and, Russon (Ch. 12)) have placed emphasis on how social learning facilitates niche construction, that is, the ability of organisms to choose, regulate, construct, and destroy important components of their environments, in the process changing the selection pressures to which they and other organisms are exposed (Laland, Odling-Smee, and Feldman, 2000; Odling-Smee, 1988; Odling-Smee, Laland, and Feldman, 1996). In species capable of social learning, tradition has greatly amplified the capacity for niche construction and the ability to modify selection pressures. Laland, Odling-Smee, and Feldman (2001) used gene–culture co-evolutionary models to explore the evolutionary consequences of culturally generated niche construction throughout hominid evolution. The analyses demonstrated that socially learned niche construction will commonly generate counter-selection that compensates for, or counteracts, a natural selection pressure in the environment (such as building a shelter to damp out temperature extremes, or storing food to compensate for seasonal fluctuations). A reasonable inference from such findings would be that competent niche constructors should be more resistant to genetic evolution in response to autonomously changing environments than less-able niche constructors. As social learning enhances the capacity of animals to alter their niches, it would seem plausible to infer that the niche construction of able social learners will be more flexible than that of other animals. This theory has been used to develop a number of predictions about human evolution. For instance, Laland *et al.* (2000, 2001) expect able social learners to show less of an evolutionary response in morphology to fluctuating climates than other animals, assuming that the latter must have been less well equipped than the former to invest in counteractive niche construction. Similarly, they expect more technologically advanced animals to exhibit less of a response to climates than less technologically advanced animals. Bergmann's and Allen's rules (Gaston, Blackburn, and Spicer, 1998) suggest that populations in warmer climates will be smaller bodied and have bigger extremities than those in cooler climates. Able social learners should show less correspondence to these rules than other animals. More generally, if sophisticated social learners have evolved more in response to self-constructed selection pressures than other animals, and less in response to selection pressures that stem from independent factors in their environment, then such populations may have become increasingly divorced from local ecological pressures. Related predictions can be made concerning the relationship between social learning, range, and dispersal (Odling-Smee, Laland, and Feldman, 2003).

As well as constructing a more stable environment, socially acquired niche-constructing behavior can also generate environmental variation. A theoretical analysis by Kendal (2002) considered the diffusion of a socially learned foraging behavior causing variation in the presence of a novel biotic resource, such as a plant species. Foraging depletes the resource, limiting further demonstration of the behavior and, therefore, the diffusion of the information. However, the resource can regrow at a rate that is dependent upon the frequency of individuals that are not performing the behavior (i.e. nonconsumers). Kendal found that such niche construction could result in individuals that have learned the behavior but are unable to perform it because they have caused resource depletion. Less intuitively, if foraging upon the resource confers a selective advantage, reflected by an increase in the birth rate, there are conditions under which the increase in informed individuals in the population can actually reduce the proportion of the population performing the behavior. By monitoring the prevalence of socially learned foraging information and behavior through the population and the prevalence of the resource, it should be possible to test the influence of this "destructive" or "negative" niche-constructing behavior upon its own rate of transmission and the associated fitness consequences.

2.3 Tractable but, as yet, neglected questions about social learning where theory and empirical work could be usefully integrated

2.3.1 How can social learning be established in the field?

Field conditions, and the nature of the data that field studies generate, do not always lend themselves to drawing clear inferences about whether particular behavior patterns are socially learned. As a consequence, intra- and interpopulation differences in behavioral repertoires resembling distinct socially transmitted traditions are frequently vulnerable to alternative "kill-joy" explanations. The primary alternative accounts are that (a) asocial learning in response to differing ecological patterns or (b) genetic differences underlie and explain much of the variation in behavior. While it is difficult to exclude these alternative explanations in absolute terms, appropriate data collection would allow the feasibility of these alternatives to be assessed and to be rejected if the probability that they can account for the data is unrealistically small.

For illustration, consider the hypothetical example of the diffusion of a novel behavior pattern depicted in Fig. 2.2. The two incidents where offspring acquire the behavior prior to their parents, and eight occasions where offspring have acquired a behavior that neither of their parents have exhibited, renders a genetic account implausible (although genetically based developmental plasticity could possibly cause the offspring to appear to exhibit the behavior before the parent). Moreover, the strong associations between the coefficient of association and time of learning ($r = 0.61$; $p = 0.046$) or (in cases where relatedness is a more tractable approximation of association) the degree of relatedness and time of learning ($r = 0.805$; $p < 0.001$) (see Fig. 2.2) are highly unlikely to arise though asocial learning.[2] In general, a high concordance between the behavior of parent and offspring might be interpreted as inconsistent with an asocial learning explanation and consistent with vertical social transmission. Here,

[2]If close associates experience similar environments and engage in the same learning as a direct result of their association, we regard this as social rather than asocial learning. However, convergent asocial learning may arise in situations where only a subsection of the population is exposed to relevant resources and hence the statistics will only be applicable in situations when all individuals are exposed to aspects of the environment that afford learning.

the failure to find a significant relationship between the behavior of parents and offspring ($r^2 = 0.98$; degrees of freedom (df) = 1, $p > 0.05$; see Fig. 2.2) suggests that a parental influence on learning does not explain the diffusion, and that horizontal or oblique processes are more important. Therefore, simply by having access to good pedigree and diffusion data, alternatives to the social learning explanation can be dismissed. Of course, genuine data will rarely be as clear cut as depicted in Fig. 2.2, but nonetheless there are likely to be occasions where such methods can be employed (see for instance Ch. 9). Experimental data from laboratory or captive studies estimating the probability of asocial learning could also be used to assess the probability that a particular pattern of diffusion or level of incidence is explicable in terms of asocial processes. For instance, if animals produce the novel behavior through asocial processes with probability 0.1, then the likelihood of finding 12 individuals in a population of 30 exhibiting the behavior as a consequence of asocial processes is estimated to be vanishingly small ($r^2 = 30$; df = 1; $p < 0.0001$; see Fig. 2.2). As the probability of asocial learning will depend on the time frame in which isolated animals are tested, we suggest that the researchers would be well advised to err on the side of longer rather than shorter time frames, such that subsequent estimates in populations are conservative. We are currently undertaking a more detailed analysis designed to establish more powerful statistical methods for distinguishing between patterns of incidence resulting from genetic inheritance, asocial learning, and social transmission.

2.3.2 Which processes of social interaction facilitate and which impede diffusion?

A feature of recent empirical work on animal social learning is the observation that learned information can be directed through populations, with diffusion dependent on the social rank, gender, age, or size of demonstrator and observer (Coussi-Korbel and Fragaszy, 1995). There is a need for models that take account of this directed social learning and population structure. The methods for developing such models are well established (Cavalli-Sforza and Feldman, 1981), but complex. As mentioned above, such models would be valuable from the perspective of making sense of patterns of diffusion. However, the models could serve other functions, for instance delineating the pathways by which information and skills spread through animal populations and predicting which variables are most likely to affect the diffusion process. Empirical scientists could

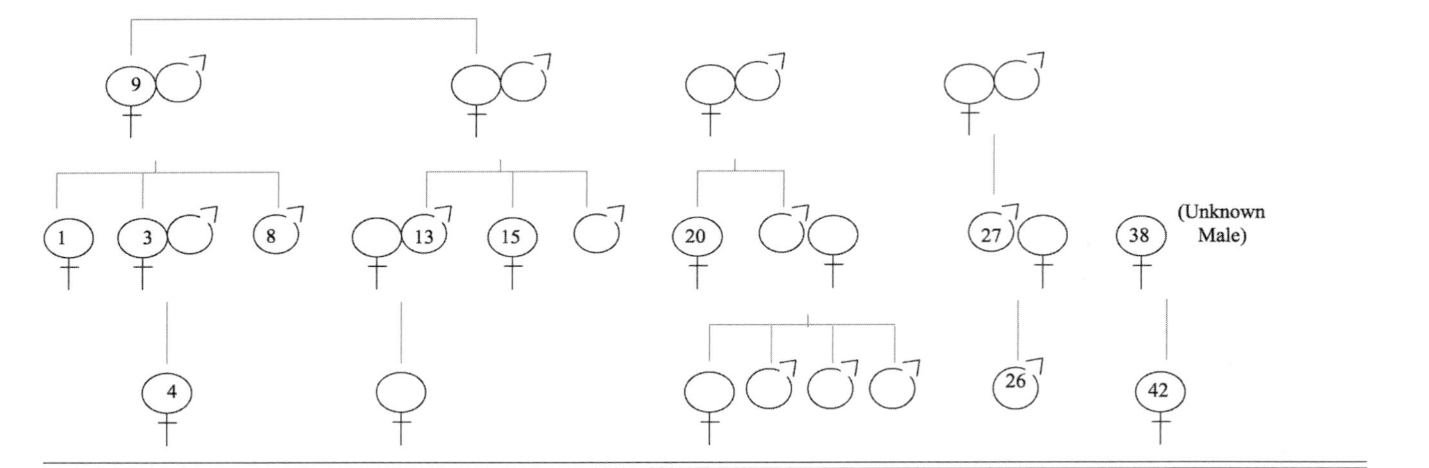

	Positions in diffusion chain										
	1–2	2–3	3–4	4–5	5–6	6–7	7–8	8–9	9–10	10–11	11–12
Time taken for diffusion of behavior	2	1	4	1	4	2	5	6	1	12	4
Degree of association	0.5	0.7	0.2	0.3	0.1	0.4	0.1	0.1	0.8	0.1	0.7
Coefficient of relatedness	0.5	0.5	0.25	0.5	0.25	0.5	0	0	0.5	0	0.5

Fig. 2.2. A hypothetical dataset, where the numbers in the pedigree indicate the days on which animals were first observed performing the novel behavior. Read down each column for the mesurements partaining to the individual that acquired the behavior on a particular day and the previous individual to have acquired the behavior. For instance, on day 1, the female on the extreme left of the figure was first observed performing the behavior, followed by her sister 2 days later. These two individuals occupy positions 1–2 in the diffusion chain. The time taken for the diffusion (i.e. for the trait to spread from the first to the second individual) is 2 days. These two individuals had been observed to have spent approximately 50% of their time together (degree of association, 0.5) and share 50% of their genes (coefficient of relatedness, 0.5). The dataset is used as follows to calculate the simple statistics referred to in the text. Directed social learning statistics: a simple regression of the time taken for diffusion of behavior upon the degree of association ($r = 0.61$; $p = 0.045$) and coefficient of relatedness ($r = 0.805$; $p < 0.001$), respectively. Vertical transmission statistic: a chi-squared test comparing observed numbers of offspring with the behavior whose parents have (six out of eight) and have not (four out of ten) performed the behavior, with the respective expected number of offspring performing the behavior (0.55×8 and 0.55×10 respectively), calculated from the overall mean fraction of offspring that have performed the behavior ($(6+4)/(8+10) = 0.55$). Asocial/social learning statistic: given that in a population of 30 animals, each animal has an independent asocial learning probability of 0.1, a chi-squared test comparing the observed incidence of animals that have and have not performed the behavior against the expected

contribute by providing data on the probability of information transmission between classes of individual.

Where animal populations have a demic structure, it would be useful to utilize and develop further models for the transmission of socially learned traits within and between populations (e.g., Cavalli-Sforza and Feldman, 1981; stepping stone models). This would allow us to explore whether the migration of individuals or the diffusion of ideas is responsible when a learned trait spreads from one population to the next. Field researchers could contribute by collating data on levels of migration, as well as sex and age differences in emigration. For instance, where there is directed social learning, if one sex is more effective at transmitting or receiving information than the other, it may make a big difference to the diffusion process if the species concerned is patrilocal or matrilocal.

2.3.3 Who are the innovators?

When a novel behavior spreads through an animal population by social learning, frequently one individual (the innovator) will have started off the process. The question of which individuals innovate to solve new problems or invent new behavior patterns is now beginning to receive some attention (Kummer and Goodall, 1985; Laland and Reader, 1999a,b; Lee, 1991; Reader and Laland, 2003). It would seem that the adage *necessity is the mother of invention* is not inappropriate. Observations of primates, birds, and fish suggest that innovators are frequently individuals of low dominance status, small size, or poor competitive ability, for whom the established risk-averse strategies are not productive and who are driven to innovate out of hunger or a lack of success in some other domain (Kummer and Goodall, 1985; Laland and Reader, 1999b; Reader, 2000). As innovation would appear to depend more on economics than genius, the findings raise the possibility that it may be possible to extend optimal foraging or state-dependent models to carry out a cost–benefit analysis of when it should pay an individual to innovate. Dewar (Ch. 5) has carried out a similar analysis to determine when it should pay an individual to attend to social aids to learning (see also Chs. 10 and 11).

2.3.4 Are the differences between the behavioral traditions of distinct populations independent of ecological constraints?

With reliable continuous data on the incidence of both relevant ecological variables and purported cultural traits within and between populations, it should be possible to use statistical analyses such as multiple regression to determine the proportion of the variance for a given trait that can be attributed to environmental factors and that which can be attributed to cultural history. This type of study has already been carried out in human populations. For instance, Guglielmino *et al.* (1995) examined variation in cultural traits among 277 contemporary African societies and found that most traits examined correlated with cultural history rather than with ecology. Such findings suggest that most human behavioral traits are maintained in populations as distinct cultural traditions rather than being evoked by the natural environment. We suggest that a similar study could be performed on nonhuman traditions.

2.3.5 Is social learning an adaptive specialization?

Lefebvre and Giraldeau (1996) have presented an excellent account on how this question may be addressed empirically. This could readily be complemented by a theoretical analysis. By collecting data on the incidence of social learning in each species across a broad taxonomic group (e.g., nonhuman primates) and plotting this against pertinent variables (e.g., group size, diet, etc.) using the relevant comparative techniques (Harvey and Pagel, 1991), it should be possible to test whether social learning is associated with particular ecological or demographic variables (see Ch. 3 for an illustration of this method).

2.4 Conclusions

There are rich opportunities for theoreticians to develop models of relevance and utility to empirical scientists in the field of animal social learning and traditions. There is also a need for researchers to collect data that can inform the development of theoretical models by testing their assumptions and predictions. We have tried to outline how these two approaches can be further integrated and hope that other researchers will take up our call for fruitful cooperation.

2.5 Acknowledgements

The research described in this chapter was supported by a Royal Society University Research Fellowship to Kevin Laland and a BBSRC studentship to Jeremy Kendal. We are grateful to Jenny Boughman, Dorothy Fragaszy, Jeff Galef, Susan Perry, Simon Reader, and the participants of the *Traditions in Nonhuman Primates* conference for helpful comments and suggestions.

References

Aoki, K. and Feldman, M. W. 1987. Toward a theory for the evolution of cultural communication: coevolution of signal transmission and reception *Proceedings of the National Academy of Sciences, USA*, **84**, 7164–7168.

Beck M. and Galef, B. G., Jr. 1989. Social influences on the selection of a protein-sufficient diet by Norway rates (Rattus norvegicus). *Journal of Comparative Psychology*, **103**, 132–139.

Bergman, A. and Feldman, M. W. 1995. On the evolution of learning: representation of a stochastic environment. *Theoretical Population Biology*, **48**, 251–276.

Boyd, R. and Richerson, P. J. 1985. *Culture and the Evolutionary Process*. Chicago, IL: University of Chicago Press.

Boyd, R. and Richerson, P. J. 1988. An evolutionary model of social learning: the effects of spatial and temporal variation. In *Social Learning: Psychological and Biological Perspectives*, ed. T. R. Zentall and B. G. Galef, Jr., pp. 29–48. Hillsdale, NY: Erlbaum.

Brown, C. and Laland, K. N. 2003. Social learning in fishes: a review. [*Learning in Fish*, ed. K. N. Laland, C. Brown and J. Krause] *Fish and Fisheries*, special edition.

Cavalli-Sforza, L. L. and Feldman, M. W. 1981. *Cultural Transmission and Evolution: A Quantitative Approach*. Princeton, NJ: Princeton University Press.

Cavalli-Sforza, L. L. and Feldman, M. W. 1983. Cultural versus genetic adaptation. *Proceedings of the National Academy of Sciences, USA*, **79**, 1331–1335.

Coussi-Korbel, S. and Fragaszy, D. M. 1995. On the relation between social dynamics and social learning. *Animal Behaviour*, **50**, 1441–1453.

Feldman M. W., and Laland, K. N. 1996. Gene–culture coevolutionary theory. *Trends in Ecology and Evolution*, **11**, 453–457.

Feldman, M. W., Aoki, K., and Kumm, J. 1996. Individual versus social learning: evolutionary analysis in a fluctuating environment. *Anthropological Science*, **104**, 209–232.

Galef, B. G., Jr. 1991. Tradition in animals: field observations and laboratory analyses. In *Interpretation and Explanation in the Study of Behaviour,* Vol. 1: *Interpretation, Intentioinality and Communication*, ed. M. Bekoff and D. Jamieson, pp. 74–95. Boulder, CO: Westview Press.

Galef, B. G., Jr. 1995. Why behaviour patterns that animals learn socially are locally adaptive. *Animal Behaviour*, **49**, 1325–1334.

Galef, B. G., Jr. 1996. The adaptive value of social learning: a reply to Laland. *Animal Behaviour*, **52**, 641–644.

Gaston, K. J., Blackburn, T. M., and Spicer, J. I. 1998. Rapoport's rule: time for an epitaph? *Trends in Ecology and Evolution,* **13**, 70–74.

Giraldeau, L.-A., Caraco, T., and Valone, T. J. 1994. Social foraging: individual learning and cultural transmission of innovations. *Behavioural Ecology,* **5**, 35–43.

Guglielmino, C. R., Viganotti, C., Hewlett, B., and Cavall-Sforza, L. L. 1995. Cultural variation in Africa: role of mechanisms of transmission and adaptation. *Proceedings of the National Academy of Sciences, USA*, **92**, 7585–7589.

Harvey, P. H. and Pagel, M. D. 1991. *The Comparative Method in Evolutionary Biology*. Oxford University Press.

Kendal J. R. 2002. An investigation into social learning: mechanisms, diffusion dynamics, functions and evolutionary consequences. PhD Thesis, University of Cambridge, UK.

Kirkpatrick, M. and Dugatkin, L. A. 1994. Sexual selection and the evolutionary effects of copying mate choice. *Behavioral Ecology and Sociobiololgy*, **34**, 443–44.

Kummer, H. and Goodall, J. 1985. Conditions of innovative behaviour in primates. *Philosophical Transactions of the Royal Society of London, Series B*, **308**, 203–214.

Lachlan R. F. and Slater, P. J. B. 1999. The maintenance of vocal learning by gene-culture interaction: the cultural trap hypothesis. *Proceedings of the Royal Society of London, Series B*, **266**, 701–706.

Laland K. N. 1994. Sexual selection with a culturally transmitted mating preference. *Theoretical Population Biology*, **45**, 1–15.

Laland K. N. 1996. Is social learning always locally adaptive? *Animal Behaviour*, **52**, 637–640.

Laland K. N. and Reader, S. 1999a. Foraging innovation in the guppy. *Animal Behaviour*, **52**, 331–340.

Laland K. N. and Reader, S. 1999b. Foraging innovation is inversely related to competitive ability in male but not female guppies. *Behavioral Ecology*, **10**, 270–274.

Laland, K. N. and Williams, K. 1998. Social transmission of maladaptive information in the guppy. *Behavioral Ecology*, **9**, 493–499.

Laland, K. N., Richerson, P. J., and Boyd, R. 1993. Animal social learning: towards a new theoretical approach. *Perspectives in Ethology*, **10**, 249–277.

Laland, K. N., Richerson, P. J., and Boyd, R. 1996. Developing a theory of animal social learning. In *Social Learning in Animals: The Roots of Culture*, ed. C. M. Heyes, and B. G. Galef Jr., pp. 129–154. New York: Academic Press.

Laland, K. N., Odling-Smee, J., and Feldman, M. W. 2000. Niche construction, biological evolution and cultural change. *Behavioural and Brain Sciences*, **23**, 131–175.

Laland, K. N., Odling-Smee, J., and Feldman, M. W. 2001. Cultural niche construction and human evolution. *Journal of Evolutionary Biology*, **14**, 22–33.

Lee, P. 1991. Adaptations to environmental change: an evolutionary perspective. In *Primate Responses to Environmental Change*, ed. H.O. Box, pp. 39–56. London: Chapman & Hall.

Lefebvre, L. 1995a. Culturally transmitted feeding behaviour in primates: evidence for accelerating learning rates. *Primates*, **36**, 227–239.

Lefebvre, L. 1995b. The opening of milk-bottles by birds: evidence for accelerating learning rates, but against the wave-of-advance model of cultural transmission. *Behavioural Processes*, **34**, 43–53.

Lefebvre, L. and Giraldeau, L. A. 1994. Cultural transmission in pigeons is affected by the number of tutors and bystanders present. *Animal Behaviour*, **47**, 331–337.

Lefebvre, L. and Giraldeau, L. A. 1996. Is social learning an adaptive specialization? In *Social Learning in Animals: The Roots of Culture*, ed. T. Zentall and B. G. Galef Jr., pp. 141–163. Hillsdale, NY: Erlbaum.

Odling-Smee, F. J. 1988. Niche constructing phenotypes. In *The Role of Behaviour in Evolution*, ed H. C. Plotkin, pp. 73–132. Cambridge, MA: MIT Press.

Odling-Smee, F. J., Laland, K. N., and Feldman, M. W. 1996. Niche construction. *American Naturalist*, **147**, 641–648.

Odling-Smee, F. J., Laland, K. N., and Feldman, M. W. 2003. *Niche Construction: The Neglected Process in Evolution*. Princeton, NJ: Princeton University Press.

Reader S. M. 2000. Social learning and innovation: individual differences, diffusion dynamics and evolutionary issues. PhD Thesis, University of Cambridge, UK.

Reader, S. M. and Laland, K. N. 2000. Diffusion of foraging innovation in the guppy. *Animal Behaviour*, **60**, 175–180.

Reader, S. M. and Laland, K. N. 2003. *Animal Innovation*. Oxford: Oxford University Press.

Rogers, A. 1988. Does biology constrain culture? *American Anthropologist*, **90**, 819–831.

Roper, T. J. 1986. Cultural evolution of feeding behaviour in animals. *Science Progress*, **70**, 571–583.

Sibley, R. M. 1999. Evolutionary biology of skill and information transfer. In *Mammalian Social Learning: Comparative and Ecological Perspectives*, ed. H. O. Box and K. R. Gibson, pp. 57–71. Cambridge: Cambridge University Press.

Stephens, D. 1991. Change, regularity and value in the evolution of learning. *Behavioral Ecology*, **2**, 77–89.

Terkel, J. 1996. Cultural transmission of feeding behavior in the black rat (*Rattus rattus*). In *Social Learning in Animals: The Roots of Culture*, ed. C. M. Heyes and B. G. Galef Jr., pp. 267–286. San Diego, CA: Academic Press.

White, D. J. and Galef, B. G., Jr. 2000. Culture in quail: social influences on mate choices of female *Coturnix japonica*. *Animal Behaviour*, **59**, 975–979.

Whiten, A., Goodall, J., McGrew, W. C., Nishida, T., Reynolds, V., Sugiyama, T., Tutin, C. E. G., Wrangham, R. W. W., and Boesch, C. 1999. Cultures in chimpanzees. *Nature*, **399**, 682–685.

Wilkinson, G. S. and Boughman, J. W. 1999. Social influences on foraging in bats. In *Mammalian Social Learning: Comparative and Ecological Perspectives*, ed. H. O. Box and K. R. Gibson, pp. 188–204. Cambridge: Cambridge University Press.

SIMON M. READER

3

Relative brain size and the distribution of innovation and social learning across the nonhuman primates

> The history of comparative learning could simply be classified as disappointing. The comparative psychologist often appears to know little more than the grade school child who would rather have a pet dog than bird, or bird than fish, or fish than worm, simply because they make better friends, as they can be taught more. This state of affairs did not arise without considerable effort.
>
> RIDDELL, 1979, P. 95

3.1 Introduction

Ecology and "intelligence" are two commonly invoked explanations for species differences in the reliance on socially learned traditions, yet we know little about how social learning evolved. Here, I examine hypotheses for the evolution and evolutionary consequences of social learning and detail possible routes to address these ideas. I will test social and ecological hypotheses for primate brain evolution to illustrate possible approaches to the study of traditions. This chapter explores cognitive, ecological, and life-history variables that may accompany a propensity for social learning, specifically, the roles of brain size and social group size. I also examine the distribution of innovations and tool use across the nonhuman primates, to determine how these aspects of behavioral plasticity are associated with social learning and to explore the relationship between asocial and social learning. Such analyses can provide important clues as to whether we can sensibly talk about the "evolution of traditions", or whether an increased reliance on social learning is simply a by-product of selection for generalized learning abilities.

3.1.1 Innovation, cultural transmission, and brain size

Links between cultural transmission, innovation, brain size, and the rate of genetic evolution have been proposed by several authors (Lefebvre *et al.*, 1997; Wilson, 1985, 1991; Wyles, Kunkel, and Wilson, 1983). Cultural transmission refers to socially learned behavior patterns, whereas behavioral innovation is the expression of a new skill in a particular individual, leading it to exploit the environment in a new way (Wyles *et al.*, 1983). Wilson (1985) developed the "behavioral drive" hypothesis, arguing that episodes of innovation and cultural transmission are more frequent in large-brained species, exposing these species to novel selective pressures and so increasing the rate of evolution in these taxa: "By suddenly exploiting the environment in a new way, a big-brained species quickly subjects itself to new selection pressures that foster the fixation of mutations complementary to the new habit" (Wilson, 1985, p. 156). Wilson thus assumed that an extensive reliance on innovation and social learning would require a large brain, and though he is not explicit on exactly why this should be the case, it seems likely he believes that innovation and social learning typically require complex cognitive processing that can only be accommodated by increases in brain size (Wilson, 1991). Let us take milk bottle opening in British birds as an example to illustrate the behavioral drive hypothesis. Milk bottle opening spread across mainland Britain and Ireland through a combination of independent innovation events and social learning processes (Fisher and Hinde, 1949; Hinde and Fisher, 1951; Lefebvre, 1995; Sherry and Galef, 1984, 1990). We could imagine that the birds were thus exposed to the novel selection pressure of digesting cream, which could have affected their subsequent evolution. In humans, the link between the cultural trait of dairy farming and expression of the gene for lactase in adults has been well established (Feldman and Laland, 1996; Holden and Mace, 1997), which supports the view that behavior can influence the course of evolution. The Baldwin effect (Baldwin, 1896; Bateson, 1988; Plotkin, 1994) is a mechanism that may also accelerate evolutionary rates in species exposing themselves to novel selection pressures (Hinton and Nolan, 1980; but see Ancel, 1999) and thus is similar in spirit to the behavioral drive hypothesis. Bateson (1988) describes other examples of the active role of behavior in evolution. However, Laland (1992 and Ch. 2) details theoretical models that show social transmission may slow evolutionary rates as well as speed them up through changing selection pressures.

The behavioral drive hypothesis is of particular interest to us here since it (a) predicts a link between brain size and social learning and (b) suggests that one consequence of an increased reliance on traditions and social learning could be changes in evolutionary rate. This second element of the hypothesis has received limited support from Wyles *et al.* (1983), who provided examples of high rates of anatomical evolution in taxonomic groups of large relative brain size (such as songbirds and *Homo sapiens*). However, it is the first element of the hypothesis that is relevant to our aim here, that is to test explanations for the distribution of social learning and innovation between taxa. This prediction that species with larger volumes of the relevant brain structures will show greater behavioral flexibility is widely held (Lefebvre *et al.*, 1997) but currently contentious. For example, Byrne and Whiten (1992, p. 609) state, "It is still a matter of dispute whether relative brain size is predictive of intelligence even in extant animals", and Byrne (1993, p. 696) laments, "We cannot yet even claim that having a larger brain gives a primate greater intelligence of any kind".

There are serious difficulties in making comparative estimates of learning and cognition. Comparative experimental studies require a test fair to all species, yet species differ widely in their reliance on different sensory modalities, in their neophobia, in their response to humans, and in innumerable other characteristics that make construction of a fair test highly problematic (Byrne, 1992; Deaner, Nunn, and van Schaik, 2000; Essock-Vitale and Seyfarth, 1987; Gibson, 1999; Lefebvre and Giraldeau, 1996). Hence the results of comparative cognitive tests are hard to interpret. Comparative tests also require the testing of a large number of species, which is difficult using traditional tests of learning ability. A novel approach is needed.

Given these problems in interpreting comparative tests of learning and cognition, Lefebvre *et al.* (1997) recently suggested, after Wyles *et al.* (1983), that behavioral innovation could be an alternative measure of behavioral plasticity. Behavioral plasticity, the capacity to modify behavior, is a type of phenotypic plasticity, that is "changeability" or the capacity of a particular genotype to produce a different phenotype in response to a change in the environment (Bateson, 1983; Bateson and Martin, 1999; Schlichting and Pigliucci, 1998). Both innovation and social learning will be components of behavioral plasticity, since both allow modification of the behavioral repertoire. Lefebvre *et al.* (1997) estimated innovation frequencies in birds by collecting published reports of opportunistic foraging innovations. Relative innovation frequency correlated with relative forebrain size, specifically the size of the hyperstriatum ventrale, the avian equivalent of the mammalian neocortex (Timmermans *et al.*, 2000). Lefebvre and coworkers thus confirmed the predicted link between innovation and brain size (Wilson, 1985; Wyles *et al.*, 1983). Sol and Lefebvre (Sol and Lefebvre, 2000; Sol, Lefebvre, and Timmermans, 2002) continued this work, demonstrating a link between forebrain size, foraging innovation frequency, and the invasion success of introduced birds, which supports the hypothesis of Lee (1991) that behavioral flexibility may radically affect the survival chances of animals under new conditions. Lefebvre *et al.* (1997, pp. 557–558) make a good case for using the innovation measure:

> Animals are judged not on their relative performance on an anthropocentric test, but on opportunistic departures from their species norm. They are not forced to perform in a captive situation that is often artificial and aversive for them, but spontaneously demonstrate changes in whatever foraging situations are relevant to them in the field. The estimate of plasticity . . . is objective, exhaustive, quantitative, ecologically relevant, non-anthropocentric [and provides] large scale data on many species.

We can make use of this methodology to examine species differences in social learning propensities. I conducted an analysis of the nonhuman primate literature to examine further links between behavioral flexibility and brain evolution, and to incorporate the role of social learning. Gibson (1999, p. 353) has reported a need for such studies, noting that, "Unfortunately, no studies have attempted to determine whether EQ [a measure of relative brain size] or absolute brain size correlates with any measures of social learning". I collected observations of social learning and tool use, as well as innovation, and, unlike Lefebvre *et al.* (1997), collected examples from all behavioral contexts rather than just foraging behaviors. I also corrected for phylogeny and the research effort into each species. However, while it seems reasonable to utilize published reports of innovation or tool use to estimate species differences in these behavior patterns, the use of social learning reports is more controversial. As several authors note in this volume (see Chs. 1, 4 and 6), social learning will often be an inferred process rather than an established fact if there is no supporting experimental evidence. If researchers' willingness to classify an observed behavior pattern as socially learned happens to correlate with some other variable of interest, such as brain size, then there is the risk of finding erroneous relationships. What, then, to do? The social learning measure used here is not perfect but is the best measure available at present, and I hope by demonstrating the utility of such estimates I will prompt the development of new, improved methods.

There are several reasons why nonhuman primates are a particularly suitable group with which to test hypotheses concerning the distribution of traditions amongst mammalian species. First, there is a large behavioral literature on this order, with a number of journals specializing in primate research. Second, the primate phylogeny is reasonably well resolved: Purvis (1995) has constructed a composite phylogeny made up of 112 previously published trees. Third, data on brain size and the volumes of various brain structures such as the neocortex and striatum are available (Stephan, Frahm, and Baron, 1981; Zilles and Rehkamper, 1988). Quantitative data on the volume of brain structures, rather than total brain size or cranial capacity, are important to the analyses for the reasons described below. Fourth, primate species show great variation in diet and social structure (Dixson, 1998), and so primatologists argue that they are particularly suitable for comparative studies of "intelligence" (Byrne, 1992).

3.1.2 Measures of brain size and intelligence

3.1.2.1 Do brains evolve as unitary structures?

What is the best measure of brain size? Several authors maintain it is the areas of the brain involved in the behaviors of interest that should be examined, rather than total brain size, since the brain has not evolved as a unitary structure (Barton and Harvey, 2000; Harvey and Krebs, 1990; Keverne, Martel, and Nevison, 1996; Purvis, 1992). Barton, an advocate of examining individual neural systems and their response to specific ecological demands, has demonstrated relationships in primates between neocortex size and frugivory and neocortex size and social group size, and has also demonstrated trade offs between visual and olfactory processing structures in primates and insectivores (Barton, 1993; Barton and Dunbar, 1997; Barton and Purvis, 1994; Barton, Purvis, and Harvey, 1995). Keverne *et al.* (1996) show trade offs between the "executive" (neocortex, striatum) and "emotional" (hypothalamus, septum) brain in primates, and food-storing birds are known to have enlarged hippocampi but similar overall brain sizes to nonstoring birds (Harvey and Krebs, 1990; Krebs *et al.*, 1989). The neocortex has received attention from researchers studying the social intelligence hypothesis (Byrne, 1993; Dunbar, 1993a; Sawaguchi and Kudo, 1990) and the role of the complexity of the ecological niche in brain evolution (Jolicoeur *et al.*, 1984). The neural processing underlying innovation and social learning is likely to reside in the neocortex and striatum, or "executive" brain (Jolicoeur *et al.*, 1984; Keverne *et al.*, 1996). It is the role of this brain region that is examined in detail here.

3.1.2.2 Brain size and body size

Large-bodied species tend to have large brains (Byrne, 1992), so it would seem important to correct for body size. However, there is controversy over whether absolute or relative measures are more appropriate and, if relative measures are used, the best method of accounting for body size (Barton, 1999). Well-known measures of relative brain size include the encephalization quotient (EQ), calculated as the residuals from a graph of the logarithm (log) of brain size against log body size (Dunbar, 1993b; Jerison, 1973) and the progression index, the ratio of neocortex or brain size to that predicted, for example, for a basal insectivore of same body size as the species of interest (Jolicoeur *et al.*, 1984). Sawaguchi and Kudo (1990) assessed the relative size of the neocortex in a given congeneric group as the difference between actual neocortical volume and the volume expected from an allometric relationship between neocortical volume and the volume of the rest of the brain. Keverne *et al.* (1996) examined the executive brain (neocortex and striatum) and the emotional brain (hypothalamus and septum), regressing these brain volumes on the brainstem and taking residuals to examine variation independent of total brain size. Lefebvre *et al.* (1997) took two measures of relative forebrain size in birds, forebrain mass divided by the brainstem mass of a galliform (the assumed primitive state in birds) of equivalent body mass, and, for a measure independent of the galliform baseline, the forebrain mass divided by the brainstem mass. These few examples illustrate the large number of competing measures of brain size.

In a number of lucid papers, Barton and his coworkers have described the various problems with relative brain measures (Barton, 1993, 1996, 1999; Barton and Dunbar, 1997; Barton and Purvis, 1994; Barton *et al.*, 1995). Since body mass is a rather inaccurate measure of body size, measures such as lifespan and metabolic rate may correlate with brain size better than body weight simply because they are more accurate indices of body size (Barton, 1999). Regression to remove body mass as a variable confounds the problem, since it adds a correlated error to each variable and so increases the chance of finding a spurious positive correlation. Barton (1999) concluded that the results of analyses including body mass should be "treated with caution". Gittleman (1986) attempted to circumvent this problem by using head and body length as a measure of body size, which avoids the problem of different gut weights, for example, but does not bypass the obstacle of measurement error. A further problem is that body size may be more evolutionary labile than brain size, so that rather than measuring encephalization we may, in fact, be measuring decreases in body size (Byrne, 1992; Dunbar, 1993a). However, Deaner and Nunn (1999) argued that there is little evidence that brain size has lagged behind body size over evolutionary time.

An alternative approach is to use the size of the brain itself as a reference variable. Including total brain weight as an independent variable is problematic where the structure of interest, such as the primate neocortex, makes up a large proportion of the brain (Byrne and Whiten, 1992). Hence using the size of the rest of the brain ('complement') is often the most appropriate technique (Barton, 1999; Purvis, 1992). Of course, this method will be unable to distinguish between an increase in neocortical size and a decrease in the size of the rest of the brain. Another possible reference variable is the size of a brain area assumed to be "primitive", such as the brainstem, on the assumption that such areas are evolutionary conservative (Barton and Dunbar, 1997; Keverne *et al.*, 1996). Again, caution is recommended, since even so-called primitive brain areas may have been subject to differential selective pressures (Barton and Dunbar, 1997). However, it is often reasonable to assume that decreases in the size of the rest of the brain or the brainstem over the course of evolution are unlikely, and techniques using these reference variables are now widely used and are often the best compromise.

The relative brain size approach treats species with identical relative brain sizes but different amounts of brain tissue as the same and so assumes that absolute volume of brain tissue is irrelevant (Byrne and Whiten, 1992). The presumption that what is important is the percentage of extra neural capacity over that "minimally required to service sensory and motor systems" (Byrne, 1992) would appear to be at odds with computing theory, where computing power is largely determined by the absolute number of computing elements (Byrne, 1992, 1993; Byrne and Whiten, 1992). If brains work like computers (and many argue they do not), absolute size may be a better measure of "computing power". Rensch (1956, 1957), for example, hypothesized that absolute brain size is positively correlated with greater learning capacities and provided evidence that elephants perform better than smaller-brained zebras and asses in discrimination learning tasks. Gibson (1999) argued that experimental primate studies indicated that absolute brain size, but not EQ, correlated with performance in learning tasks measuring mental flexibility. By this logic, the learning abilities of whales or elephants should be greater than that of humans.

Perhaps more reasonably, Byrne (1992) argued that what matters most is the absolute volume of neural tissue free for computation, suggesting as a suitable measure the ratio of neocortex to the rest of the brain: Dunbar's "neocortical ratio" (Aiello and Dunbar, 1993; Dunbar, 1992, 1993b). Dunbar (1992, 1993b) noted that this brain measure provided the best fit to the data on primate group sizes and advocated the use of this measure for tests of alternative hypotheses regarding brain evolution in primates. Neocortex ratio is also related to the frequency of tactical deception in primates (Byrne 1992, 1993; Byrne and Whiten, 1992). Hence neocortex ratio seems to work as a correlate of hypothesized indicators of "intelligence". However, neocortex ratio has been accused of obscuring or "smuggling in" body size confounds as a dimensionless variable (Deacon, 1993). The neocortex ratio does not completely remove the effect of body size, because neocortex size increases with body size more rapidly than the rest of the brain (Barton, 1993; Byrne, 1993).

In summary, a large number of measures have been used to compare brain size. Virtually all of these measures have methodological weaknesses, and all make different assumptions about the most appropriate way to measure the brain; consequently, the brain measure chosen will reflect a hypothesis concerning what underlies intelligence. In the studies described below, I have used three measures to reflect the different hypotheses described above. These are the executive brain ratio (executive brain volume over brainstem), the absolute executive brain volume, and what I term residual executive brain volume, which can be visualized as the residuals from a natural log–log plot of executive brain volume against brainstem. The residual descriptor is a convenience of terminology, as residuals were not calculated explicitly but instead brainstem was included in a multiple regression with executive brain volume and the behavioral measure of interest. The latter measure could be considered the most stringent method of accounting for body size.

3.1.3 Correcting for phylogeny

Species may show similar characteristics simply because they are closely related rather than because they have evolved independently under similar selection pressures. For example, imagine that new genetic data had resulted in reclassification of the common chimpanzee, and this species was now considered to be 10 species. The common chimpanzee has a large relative brain size, and more reports of social learning, innovation, and tool use (relative to research effort) are made for this species than for any other nonhuman primate (see Section 3.3.3). The reclassification of the chimpanzee would result in a cluster of points in the top right of a plot of relative brain size against relative social learning frequency. This cluster of points would not represent 10 independent cases of the co-evolution of large brains and social learning but, more likely, would represent one evolutionary event in the ancestor of the 10 species. Treating species as independent data points can reduce the chances of finding the true evolutionary relationship between brain size and social learning frequency. Hence it is essential to consider phylogeny when conducting comparative analyses of this type (Barton, 1999; Harvey and Pagel, 1991).

A number of techniques have been developed to incorporate phylogeny into comparative analyses and to account for the fact that species can often not be treated as independent data points, since this would overestimate the degrees of freedom (Harvey and Pagel, 1991). Some studies cope with this demand by simply conducting the analysis at a higher taxonomic level, such as the genus (e.g., Dunbar, 1992), the subfamily (e.g., Harvey, Martin, and Clutton-Brock, 1987), or the order (e.g., Lefebvre *et al.*, 1997). Choosing the appropriate taxonomic level is often a rather ad hoc process, though statistical techniques are available (Harvey and Pagel, 1991). Such procedures reduce rather than solve the problem and are vulnerable to both type I and type II errors.

An alternative approach is to take independent contrasts, now often the method of choice for comparative studies (Harvey and Pagel, 1991; Harvey and Purvis, 1991; but see Harvey and Rambaut, 2000; Pagel, 1999). For each variable of interest, comparisons are made at each pair of nodes in a bifurcating phylogeny. While the character at each pair of nodes may not be independent of common ancestry, the difference between them is assumed to be (Felsenstein, 1985). For example, if two sister species had relative brain sizes of 10.7 and 6.7, we would assume the difference between them (4.0) was the result of independent evolution in the two lineages subsequent to a speciation event. This difference score, standardized, would be one contrast, one piece of information in our analysis. The sets of independent comparisons can be correlated with each other by regression through the origin to determine whether the two variables have evolved together. The CAIC (comparative analysis by independent contrasts) computer program (Purvis and Rambaut, 1995) is one widely used implementation of this technique. Uncertainties about branch lengths or phylogeny are not sufficient justification for treating species as independent data points since studies show that CAIC performs reasonably well under these conditions and makes fewer type I errors than across-species analyses, even when the phylogeny is very inaccurate (Martins, 1993; Purvis, Gittleman, and Luh, 1994; Purvis and Rambaut, 1995). Purvis and Webster (1999) give a highly readable, primate-orientated description of the logic behind independent contrast analyses that is recommended to readers left unsatisfied by the necessarily short description given here.

3.1.4 Innovation, asocial and social learning

It is still an open question as to whether a binary distinction can be made between social learning and asocial learning. Some authors view social learning as a subcategory of asocial learning, predicting that social learning will covary across species with general behavioral flexibility (Galef, 1992; Laland and Plotkin, 1992; Lefebvre and Giraldeau, 1996). Heyes (1994), for example, argues that social and asocial learning processes share similar mechanisms, and Fragaszy and Perry take a similar stance in Ch. 1. These authors would predict a positive correlation between asocial and social learning competence. This would mean that hypotheses regarding the distribution of asocial learning or behavioral plasticity across taxa may also be applicable to the distribution of social learning propensities.

Other authors suggest or assume that there may be a trade off between individual learning and social learning abilities (Boyd and Richerson, 1985, 2000; Richerson and Boyd, 2000; Rogers, 1988), predicting a negative correlation between the two, rather like the negative relationship found between spatial and nonspatial learning competence in food-caching birds (Lefebvre and Giraldeau, 1996). Asocial learning and social learning are viewed by some as different, domain-specific, special-purpose adaptive mechanisms (Giraldeau, Caraco, and Valone, 1994; Tooby and Cosmides, 1989). Sometimes implicit in this view is the assumption that social learning is dependent on a specialized neural substrate at least partly separate from that required for asocial learning. There is currently little comparative evidence that social learning is an adaptive specialization to particular environmental demands (Lefebvre and Giraldeau, 1996). The results described here are based on the assumption that innovation is, at least partly, a manifestation of asocial learning, since in many cases innovations reported in the literature will be the result of an initial discovery process (which may in itself require learning about the affordances of an object, for example) and subsequent learning. Therefore, the index of innovation frequencies can be used to test hypotheses regarding the co-evolution of asocial and social learning.

3.1.5 Tool use and social learning

Tool use has traditionally been defined as the use of an external object that is detached from the substrate and held in the hand or mouth to obtain an immediate goal (Beck, 1980; van Lawick-Goodall, 1970). Commonly, but by no means universally, tool use has been considered as requiring complex cognitive abilities (Beck, 1980; Shettleworth, 1998; van Schaik, Deaner, and Merrill, 1999). "Technical intelligence" hypotheses that argue that technology or technical skills drove brain evolution would suggest that a large brain would be associated with tool use (Byrne, 1992, 1997; Passingham, 1982). Lefebvre, Nicolakakis, and Boire (2002) have presented evidence in support of this view, with avian taxa that frequently use tools having larger relative brain sizes than taxa that use tools less often. The frequency of reported tool use may be a useful measure of cognitive ability. Often a behavior pattern involving a tool will be novel, in which case it can also be classified as an innovation, but in many cases this will not be the case and so tool-use frequency can be regarded as a separate index of behavioral flexibility.

Van Schaik and coworkers (1999; see Ch. 11) have argued that social learning abilities are amongst the key determinants of primate tool use in the wild. So, we would expect species that exhibit a high incidence of social learning to also show high tool-using frequencies, an idea that can be easily tested. Van Schaik *et al.* (1999) also cite invention as a likely covariable of tool use that, like social learning, allows the rapid acquisition of complex technical skills. Again, we can test this idea here and also examine the relationship between brain size and tool use in primates.

3.1.6 Social learning, group size, and social intelligence hypotheses

Species that live a gregarious lifestyle have frequently been predicted to rely more on social learning processes than solitary species (Lee, 1991; Lefebvre and Giraldeau, 1996; Lefebvre *et al.*, 1996; Reader and Lefebvre, 2001; Roper, 1986). There have been few large-scale comparative tests of this theory, indeed Lefebvre *et al.* (1996) noted that only three research programs have tested for species differences in social learning (Cambefort, 1981; Jouventin, Pasteur, and Cambefort, 1976; Klopfer, 1961; Sasvàri, 1979, 1985). Relevant to this are the social (or Machiavellian) intelligence hypotheses (Byrne and Whiten, 1988; Flinn, 1997; Humphrey, 1976; Jolly, 1966; Whiten and Byrne, 1997), which argue that the large brains of primates evolved as an adaptation to living in large, complex social groups. Byrne and Whiten (1997) distinguished "narrow Machiavellianism", the idea that it is selection for strategies of social manipulation or deception that has driven primate brain evolution, from their own broader use of the term "Machiavellian intelligence", which includes all forms of social intelligence. There is considerable evidence that relative neocortex size is positively correlated with social group size in primates, carnivores, and cetaceans (Barton, 1999; but see Connor, Mack, and Tyack, 1998 on the cetacean data) so it is important to take brain size into account when testing relationships between social learning frequency and social group size.

In summary, the following sections investigate (a) whether the relative frequencies of social learning, innovation, and tool use are related to executive brain size; (b) whether rates of innovation and social learning covary; and (c) whether the incidence of social learning correlates with social group size.

3.2 Methods and data analysis

3.2.1 Overview

Back issues of primate journals and the social learning literature were searched for examples of innovation, social learning, and tool use. Lefebvre *et al.* (1997) used keywords to define examples of foraging behaviors as innovations and a similar approach was used here. For example, if the author or editor classified a behavior as "opportunistic" or "never seen before" in that species, this behavior pattern was scored as an innovation. By leaving such judgments to the authors, this approach aims to avoid any subjective bias imposed through data collection. Data on the identity of the individual(s) performing the behavior pattern and the circumstances of the behavior are described in Reader and Laland (2001). For each article examined in the four primate journals, the species studied was noted, regardless of whether that article contained an example of one of the behavior patterns of interest. This count of the number of studies on

Table 3.1. *Journals examined in the primate study*

Journal	Volumes	Years covered
American Journal of Primatology	39–41	1996–1997
International Journal of Primatology	17–18	1996–1997
Folia Primatologica	15–68	1971–1997
Primates	32–38	1991–1997

each species allowed an estimate of research effort to be made. Whiten and Byrne collected "opportunistic observations" of tactical deception (Byrne, 1993; Byrne and Whiten, 1992; Whiten and Byrne, 1988), advocating such an approach when the behavior of interest is rarely performed and emphasizing that the reports were not uninformed casual observations (or "anecdotes") but come from experienced scientists familiar with their subjects. A similar assertion applies here. However, Byrne and Whiten (1992) cautioned that in any such exercise there is no way of dissociating the tendency for scientists with particular interests, such as deception, to study species they consider "appropriate". Consequently, the collected data may still be vulnerable to biases.

3.2.2 Data sources

3.2.2.1 Literature

Approximately 1000 articles in four primate journals (*Primates*, *American Journal of Primatology*, *Folia Primatologica* and the *International Journal of Primatology*) were searched for examples of innovation, social learning, and tool use. Examples were also taken from relevant literature. Examples cited in the text of these articles were included, with the final database carefully checked to remove any repeated examples. The volumes examined are indicated in Table 3.1.

3.2.2.2 Phylogeny

The composite primate phylogeny used covers 203 species of primate and is relatively well resolved, containing 160 nodes (Purvis, 1995). Of these 160 nodes, 90 are dated, but dated nodes are not spread evenly over the tree and seven date estimates imply that a node is older than an ancestral node. Hence for the purposes of this analysis, an assumption of equal distances between phylogenetic nodes was made. That is, branch lengths were all assigned the same value. CAIC is reported to be robust to such assumptions

(Purvis *et al.*, 1994; Purvis and Rambaut, 1995). Where Purvis indicated simply the number of species in a genus (e.g., *Saimiri* and *Pithecia*), species names were taken from Rowe (1996) and these genera were assumed to be monophyletic groups. Purvis (1995) only include two species of *Aotus*, but recent classifications (Rowe, 1996) have documented 10 species. The two species in the original phylogeny were replaced with two monophyletic groups (*A. nigriceps* group: *A. azarai, A. infulatus, A. miconax, A. nancymaae,* and *A. nigriceps*; *A. trivirgatus* group: *A. brumbacki, A. hershkovitzi, A. lemurinus, A. trivirgatus,* and *A. vociferans*).

3.2.2.3 Group size, body weight, and life history data

Data on group sizes were taken from Rowe (1996) and missing data from Smuts *et al.* (1987) or Dixson (1998). These group sizes represent spatially and temporally cohesive associations (Dunbar, 1991). In complex social systems, such as those of the common chimpanzee, the group is defined as the number of individuals that an animal "knows and interacts regularly with" (Dunbar, 1991, 1992). Where a range of group sizes was indicated, the mean was taken. In some species of bushbaby, matriarchies are present where related adult females have overlapping ranging areas (Bearder, 1987). Matriarchy size was used as an estimate of group size in these species if no other group size data were available. Body weight, from Rowe (1996), was taken as the mean for the two sexes. If no body weight data were available, or if Rowe (1996) gave a figure for only one sex, data were taken from Harvey *et al.* (1987). There are possible errors in these measures, but these are the best estimates that are possible at this time.

3.2.2.4 Brain size

Analysis was conducted using three measures of brain size, the executive brain ratio, the absolute executive brain volume, and the residual executive brain volume, as described in Section 3.1.2. Information on the volume of the relevant brain regions was taken from Stephan *et al.* (1981), who detailed the brain sizes of 46 primate species and three tree shrews, which are no longer considered as primates (Martin, 1990). Stephan *et al.* (1981) listed data for a species not listed in current phylogenies, *Saguinus tamarin,* and indicate the genus, but not the species, in two cases. It was possible to identify the species involved as *Saguinus midas, Alouatta palliata,* and *Cebus apella,* respectively, by using total brain weight, which matched the figures given in Harvey *et al.* (1987). Stephan *et al.* (1981) are the only source of data on the volumes of primate brain structures available, apart from the more recent publication of similar data for the orangutan (Zilles and Rehkamper, 1988). Neocortex sizes were not estimated from cranial capacities or total brain volumes as in some studies (Aiello and Dunbar, 1993; Dunbar, 1995). Such an estimate would increase the size of the dataset but would compromise the specific hypotheses examining the deviation of executive brain size from that expected by allometry. The executive brain volume was calculated as the sum of neocortex and striatum volumes, and the brainstem as the sum of mesencephalon and medulla oblongata volumes. Executive brain ratio was the executive divided by brainstem volume. Stephan *et al.* (1981) corrected brain sizes to species means, which reduces the problem of accounting for sex differences in brain size.

3.2.3 Data collection

3.2.3.1 Procedure

Examples of social learning, innovation, and tool use were collected, with the species of the individual performing the behavior recorded in each case. Note that tool use is not a subset of the innovation data, since all instances of tool use are collated, not only novel ones. Where several species were noted as performing the same behavior, the behavior was scored for each species. For each journal article searched, whether or not it contained an example of the behavior patterns of interest, the species studied was noted. This allowed an estimate of research effort to be made in terms of the number of studies on each species. Theoretical articles, papers on extinct or fossil primates, and papers on several (three or more) species were not counted for the estimate of research effort. *Homo sapiens* was excluded from the analysis, since this species is often an outlier, and the rapid evolution of the human brain violates the assumptions of CAIC (Harvey *et al.*, 1987; Purvis, 1992).

All episodes were recorded, whether they occurred in captivity or in the field, as a result of experimental manipulations or as a result of human intervention such as provisioning or habitat degradation. Unusual behaviors that were described as pathological were not included in the analysis. "Questionable" examples, where, for example, social learning was implied rather than explicitly stated, were initially included in the analysis. Statistical analyses were conducted removing examples that were questionable, that occurred in captivity, under experimental manipulation, or under human intervention to check that the inclusion of these data did not produce artefactual results (see below).

3.2.4 Analysis

3.2.4.1 Research effort

There were huge differences in research effort among species, with a large number of studies conducted on, for example, common chimpanzees (*Pan troglodytes*), tufted capuchins (*C. apella*), Japanese macaques (*Macaca fuscata*), gorillas (*Gorilla gorilla*), and common marmosets (*Callithrix jacchus*) compared with, for instance, the relatively understudied bushbabies and gibbons. Studies covered 116 species, in comparison with the 203 species of living nonhuman primates (Purvis, 1995).

The frequencies of social learning, innovation, and tool use were corrected for research effort. There are several methods of calculating the difference between the number of observations and the number of observations expected from the research effort on each species. Lefebvre *et al.* (1997) and Byrne and Whiten (1992; Byrne, 1993) used a formula derived from the chi-square to make similar corrections. They calculated the observed value minus the expected, divided by the square root of the expected value (i.e., the square root of the chi-square: a chi-square would not differentiate between deviations above and below the expected value). This measure assumes that the expected value is directly proportional to the number of studies conducted on that species. This may not be the case. For example, well-studied species may attract specialists looking for examples of social transmission; as a result, more observations are made per unit of research effort than in other species. A superior method is to use the observed relationship between research effort and observation frequency to estimate expected values. Such a measure is the residual from a natural log–log plot of observation frequency against research effort. This technique is used here because it makes fewer assumptions about the relationship between the number of studies and the expected number of observations.

3.2.4.2 Comparative analysis

The terms "across-species" and "independent contrast" analysis are used here to refer to comparative methods that do or do not treat species as independent data points, respectively. Analysis by independent contrasts is widely recommended, but interpretation of graphs of contrasts can be less intuitive than those for across-species analyses, where each datum represents a single species. Graphs of species data may be especially informative if one is interested in the relative position of a particular species. Stephan *et al.* (1981) generally chose a single representative from each genus for brain volume measurement. Hence, across-species analyses of brain data are similar to a genus-level analysis, because each datum will be a species from a different genus. Analysis at a higher taxonomic level, such as the genus, is sometimes utilized as a partial solution to the problem of accounting for the effects of phylogeny (Dunbar, 1992). Additionally, recent developments in phylogenetic analysis have suggested that across-species analyses may occasionally be more appropriate than independent contrasts (Harvey and Rambaut, 2000). For these reasons, the results of both across-species and independent contrast analyses are of interest, and so data from both are presented. In general, the across-species and independent contrast analyses give a similar pattern of results, but where across-species analyses provide a significant result and independent contrasts do not, it cannot be excluded that the significant relationship is a result of the confounding effects of phylogeny.

Independent contrasts were calculated using CAIC version 2.0.0 (Purvis and Rambaut, 1995). Observation frequency data were corrected for research effort and natural logarithm (ln) transformed before taking contrasts. All brain volumes and body weights, apart from the executive brain ratio, were natural log transformed before taking contrasts since CAIC assumes that different lineages are equally likely to make the same proportional change in size. Independent contrasts were regressed through the origin using least-squares regression (Purvis, 1992; Purvis and Rambaut, 1995).

3.2.5 Interobserver reliabilities

A second observer coded previously examined issues of the journals *Folia Primatologica* and *Primates* using the definitions of social learning, innovation, and tool use given above. Interobserver reliabilities were calculated for 241 records: approximately 10% of the total number of records examined. Two points were clarified once coding began. Geophagy was not considered as a innovation unless the paper specifically stated that the behavior was novel, and only novel tool use or tool use in a novel context were classified as innovations: that is, not all cases of tool use were termed innovations. Agreement between the two observers was calculated as an index of concordance (Martin and Bateson, 1986). The interobserver reliability for social learning was 0.95, for innovation 0.83, and for tool use 0.94.

3.3 Results and discussion

In total, 533 instances of innovation, 445 observations of social learning, and 607 episodes of tool use were recorded from a total of approximately 2000 records and 1000 articles searched. The results section is divided into four sections. Section 3.3.1 addresses the relationship between brain size and social learning, innovation, and tool use frequencies. Section 3.3.2 examines the links between social learning, innovation, and tool use frequencies. Section 3.3.3 looks at group size, and section 3.3.4 examines whether the analyses are robust. Reader and Laland (2002) have presented a concise discussion of some of these data, with an emphasis on innovation and brain evolution.

3.3.1 Innovation, social learning, tool use, and brain size

The results for the three alternative brain measures are presented in turn and are summarized in Table 3.2.

3.3.1.1 Executive brain ratio

There was a significant positive correlation between social learning frequency and executive brain ratio, both across-species and for independent contrasts (Fig. 3.1). Similarly, there was a significant positive correlation between executive brain ratio and innovation frequency, and executive brain ratio and tool use frequency, both across-species and for independent contrasts (Fig. 3.1).

The executive brain ratio measure partially controls for differences in body size by dividing executive brain volume by brainstem volume. However, ratio measures have been criticized because they do not completely remove the effect of body size (see Section 3.1.2). Therefore, body weight was subsequently included as an independent variable in the analyses. Across-species, factoring out body weight resulted in similar results for all three measures of behavioral plasticity. Using independent contrasts and including body weight in the multiple regression with executive brain ratio resulted in a nonsignificant correlation between executive brain ratio and social learning frequency, and correlations with tool use and innovation frequencies that approached significance. Using body weight rather than a brain size measure to control for differences in body size is problematic (see Section 3.1.2). However, the fact that including body weight in the analyses resulted in similar observed patterns for five of the six analyses should

Table 3.2. *Summary of brain size findings*

Brain measure	Across species correlation			Independent contrasts correlation		
	Innovation	Social learning	Tool use	Innovation	Social learning	Tool use
Executive: brainstem ratio						
r^2_{adj}	0.34	0.48	0.40	0.18	0.13	0.17
F^a	16.70	29.49	21.46	7.66	5.55	7.28
p value[b]	**<0.0005**	**<0.0001**	**<0.0001**	**<0.01**	**<0.05**	**<0.05**
Executive brainstem ratio controlling for body weight						
Partial r^c	0.38	0.46	0.46	0.34	0.21	0.35
t^d	2.22	2.77	2.77	1.97	1.18	1.98
p value[b]	**<0.05**	**<0.01**	**<0.01**	0.06	>0.1	0.06
Absolute executive volume						
r^2_{adj}	0.24	0.34	0.27	0.14	0.13	0.14
F^a	10.95	17.00	12.17	5.92	5.41	5.99
p value[b]	**<0.005**	**<0.0005**	**<0.005**	**<0.05**	**<0.05**	**<0.05**
Executive brain volume controlling for brainstem volume						
Partial r^c	0.14	0.24	0.19	0.14	0.06	0.13
t^d	0.73	1.32	1.03	1.13	0.37	0.88
p value[b]	>0.1	>0.1	>0.1	>0.1	>0.1	>0.1

r, correlation coefficient; F, variance ratio; t Student distribution (with the number of degrees of freedom given as a subscript).
[a] $F_{1,30}$ for across species and $F_{1,29}$ for independent contrasts.
[b] Bold indicates significant correlations ($p < 0.05$).
[c] Where multiple regressions were used to control for the effect of a potential confounding variable (such as brainstem volume, an index of body size), the partial correlation coefficient (r) is given (Howell, 1997).
[d] t_{29} for across species and t_{28} for independent contrasts.

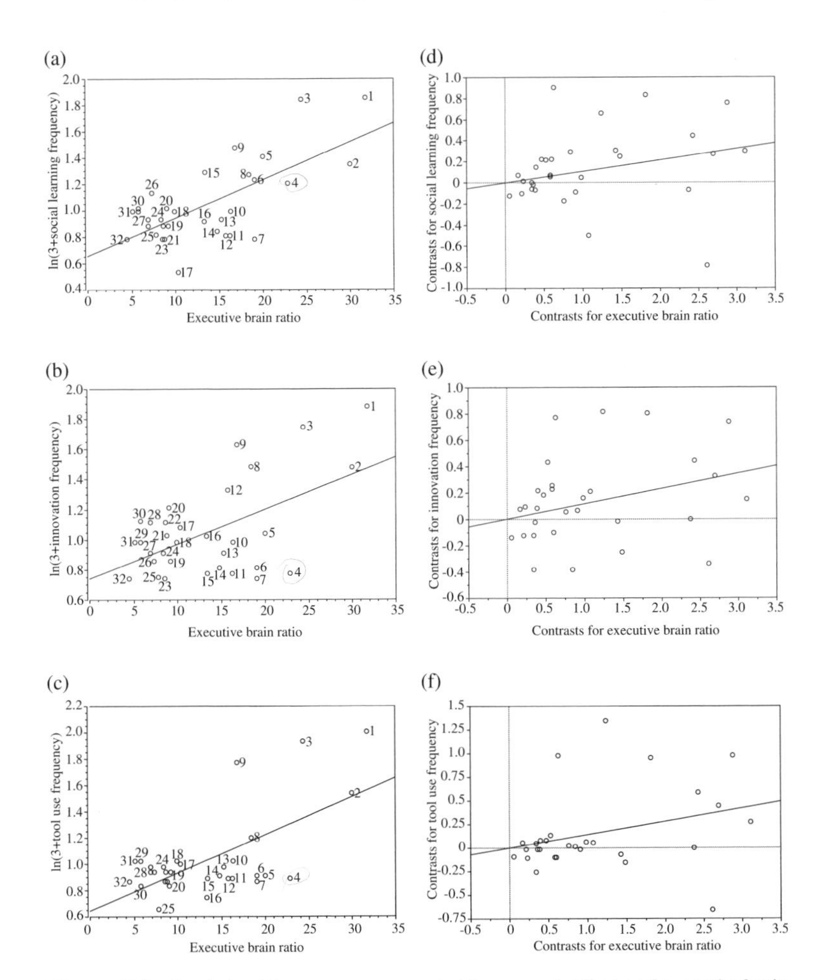

Fig. 3.1. Behavioral plasticity measures, corrected for research effort, and executive brain ratio. (a–c) Across-species analyses, with each point representing one species and (d–f) independent contrast analyses for (a,d) social learning frequency, (b,e) innovation frequency, and (c,f) tool-use frequency. Frequencies were corrected for research effort by taking residuals from a plot through the origin of natural logarithm (ln) frequency against ln research effort. Species are as follows, in descending order of executive brain ratio: 1, *Pan troglodytes*; 2, *Gorilla gorilla*; 3, *Pongo pygmaeus*; 4, *Ateles geoffroyi*; 5, *Macaca mulatta*; 6, *Erythrocebus patas*; 7, *Hylobates lar*; 8, *Papio anubis*; 9, *Cebus apella*; 10, *Cercocebus albigena*; 11, *Colobus badius*; 12, *Cercopithecus mitis*; 13, *Miopithecus talapoin*; 14, *Nasalis larvatus*; 15, *Alouatta palliata*; 16, *Saimiri sciureus*; 17, *Daubentonia madagascariensis*; 18, *Aotus trivirgatus*; 19, *Callicebus moloch*; 20, *Petterus fulvus*; 21, *Callimico goeldii*; 22, *Loris tardigradus*; 23, *Saguinus oedipus*; 24, *Saguinus midas*; 25, *Callithrix jacchus*; 26, *Propithecus verreauxi*; 27, *Varecia variegata*; 28, *Cebuella pygmaea*; 29, *Galagoides demidoff*; 30, *Otolemur crassicaudatus*; 31, *Galago senegalensis*; 32, *Microcebus murinus*.

increase confidence that the results are not a consequence of a body size confound.

3.3.1.2 Absolute executive brain volume

There was a significant positive correlation between corrected social learning frequency and absolute executive brain volume, both across-species and for independent contrasts. There were similar results for innovation frequencies and tool use frequencies.

3.3.1.3 Residual executive brain volume

A multiple regression with social learning frequency as the dependent variable, executive brain volume as the predictor variable, and brainstem volume as a covariate revealed no significant correlation between executive brain volume and social learning frequency, neither across-species nor for independent contrasts, once brainstem volume had been accounted for. Similar results were found for innovation frequencies and tool use frequencies.

3.3.1.4 Summary of links with brain size

In summary, there was not a significant correlation between the behavioral measures chosen and every measure of brain size (Table 3.2). There were significant positive correlations, both across-species and for independent contrasts, between executive brain ratio and rates of social learning, innovation, and tool use, and between absolute executive brain volume and rates of social learning, innovation, and tool use. However, no significant relationships were found using residual executive brain volume. The disparities between different brain size measures suggest that either the three measures gauge different things or some measures are more susceptible to type I or type II errors. Deaner *et al.* (2000) reviewed various relative brain size measures and found no reasonable basis to prefer one measure over another, so it is pertinent to discuss which measure may be most relevant.

Because few data on brain size are available, analyses were typically performed on a small number (30 to 32) of data points. The techniques used by Stephan *et al.* (1981) to determine brain volumes are highly labor intensive, which means only a small proportion of primate brains have been measured, and in the majority of species where data are available the figures are based upon measurements of only one or two individuals. Until more brain data become available, conclusions are necessarily tentative. CAIC analysis seemed peculiarly vulnerable to the exclusion of individual species. For example, removal of the gorilla from the analysis strengthened the correlations considerably. Primatologists have noted that gorilla tool use is less frequent than might be expected, but that gorillas make use of presumably cognitively complex hierarchical food-processing techniques (Byrne, 1997). Excluding younger nodes (e.g., nodes less than 5 million years old in primates) may improve the analysis, since error variance tends to be amplified at contrasts at younger nodes (Barton, 1999). Alternatively, improved comparative techniques that use a maximum likelihood framework, such as generalized least squares models, are becoming available and could be implemented (Pagel, 1999).

The finding that absolute executive brain volume correlated with innovation, social learning, and tool use frequencies supports the hypothesis that absolute brain size is positively correlated with greater learning capacities (Gibson, 1999; Rensch, 1956). However, this finding could be the result of a confound with body size. The use of executive brain ratio has received more theoretical and empirical support as an appropriate measure of relative brain size and cognitive ability (Barton and Dunbar, 1997; Byrne, 1992; see Section 3.1.2). This, combined with the finding that the correlations between executive brain ratio and innovation and tool use frequencies remained significant or approached significance when body weight is factored out, gives reasonable confidence that what is being measured is some index of brain size rather than simply a body size confound.

The detection, in nonhuman primates, of positive correlations between executive brain ratio and social learning, innovation, and tool-use frequencies confirms predicted trends linking innovation, cultural transmission, and brain size. That is, large brained species are reported to learn socially and innovate more, as assumed by the behavioral drive hypothesis (Wilson, 1985; Wyles *et al.*, 1983). There are at least two explanations for these relationships, which are not mutually exclusive and may work in concert (Lefebvre *et al.*, 1997). First, selection has favored individuals with large executive brain ratios because they have greater innovative, social learning, or tool-using capacities or propensities. That is, there has been direct selection for an increase in executive brain ratio in these animals. Second, animals may make opportunistic use of information-processing capabilities afforded by a large executive brain, which has evolved for some other reason, to cope with challenges in new flexible ways, through social learning or by using tools. The results fit with similar results linking relative brain size and deception (Byrne, 1992; Byrne and Whiten, 1992), mating competition (Sawaguchi, 1997), environmental

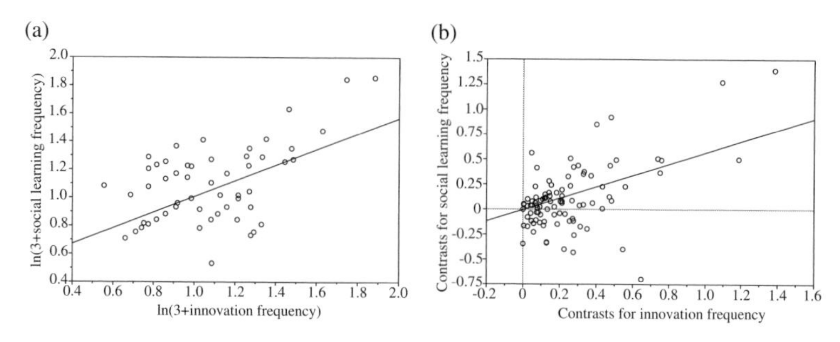

Fig. 3.2. Innovation and social learning frequencies (corrected for research effort): (a) across-species analysis, with each point representing one species and (b) independent contrasts.

complexity (Jolicoeur *et al.*, 1984), social group size (Dunbar, 1992), and frugivory (Barton, 1999). The findings also support a number of hypotheses concerning primate brain evolution (Byrne and Whiten, 1997), which are discussed at the end of this chapter.

3.3.2 Innovation and social learning covary

Figure 3.2 shows that innovation and social learning frequencies are positively correlated, across-species and for independent contrasts ($r^2_{adj} = 0.48$, $F_{1,114} = 108.38$, $p < 0.0001$; $r^2_{adj} = 0.35$, $F_{1,100} = 55.47$, $p < 0.0001$, respectively where r is the correlation coefficient and F the variance ratio). Of the available brain measures, executive brain ratio explains most variance in innovation and social learning frequencies (see above), so controlling for this brain measure is the more conservative analysis. This result was unaffected by the inclusion of executive brain ratio as an independent variable (multiple regression across-species: partial $r = 0.69$, controlling for relative executive brain size, $t_{29} = 4.87$, $p < 0.0001$; multiple regression independent contrasts: partial $r = 0.69$, $t_{28} = 5.07$, $p < 0.0001$).

There was a similar positive correlation between tool use and innovation frequencies (Fig. 3.3), statistically significant both across-species and for independent contrasts ($r^2_{adj} = 0.63$, $F_{1,114} = 198.96$, $p < 0.0001$, and $r^2_{adj} = 0.54$, $F_{1,100} = 118.89$, $p < 0.0001$, respectively). This result was unaffected by the inclusion of executive brain ratio as an independent variable (multiple regression across-species: partial $r = 0.85$, controlling for relative executive brain size, $t_{29} = 8.84$, $p < 0.0001$; multiple regression independent contrasts: partial $r = 0.88$, $t_{28} = 9.86$, $p < 0.0001$). There was also a positive correlation between tool-use and social learning frequencies (Fig. 3.3), supporting the predictions of van Schaik

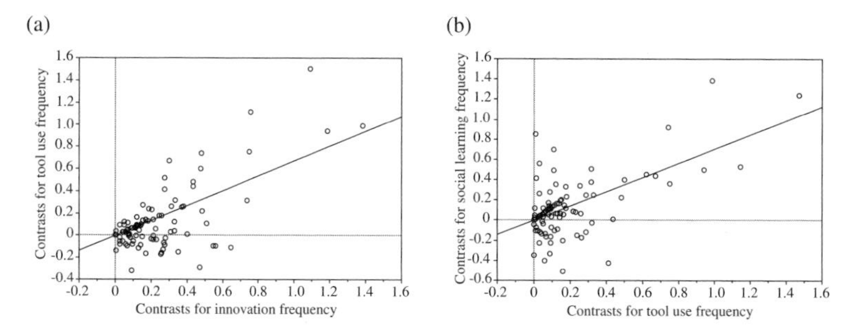

Fig. 3.3. Independent contrasts analysis for (a) innovation and tool use and (b) social learning and tool-use frequencies. Frequencies are corrected for research effort.

(Ch. 11). This correlation was significant both across-species and for independent contrasts ($r^2_{adj} = 0.57$, $F_{1,114} = 154.75$, $p < 0.0001$, and $r^2_{adj} = 0.45$, $F_{1,100} = 84.65$, $p < 0.0001$, respectively). This result was unaffected by the inclusion of executive brain ratio as an independent variable (multiple regression across-species: partial $r = 0.78$, controlling for relative executive brain size, $t_{29} = 6.71$, $p < 0.0001$; multiple regression independent contrasts: partial $r = 0.83$, $t_{28} = 7.69$, $p < 0.0001$).

It would be useful to demonstrate that species covary in their propensities to perform these three kinds of behavior regardless of the opportunities afforded by a propensity to perform one of these behavior types. For example, a high propensity to innovate could result in a high incidence of social learning because there are plenty of innovations to be learned. Similar arguments apply to tool use. Ideally, the relationship between the propensity for social learning and the propensity for innovation would be determined, regardless of the number of opportunities for social learning afforded by a large number of innovations. Though Kummer and Goodall (1985) noted that the majority of innovations in primate populations do not appear to spread, these variables may be partly confounded in the field data, and controlled studies may be the only route to resolving these confounds (Lefebvre and Giraldeau, 1996). However, re-analysis may go some way towards a solution. In the present study, examples included in more than one of the three categories of innovation, social learning, and tool use (for example, a novel tool use) were removed from the analysis. This gave a restricted dataset including only observations of tool use that were not innovations or learned from others, and only cases of social learning that did not involve learning an innovation, or at least observations that were not recorded as such. Note that learning an innovation is not a defining feature of social learning, and it is possible to learn a behavior well established in the population repertoire and novel only to the individual who is learning.

Re-analysis of this restricted dataset gave similar results to previous analyses. There was a significant positive correlation between social learning and innovation frequencies (across-species: $r^2_{adj} = 0.28$, $F_{1,113} = 44.63$, $p < 0.0001$; independent contrasts: $r^2_{adj} = 0.22$, $F_{1,99} = 28.95$, $p < 0.0001$). There was also a significant positive correlation between tool use and innovation frequencies (across-species: $r^2_{adj} = 0.27$, $F_{1,113} = 43.19$, $p < 0.0001$; independent contrasts: $r^2_{adj} = 0.15$, $F_{1,99} = 19.01$, $p < 0.0001$). Again, a significant positive correlation between tool-use and social learning frequencies was found (across-species: $r^2_{adj} = 0.31$, $F_{1,113} = 51.41$, $p < 0.0001$; independent contrasts: $r^2_{adj} = 0.23$, $F_{1,99} = 30.35$, $p < 0.0001$).

It was not possible to account for reporting biases in the data. There was no evidence for such biases, but it was possible that researchers were more likely to score behaviors as socially transmitted in species they considered to be innovatory, for example. It is also possible that socially learned behaviors may be recorded instead as innovations if the transmission episode is unobserved or undetermined. Theoretically, this should not bias the results since both an innovation and a social transmission episode have occurred. However, in species that socially learn innovations, there will be more individuals performing these new behavior patterns and so a greater chance of these behavior patterns being observed. In an equally innovatory species where individuals rarely socially learn, fewer innovations may be recorded.

With these issues in mind, the view that social learning capacities covary with general behavioral plasticity is supported by a number of facts: the results discussed here are robust to the re-analysis, the frequency of social learning not only correlates with innovation but also with tool-use frequency, and rates of tool use and innovation covary together, even with executive brain ratio controlled for in the analysis. The finding that innovation and social learning frequencies covary is important since it provides the first large-scale comparative evidence consistent with social learning being a component of general learning abilities. The data are also consistent with the hypothesis that social learning and asocial learning are separate, domain-specific capacities, but that correlated evolution of these two traits has been favored by one or more selection pressures. The results are not consistent with a third hypothesis, that there has been a trade off between asocial learning and social learning capacities over evolutionary time. For the moment, there is little evidence that social learning is an adaptive specialization beyond that for a general selection pressure favoring behavioral plasticity.

3.3.3 Group size and social learning

It has been previously reported that social group size and social learning frequency correlated weakly across species but this relationship was no longer significant when phylogeny was taken into account by taking independent contrasts (Reader and Lefebvre, 2001). The additional analyses described below test the robustness of this claim.

Since group size and neocortex ratio have been shown to covary (Dunbar, 1992), executive brain ratio was included as an independent variable. No significant relationships were found between group size and social learning frequencies (across-species: partial $r = -0.22$, $t_{28} = 1.19$, $p > 0.1$; independent contrasts: partial $r = -0.18$, $t_{27} = 0.90$, $p > 0.1$). Orangutans are unusual in that they have a very much smaller group size than would be expected from their brain size, and some authors argue they may have a more complex social life than their supposed group size would suggest (Dunbar, 1992; see also Ch. 11). However, exclusion of orangutans from the independent contrast analyses did not affect the results. Similarly, the results were not a consequence of the inclusion of matriarchy data for bushbabies (see Section 3.2.2), since exclusion of these data gave similar results.

It could be argued that the dataset included several examples of social learning from groups of primates living in artificially large groups, such as captive or provisioned populations. Groups including individuals who learn from humans could also be considered to be artificially large, since the humans can be counted as potential demonstrators. Therefore, a re-analysis was conducted excluding captive studies and data where a human influence was stated or suggested. Similar results were found. Group size and social learning frequency appeared to correlate across-species ($r^2_{adj} = 0.03$, $F_{1,103} = 4.18$, $p < 0.05$) but this relationship was weak and was not significant after taking independent contrasts ($r^2_{adj} = 0.00$, $F_{1,92} = 0.99$, $p > 0.1$). Including executive brain ratio as an independent variable had similar effects on this restricted dataset as on the full social learning measure.

Hence, contrary to predictions, there was no significant relationship between group size and social learning frequency after taking phylogeny or executive brain ratio into account. This finding would seem to be inconsistent with "broad" Machiavellian intelligence hypotheses, which argue that living in complex social groups has favored the evolution of all forms of social intelligence in primates (Whiten and Byrne, 1997). This may suggest social learning capacities are not aspects of a general social cognitive ability or general social "intelligence". However, it is also plausible that social group size may be a poor or inexact measure of social complexity, and that a better measure of social complexity would reveal an association with social learning.

3.3.4 How robust is the analysis?

The data were re-analyzed, taking only examples from the field and excluding questionable examples (e.g., where social learning was implied rather than explicitly stated), experimental manipulations, and cases where human intervention was stated or implied. Analyses of the relationships between social learning, innovation, and tool-use frequencies were unaffected by this procedure (see Table 3.3), both across-species and for independent contrasts. Similarly, the across-species relationships between social learning, innovation, and tool-use frequency and executive brain ratio were unaffected. However, after taking independent contrasts, no relationship was found between executive brain ratio and innovation frequency or tool-use frequency. The tool-use result probably reflects the loss of power associated with the relatively small number of species that have been observed using tools in the wild compared with tool use in captivity (Byrne, 1997). However, the fact that the vast majority of the results are robust to the extremely conservative nature of the re-analysis suggests reasonable confidence in the results.

3.4 Conclusions

The principal findings of this study are that, once research effort and phylogenetic relationships have been taken into account, (a) executive brain ratio and absolute executive brain volume correlate with social learning, innovation, and tool-use frequencies; (b) incidence of social learning covaries with that of innovation; and (c) there is no evidence for a relationship between social learning frequency and social group size. These findings and possible confounding factors have been discussed in the relevant sections, so here only the major conclusions are summarized.

First, there is now evidence that members of large-brained nonhuman primate species learn from others and innovate more frequently than

Table 3.3. *Re-analysis using the most conservative dataset*

Analysis	Correlation coefficient (r)		Variance ratio (df)	t(df) controlling for executive brain ratio	p value	p value controlling for executive brain ratio
	r^2_{adj}	Partial r controlling for executive brain ratio				
Across-species						
Social learning against innovation	0.51	*0.74*	121.42 (1,114)	*5.97 (29)*	**<0.0001**	***<0.0001***
Tool use against innovation	0.49	*0.89*	111.24 (1,114)	*10.31 (29)*	**<0.0001**	***<0.0001***
Tool use against social learning	0.47	*0.82*	103.91 (1,114)	*7.66 (29)*	**<0.0001**	***<0.0001***
Innovation against executive brain ratio	0.26	–	11.83 (1,30)	–	**<0.005**	–
Social learning against executive brain ratio	0.53	–	35.67 (1,30)	–	**<0.0001**	–
Tool use against executive brain ratio	0.29	–	13.90 (1,30)	–	**<0.001**	–
Independent Contrasts						
Social learning against innovation	0.61	*0.86*	159.32 (1,100)	*8.63 (28)*	**<0.0001**	***<0.0001***
Tool use against innovation	0.54	*0.96*	118.92 (1,100)	*17.41 (28)*	**<0.0001**	***<0.0001***
Tool use against social learning	0.35	*0.87*	56.00 (1,100)	*8.86 (28)*	**<0.0001**	***<0.0001***
Innovation against executive brain ratio	0.00	–	0.72 (1,29)	–	>0.1	–
Social learning against executive brain ratio	0.21	–	8.94 (1,29)	–	**<0.01**	–
Tool use against executive brain ratio	0.02	–	1.54 (1,29)	–	>0.1	–

df, degrees of freedom; t_1 Student distribution.
Figures in italics indicate the partial r, t value and probability level after executive brain ratio was controlled for using multiple regression (Howell, 1997).
Bold indicates significant correlations ($p < 0.05$).

members of small-brained primate species. Macphail (1982) has argued that evidence that brain-size measures predict intellectual capacity is lacking. This study provides evidence to the contrary, if the reasonable assumption is made that the reported incidence of innovation and social learning correlates with ability or capacity. Furthermore, the results support the argument that an increase in brain size and complexity is one cost of a reliance on learning (Johnston, 1982). Moreover, the findings presented here provide support for the behavioral drive hypothesis (Wilson, 1985). Therefore, brain size measures appear a valuable tool in explaining species differences in social learning.

Second, social learning frequencies appear to correlate with general behavioral flexibility, to the extent that innovation and tool-use frequencies are measures (Lefebvre *et al.*, 1997). This is an important finding since it suggests that social learning is not independent of asocial learning. The same selection pressures may favor both asocial and social learning. The correlation between rates of social learning and innovation is also consistent with the view that social learning and asocial learning share similar mechanisms (Heyes, 1994) and perhaps share similar neural substrates. However, the possibility cannot be ruled out that some social learning processes, such as imitation, may rely on different brain systems. Like the findings of Lefebvre and Giraldeau (1996), the results presented here do not support the idea that social learning is an adaptive specialization. If social learning and asocial learning propensities or capacities are closely tied, the relationship between life-history variables and rates of social learning may be rather uninformative. Instead, it will be the deviations from the relationship between innovation and social learning that are interesting and instructive (Lefebvre and Giraldeau, 1996). The correlations between innovation, social learning, and tool-use frequencies suggest that either these processes are part of one "domain-general" "intelligence" or that one or more selection pressures, either acting consecutively or concurrently, have favored the correlated evolution of several, domain-specific "intelligences" (Byrne and Whiten, 1997; Tooby and Cosmides, 1989).

Third, the findings are consistent with several hypotheses concerning primate brain evolution (Byrne and Whiten, 1988; Whiten and Byrne, 1997). Rates of tool use were found to correlate with executive brain ratio, which is consistent with technical intelligence hypotheses, which argue that technology or technical skills drove brain evolution

(Byrne, 1992, 1997; Passingham, 1982; Wynn, 1988). Rates of innovation and social learning were also found to correlate with executive brain ratio. Taken with the fact that most of the recorded socially learned behaviors and innovations were in the foraging context (Reader and Laland, 2001), the results seem consistent with hypotheses suggesting ecological function as important in the evolution of primate intelligence. Examples of such arguments are the extractive foraging hypothesis (Parker and Gibson, 1977) and Milton's (1988) cognitive mapping hypothesis, the idea that intelligence developed as a response to the challenge of locating patchily distributed, but potentially predictable, food sources. Further, the results described here may suggest an alternative social intelligence hypothesis. Social (or Machiavellian) intelligence hypotheses argue that the complex cognitive demands of living in social groups promote the evolution of a larger brain (Byrne and Whiten, 1988; Whiten and Byrne, 1997). Johnston (1982) has suggested that one cost of learning is that of parental care for a dependant infant. Complex supportive social systems may reduce this cost of learning and simultaneously increase selective pressures for the development of learning abilities (Johnston, 1982). If social learning is causal in the relationship between executive brain size and rates of social learning, then social learning, a manifestation of "social intelligence", may be an additional driving force behind the evolution of large brains in primates.

The fact that the findings described above are consistent with a number of competing theories supports the contention that several selective pressures are responsible for the development of large relative brain sizes and intelligence in primates (Byrne and Whiten, 1997). An alternative view is that one factor is driving brain evolution, but that the cognitive abilities afforded by a large brain are applied to other domains. An interesting extension to the study would be to examine taxon-level differences in the relationships between brain size and innovation, social learning, and tool use. Such analyses may provide useful information on the relative importance of the effect of different selection pressures on brain size evolution in different primate taxa. For example, Barton (1993) finds a correlation between group size and neocortex size in haplorhines, but not strepsirhines, which may indicate that group living favored brain size evolution amongst haplorhines only. In conclusion, frequencies of reports of social learning and innovation gathered from published literature seem likely to be both ecologically relevant and measurable indices of learning propensities (Lefebvre *et al.*, 1997) that allow the various hypothesis concerning the evolution of social learning to be tested.

3.4.1 Implications for the study of traditions

Comparative studies across all primates of the kind described here seem a powerful tool for testing hypotheses about the evolution of a reliance on traditional behaviors. What do the alternative comparative approaches described elsewhere in this volume have to offer and gain from such all-primate studies? There are a number of obvious avenues by which observational and experimental studies in the field and laboratory could improve the power of analyses such as those detailed here. First, the method depends upon the accurate recording of innovation and social learning across a wide range of species, ideally in unprovisioned field populations. It would also be helpful if researchers could note such characteristics as the age, sex, and social rank of individuals innovating, socially learning or using tools. Such data would allow hypotheses regarding the distribution of social learning between age or sex classes to be tested (e.g., Reader and Laland, 2001; see Ch.5) and may also allow true innovations and rarely observed conditional strategies to be distinguished. Second, in many field studies, circumstantial evidence is used to identify socially learned behavior patterns, and it is by no means certain that social learning is actually involved. Methods such as those described by Dewar (Ch. 5) will be helpful to determine the true reliance on social learning in field populations. Third, the comparative methods described here rely upon accurate estimates of brain size, group size, and other ecological or social variables proposed to be linked to traditions. Again, field researchers can help by gathering accurate data, and students of brain evolution should prioritize the gathering of precise brain size data for the 156 species where no data are available.

Comparative methods also have much to offer those interested in studying traditions in the field. For example, they suggest species particularly likely to rely on traditions and so suitable for future study, indicate less-studied species, and suggest hypotheses that can be tested experimentally in the field or in captivity. The conclusions drawn at the species level may also apply to population differences, which means that population level comparisons will also be valuable. Comparative studies and models that make predictions for primates in general (see also Chs. 2, 5, and 11) should be integrated with population-level comparative field studies such as those described in Chs. 10, 13, and 14. This would make a powerful combination with which to study the evolution and function of traditional behaviors.

3.4.2 Summary

A comparative study of social learning, innovation, and tool use in nonhuman primates was conducted by collecting over 450 reports of such behaviors from the primate and social learning literature. Comparative studies of learning and behavioral flexibility require a test fair to all species and data on large numbers of species, which this method provides by measuring the tendency to discover or learn novel solutions to environmental or social problems relevant to the animal. Social learning, innovation, and tool-use frequencies, corrected for research effort and phylogeny, were positively correlated with two brain measures, absolute executive brain volume and the ratio of executive brain over brainstem, confirming predicted trends linking innovation and brain size. These findings are consistent with several hypotheses regarding brain evolution, and, if social learning is causal in brain size evolution, suggest an alternative, complementary, social intelligence hypothesis. Moreover, innovation and social learning frequencies were found to covary, which is consistent with social learning capacities correlating with general behavioral flexibility. Contrary to predictions, the results do not support a relationship between social learning frequencies and social group size. The results have a number of implications for the future study of traditions in the wild.

3.5 Acknowledgements

I am indebted to Rachel Day for collecting data for interobserver reliability calculations, to Phyllis Lee, Alan Dixson, and Barry Keverne for pointing out a number of essential data sources, and to Kevin Laland, Gillian Brown, Gwen Dewar, Melissa Panger, and Robert Boyd for helpful comments on the manuscript. Funding for this research was provided by the BBSRC and the Royal Society.

References

Aiello, L. C. and Dunbar, R. I. M. 1993. Neocortex size, group size and the evolution of language. *Current Anthropology*, **34**, 184–193.

Ancel, L. W. 1999. A quantitative model of the Simpson–Baldwin effect. *Journal of Theoretical Biology*, **196**, 197–209.

Baldwin, J. M. 1896. A new factor in evolution. *American Naturalist*, **30**, 441–451, 536–553.

Barton, R. A. 1993. Independent contrasts analysis of neocortical size and socioecology in primates. *Behavioral and Brain Sciences*, **16**, 694–695.

Barton, R. A. 1996. Neocortex size and behavioural ecology in primates. *Proceedings of the Royal Society of London, Series B*, **263**, 173–177.

Barton, R. 1999. The evolutionary ecology of the primate brain. In *Comparative Primate Socioecology*, ed. P. C. Lee, pp. 167–194. Cambridge: Cambridge University Press.

Barton, R. A. and Dunbar, R. I. M. 1997. Evolution of the social brain. In *Machiavellian Intelligence II*, ed. A. Whiten, and R. W. Byrne, pp. 240–263. Cambridge: Cambridge University Press.

Barton, R. A. and Harvey, P. H. 2000. Mosaic evolution of brain structure in mammals. *Nature*, **405**, 1055–1058.

Barton, R. A. and Purvis, A. 1994. Primate brains and ecology: looking beneath the surface. In *Current Primatology: Proceedings of the XIVth Congress of the International Primatological Society*, ed. J. R. Anderson, B. Thierry, and N. Herrnschmidt, pp. 1–11. Strasbourg: Université Louis Pasteur.

Barton, R. A., Purvis, A., and Harvey, P. H. 1995. Evolutionary radiation of visual and olfactory brain systems in primates, bats and insectivores. *Philosophical Transactions of the Royal Society of London, Series B*, **348**, 381–392.

Bateson, P. P. G. 1983. Genes, environment and the development of behavior. In *Animal Behavior*, Vol. 3, *Genes, Development and Learning*, ed. T. R. Halliday, and P. J. B. Slater, pp. 52–81. Oxford: Blackwell.

Bateson, P. G. 1988. The active role of behavior in evolution. In *Evolutionary Processes and Metaphors*, ed. M.-W. Ho, and S. W. Fox, pp. 191–207. Chicester: Wiley.

Bateson, P. P. G. and Martin, P. 1999. *Design for a Life: How Behavior Develops*. London: Jonathan Cape.

Bearder, S. K. 1987. Infants and adult males. In *Primate Societies*, ed. B. B. Smuts, D. L. Cheney, R. M. Seyfarth, R. W. Wrangham, and T. T. Struhsaker, pp. 11–24. Chicago, IL: University of Chicago Press.

Beck, B. B. 1980. *Animal Tool Behavior: The Use and Manufacture of Tools by Animals*. New York: Garland.

Boyd, R. and Richerson, P. J. 1985. *Culture and the Evolutionary Process*. Chicago, IL: University of Chicago Press.

Boyd, R. and Richerson, P. J. 2000. Memes: universal acid or a better mousetrap. In *Darwinizing Culture: The Status of Memetics as a Science*, ed. R. Aunger, pp. 143–162. Oxford: Oxford University Press.

Byrne, R. W. 1992. The evolution of intelligence. In *Behaviour and Evolution*, ed. P. J. B. Slater and T. R. Halliday, pp. 223–265. Cambridge: Cambridge University Press.

Byrne, R. W. 1993. Do larger brains mean greater intelligence? *Behavioral and Brain Sciences*, **16**, 696–697.

Byrne, R. W. 1997. The technical intelligence hypothesis: an additional evolutionary stimulus to intelligence? In *Machiavellian Intelligence II*, ed. A. Whiten and R. W. Byrne, pp. 289–311. Cambridge: Cambridge University Press.

Byrne, R. W. and Whiten, A. 1992. Cognitive evolution: evidence from tactical deception. *Man*, **27**, 609–627.

Byrne, R. W. and Whiten, A. 1997. Machiavellian intelligence. In *Machiavellian Intelligence II*, ed. A. Whiten and R. W. Byrne, pp. 1–23. Cambridge: Cambridge University Press.

Byrne, R. W. and Whiten, A. 1988. *Machiavellian Intelligence: Social Expertise and the Evolution of Intellect in Monkeys, Apes and Humans*. Oxford: Oxford University Press.

Cambefort, J. P. 1981. A comparative study of culturally transmitted patterns of feeding habits in the chacma baboon *Papio ursinus* and the vervet monkey *Cercopithecus aethiops*. *Folia Primatologica*, **36**, 243–263.

Connor, R. C., Mann, J., and Tyack, P. L. 1998. Reply to Marino. *Trends in Ecology and Evolution*, **13**, 408.

Deacon, T. W. 1993. Confounded correlations, again. *Behavioral and Brain Sciences*, **16**, 698–699.

Deaner, R. O. and Nunn, C. L. 1999. How quickly do brains catch up with bodies? A comparative method for detecting evolutionary lag. *Proceedings of the Royal Society of London, Series B*, **266**, 687–694.

Deaner, R. O., Nunn, C. L., and van Schaik, C. P. 2000. Comparative tests of primate cognition: different scaling methods produce different results. *Brain, Behavior and Evolution*, **55**, 44–52.

Dixson, A. F. 1998. *Primate Sexuality: Comparative Studies of the Prosimians, Monkeys, Apes and Human Beings*. Oxford: Oxford University Press.

Dunbar, R. I. M. 1991. Functional significance of social grooming in primates. *Folia Primatologica*, **57**, 121–131.

Dunbar, R. I. M. 1992. Neocortex size as a constraint on group size in primates. *Journal of Human Evolution*, **20**, 469–493.

Dunbar, R. I. M. 1993a. On the origins of language: a history of constraints and windows of opportunity. *Behavioral and Brain Sciences*, **16**, 721–729.

Dunbar, R. I. M. 1993b. Coevolution of neocortical size, group size and language in humans. *Behavioral and Brain Sciences*, **16**, 681–735.

Dunbar, R. I. M. 1995. Neocortex size and group size in primates: a test of the hypothesis. *Journal of Human Evolution*, **28**, 287–296.

Essock-Vitale, S. and Seyfarth, R. M. 1987. Intelligence and social cognition. In *Primate Societies*, ed. B. B. Smuts, D. L. Cheney, R. M. Seyfarth, R. W. Wrangham, and T. T. Struhsaker, pp. 452–461. Chicago, IL: University of Chicago Press.

Feldman, M. W. and Laland, K. N. 1996. Gene–culture coevolutionary theory. *Trends in Ecology and Evolution*, **11**, 453–457.

Felsenstein, J. 1985. Phylogenies and the comparative method. *American Naturalist*, **125**, 1–15.

Fisher, J. and Hinde, R. A. 1949. The opening of milk bottles by birds. *British Birds*, **42**, 347–357.

Flinn, M. V. 1997. Culture and the evolution of social learning. *Evolution and Human Behaviour*, **18**, 23–67.

Galef, B. G., Jr. 1992. The question of animal culture. *Human Nature*, **3**, 157–178.

Gibson, K. R. 1999. Social transmission of facts and skills in the human species: neural mechanisms. In *Mammalian Social Learning: Comparative and Ecological Perspectives*, ed. H. O. Box, and K. R. Gibson, pp. 351–366. Cambridge: Cambridge University Press.

Giraldeau, L.-A., Caraco, T., and Valone, T. J. 1994. Social foraging: individual learning and cultural transmission of innovations. *Behavioural Ecology*, **5**, 35–43.

Gittleman, J. L. 1986. Carnivore brain size, behavioural ecology, and phylogeny. *Journal of Mammology*, **67**, 23–36.

Harvey, P. H. and Krebs, J. R. 1990. Comparing brains. *Science*, **249**, 140–146.

Harvey, P. H., and Pagel, M. D. 1991. *The Comparative Method in Evolutionary Biology*. Oxford: Oxford University Press.

Harvey, P. H. and Purvis, A. 1991. Comparative methods for explaining adaptations. *Nature*, **351**, 619–624.

Harvey, P. H. and Rambaut, A. 2000. Comparative analyses for adaptive radiations. *Philosophical Transactions of the Royal Society of London, Series B*, **355**, 1599–1605.

Harvey, P. H., Martin, R. D., and Clutton-Brock, T. H. 1987. Life histories in comparative perspective. In *Primate Societies*, ed. B. B. Smuts, D. L. Cheney, R. M. Seyfarth, R. W. Wrangham, and T. T. Struhsaker, pp. 181–196. Chicago, IL: University of Chicago Press.

Heyes, C. M. 1994. Social learning in animals: categories and mechanisms. *Biological Reviews*, **69**, 207–231.

Hinde, R. A. and Fisher, J. 1951. Further observations on the opening of milk bottles by birds. *British Birds*, **44**, 393–396.

Hinton, S. J. and Nolan, G. E. 1980. How learning can guide evolution. *Complex Systems*, **1**, 495–502.

Holden, C. and Mace, R. 1997. A phylogenetic analysis of the evolution of lactose digestion. *Human Biology*, **69**, 605–628.

Howell, D. C. 1997. *Statistical Methods for Psychology, 4th edn*. Belmont: Duxbury.

Humphrey, N. K. 1976. The social function of intellect. In *Growing Points in Ethology*, ed. P. P. G. Bateson and R. A. Hinde, pp. 303–317. Cambridge: Cambridge University Press.

Jerison, H. J. 1973. *Evolution of the Brain and Intelligence*. New York: Academic Press.

Johnston, T. D. 1982. The selective costs and benefits of learning: an evolutionary analysis. *Advances in the Study of Behaviour*, **12**, 65–106.

Jolicoeur, P., Pirlot, P., Baron, G., and Stephan, H. 1984. Brain structure and correlation patterns in Insectivora, Chiroptera, and primates. *Systematic Zoology*, **33**, 14–29.

Jolly, A. 1966. Lemur social behavior and primate intelligence. *Science*, **153**, 501–506.

Jouventin, P., Pasteur, G., and Cambefort, J. P. 1976. Observational learning of baboons and avoidance of mimics: exploratory tests. *Evolution*, **31**, 214–218.

Keverne, E. B., Martel, F. L., and Nevison, C. M. 1996. Primate brain evolution: genetic and functional considerations. *Proceedings of the Royal Society of London, Series B*, **262**, 689–696.

Klopfer, P. H. 1961. Observational learning in birds: the establishment of behavioral modes. *Behaviour*, **17**, 71–80.

Krebs, J. R., Sherry, D. F., Healy, S. D., Perry, H., and Vaccerino, A. L. 1989. Hippocampal specialization of food-storing birds. *Proceedings of the National Academy of Sciences, USA*, **86**, 1388–1392.

Kummer, H. and Goodall, J. 1985. Conditions of innovative behaviour in primates. *Philosophical Transactions of the Royal Society of London, Series B*, **308**, 203–214.

Laland, K. N. 1992. A theoretical investigation of the role of social transmission in evolution. *Ethology and Sociobiology*, **13**, 87–113.

Laland, K. N. and Plotkin, H. C. 1992. Further experimental analysis of the social learning and transmission of foraging information amongst Norway rats. *Behavioural Processes*, **27**, 53–64.

Lee, P. 1991. Adaptations to environmental change: an evolutionary perspective. In *Primate Responses to Environmental Change*, ed. H. O. Box, pp. 39–56. London: Chapman & Hall.

Lefebvre, L. 1995. The opening of milk-bottles by birds: evidence for accelerating learning rates, but against the wave-of-advance model of cultural transmission. *Behavioural Processes*, **34**, 43–53.

Lefebvre, L. and Giraldeau, L.-A. 1996. Is social learning an adaptive specialization? In *Social Learning in Animals: The Roots of Culture*, ed. C. M. Heyes and B. G. Galef Jr., pp. 107–128. London: Academic Press.

Lefebvre, L., Palameta, B., and Hatch, K. K. 1996. Is group-living associated with social learning? A comparative test of a gregarious and a territorial columbid. *Behaviour*, **133**, 1–21.

Lefebvre, L., Whittle, P., Lascaris, E., and Finkelstein, A. 1997. Feeding innovations and forebrain size in birds. *Animal Behaviour*, **53**, 549–560.

Lefebvre, L., Nicolakakis, N., and Boire, D. 2002. Tools and Brains in Birds. *Behaviour*, **139**, 939–973.

Macphail, E. M. 1982. *Brain and Intelligence in Vertebrates*. Oxford: Clarendon Press.

Martin, R. D. 1990. *Primate Origins and Evolution: A Phylogenetic Reconstruction*. London: Chapman & Hall.

Martin, P. R. and Bateson, P. 1986. *Measuring Behaviour: An Introductory Guide*. Cambridge: Cambridge University Press.

Martins, E. P. 1993. Comparative studies, phylogenies and predictions of coevolutionary relationships. *Behavioral and Brain Sciences*, **16**, 714–716.

Milton, K. 1988. Foraging behaviour and the evolution of primate intelligence. In *Machiavellian Intelligence: Social Expertise and the Evolution of Intellect in Monkeys, Apes and Humans*, ed. R. W. Byrne, and A. Whiten, pp. 271–284. Oxford: Oxford University Press.

Pagel, M. 1999. Inferring the historical patterns of biological evolution. *Nature*, **401**, 877–884.

Parker, S. T. and Gibson, K. R. 1977. Object manipulation, tool use and sensorimotor intelligence as feeding adaptations in Cebus monkeys and great apes. *Journal of Human Evolution*, **6**, 623–641.

Passingham, R. E. 1982. *The Human Primate*. Oxford: W. H. Freeman.

Plotkin, H. C. 1994. *Darwin Machines and the Nature of Knowledge*. London: Penguin.

Purvis, A. 1992. Comparative Methods: Theory and Practice. DPhil Thesis, University of Oxford.

Purvis, A. 1995. A composite estimate of primate phylogeny. *Philosophical Transactions of the Royal Society of London, Series B*, **348**, 405–421.

Purvis, A. and Rambaut, A. 1995. Comparative analysis by independent contrasts (CAIC): an Apple Macintosh application for analysing comparative data. *Computer Applications in the Biosciences*, **11**, 247–251.

Purvis, A. and Webster, A. J. 1999. Phylogentically independent comparisons and primate phylogeny. In *Comparative Primate Socioecology*, ed. P. C. Lee, pp. 44–70. Cambridge: Cambridge University Press.

Purvis, A., Gittleman, J. L., and Luh, H.-K. 1994. Truth or consequences: effects of phylogenetic accuracy on two comparative methods. *Journal of Theoretical Biology*, **167**, 293–300.

Reader, S. M. and Laland, K. N. 2001. Primate innovation: sex, age and social rank differences. *International Journal of Primatology*, **22**, 787–805.

Reader, S. M. and Laland, K. N. 2002. Social intelligence, innovation and enhanced brain size in primates. *Proceedings of the National Academy of Sciences, USA*, **99**, 4436–4441.

Reader, S. M. and Lefebvre, L. 2001. Social learning and sociality. *Behavioral and Brain Sciences*, **24**, 353–355.

Rensch, B. 1956. Increase of learning capability with increase of brain size. *American Naturalist*, **90**, 81–95.

Rensch, B. 1957. The intelligence of elephants. *Scientific American*, **196**, 44–49.

Richerson, P. J. and Boyd, R. 2000. Climate, culture and the evolution of cognition. In *The Evolution of Cognition*, ed. C. Heyes, and L. Huber, pp. 329–346. Cambridge, MA: MIT Press.

Riddell, W. I. 1979. Cerebral indices and behavioral differences. In *Development and Evolution of Brain Size: Behavioral Implications*, ed. M. E. Hahn, C. Jenson, and B. C. Dudek, pp. 89–109. New York: Academic Press.

Rogers, A. R. 1988. Does biology constrain culture? *American Anthropologist*, **90**, 819–831.

Roper, T. J. 1986. Cultural evolution of feeding behaviour in animals. *Science Progress*, **70**, 571–583.

Rowe, N. 1996. *The Pictorial Guide to the Living Primates*. New York: Pogonias Press.

Sasvàri, L. 1979. Observational learning in great, blue and marsh tits. *Animal Behaviour*, **27**, 767–771.

Sasvàri, L. 1985. Different observational learning capacity in juvenile and adult individuals of congeneric bird species. *Zietschrift für Tierpsychologie*, **69**, 293–304.

Sawaguchi, T. 1997. Possible involvement of sexual selection in neocortical evolution of monkeys and apes. *Folia Primatologica*, **65**, 95–99.

Sawaguchi, T. and Kudo, H. 1990. Neocortical development and social structure in primates. *Primates*, **31**, 283–290.

Schlichting, C. D. and Pigliucci, M. 1998. *Phenotypic Evolution: A Reaction Norm Perspective*. Sunderland, MA: Sinauer.

Sherry, D. F. and Galef, B. G., Jr. 1984. Cultural transmission without imitation: milk bottle opening by birds. *Animal Behaviour*, **32**, 937–938.

Sherry, D. and Galef, B. G., Jr. 1990. Social learning without imitation: more about milk bottle opening by birds. *Animal Behaviour*, **40**, 987–989.

Shettleworth, S. J. 1998. *Cognition, Evolution, and Behaviour*. Oxford: Oxford University Press.

Smuts, B. B., Cheney, D. L., Seyfarth, R. M., Wrangham, R. W., and Struhsaker, T. T. (ed.) 1987. *Primate Societies*. Chicago, IL: University of Chicago Press.

Sol, D. and Lefebvre, L. 2000. Forebrain size and foraging innovations predict invasion success in birds introduced to New Zealand. *Oikos*, **90**, 599–605.

Sol, D., Lefebvre, L., and Timmermans, S. 2002. Behavioural flexibility and invasion success in birds. *Animal Behaviour*, **63**, 495–502.

Stephan, H., Frahm, H., and Baron, G. 1981. New and revised data on volumes of brain structure in insectivores and primates. *Folia Primatologica*, **35**, 1–29.

Timmermans, S., Lefebvre, L., Boire, D., and Basu, P. 2000. Relative size of the hyperstriatum ventrale is the best predictor of feeding innovation rate in birds. *Brain, Behavior and Evolution*, **56**, 196–203.

Tooby, J. and Cosmides, L. 1989. Evolutionary psychology and the generation of culture, part I. *Ethology and Sociobiology*, **10**, 29–49.

van Lawick-Goodall, J. 1970. Tool-using in primates and other vertebrates. *Advances in the Study of Behaviour*, **3**, 95–249.

van Schaik, C. P., Deaner, R. O., and Merrill, M. Y. 1999. The conditions for tool use in primates: implications for the evolution of material culture. *Journal of Human Evolution*, **36**, 719–741.
Whiten, A. and Byrne, R. 1988. Taking (Machiavellian) intelligence apart: editorial. In *Machiavellian Intelligence: Social Expertise and the Evolution of Intellect in Monkeys, Apes and Humans*, ed. R. W. Byrne, and A. Whiten, pp. 50–65. Oxford: Oxford University Press.
Whiten, A. and Byrne, R. W. 1997. *Machiavellian Intelligence II: Extensions and Evaluations*. Cambridge: Cambridge University Press.
Wilson, A. C. 1985. The molecular basis of evolution. *Scientific American*, **253**, 148–157.
Wilson, A. C. 1991. From molecular evolution to body and brain evolution. In *Perspectives on Cellular Regulation: From Bacteria to Cancer*, ed. J. Campisi, D. Cunningham, M. Inouye, and M. Riley, pp. 331–340. New York: Wiley-Liss.
Wyles, J. S., Kunkel, J. G., and Wilson, A. C. 1983. Birds, behaviour, and anatomical evolution. *Proceedings of the National Academy of Sciences, USA*, **80**, 4394–4397.
Wynn, T. 1988. Tools and the evolution of human intelligence. In *Machiavellian Intelligence*, ed. R. W. Byrne, and A. Whiten, pp. 271–284. Oxford: Oxford University Press.
Zilles, K. and Rehkamper, G. 1988. The brain, with special reference to the telencephalon. In *Orang-Utan Biology*, ed. J. H. Schwartz, pp. 157–176. Oxford: Oxford University Press.

LOUIS LEFEBVRE AND JULIE BOUCHARD

4

Social learning about food in birds

4.1 Introduction

Since the classic studies on potato and wheat washing in Japanese macaques (Kawai, 1965), traditions have often been studied in nonhuman animals because they represent an important precursor to human culture. This anthropocentric program has led many researchers to study primates and to focus on cognitive traits that are associated with human culture, for example imitation, language, tool use, and theory of mind. In this perspective, the study of nonhuman culture has recently culminated in the demonstration that wild chimpanzees in seven African populations show as many as 39 behavioral variants that may be attributed to "culture" (Whiten *et al.*, 1999). For psychologists and anthropologists, the concern with precursors of human behavior in the closest relatives of *Homo sapiens* is perfectly justified. For biologists, however, the evolution of cognition must be studied on a much broader and phylogenetically distant set of taxa; in comparative biology (Harvey and Pagel, 1991), one of the goals is to remove phylogenetic influences from taxonomic data and to look for independent evolution of traits as adaptations to particular ecological and life-history conditions.

In this chapter, we compare the origin and diffusion of new feeding behaviors in birds and mammals. We begin by explaining why birds are particularly suitable to a comparison with mammals, and we discuss the use of anecdotal reports in the study of cognition. We then highlight three features by which the current literature on birds appears to differ from that on mammals and propose hypotheses to explain the differences. If this literature is an unbiased estimate of real differences between birds and mammals, the differences raise important questions on the evolution of social learning and innovations. If current trends are a consequence of research biases, the apparent differences between birds and mammals point to gaps in our knowledge that need to be filled.

4.2 Why birds are important for the study of cognition and traditions

In a comparative approach centered on independent evolution, birds are a particularly interesting group for the study of social learning and cognition. The ancestors of modern-day birds and mammals diverged more than 300 million years ago (Hedges *et al.*, 1996). Current avian orders are thought to have appeared 100 to 150 million years ago (Cooper and Penny, 1997; Cracraft, 2001; Hedges *et al.*, 1996). If similar cognitive traits are found in some mammalian and avian taxa, it is unlikely that common ancestry could be behind the similarity. The molecular relationships between modern bird taxa have been worked out for the entire class (approximately 10 000 species; Sibley and Ahlquist, 1990; Sibley and Monroe, 1990), so phylogenetic confounds can be removed from any comparative study. At least seven avian taxa appear to have independently evolved large brains (Fig. 4.1; based on data for 737 species in Mlíkovský, 1989a,b,c, 1990; see Nicolakakis, Sol, and Lefebvre, 2002 for details): Piciformes (woodpeckers), Bucerotiformes (hornbills), Psittaciformes (parrots), Strigi (owls), Accipitrida and Falconida (hawks, eagles, and falcons), Ciconiida (herons and penguins) and Passeriformes (suboscines and oscines, especially corvids). These taxa represent a wide range of ecological adaptations, from tropical nut eating in parrots to nocturnal carnivory in owls, polar piscivory in penguins, insect eating in woodpeckers and carrion eating in corvids. Based on embryological, neuromorphological, cytoarchitectonic, and cytochemical evidence, Dubbeldam (1998), Karten (1991) and Rehkämper and Zilles (1991) have underlined the similarities between the mammalian neocortex and parts of the avian telencephalon like the hyperstriatum ventrale and neostriatum. In large-brained taxa, these are the structures that show the largest relative increase in size (Boire, 1989; Rehkämper, Frahm, and Zilles, 1991), just as the neocortex does in mammals (Stephan, Baron, and Frahm, 1988).

Birds occupy environments that range from polar landmasses to open seas and deserts. Ecological and life-history variables thought to be associated with complex cognition (e.g., generalism, group living, slow development) show large variation within the class Aves. There are small, rapidly

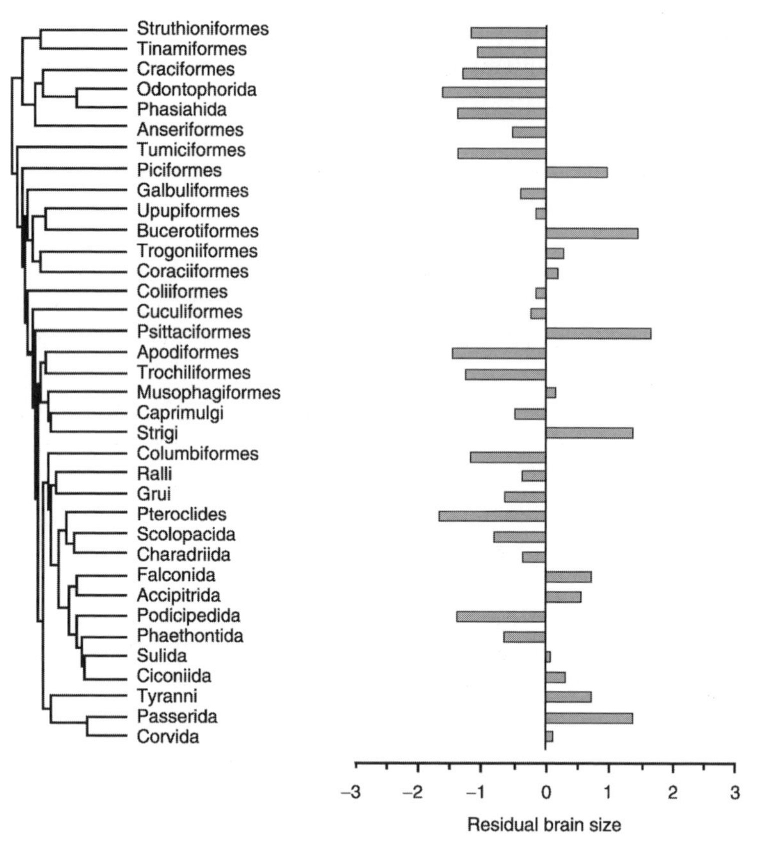

Fig. 4.1. Mean residual brain size (regressed against body weight) for avian orders and parvorders, based on data for 737 species in Mlíkovský (1989a–c; 1990). Phyletic tree and branch lengths based on Sibley and Ahlquist (1990).

developing species like quail and large, slowly developing species like parrots. Some birds, such as Florida scrub jays, live in cohesive groups with individual recognition and complex communication, while species like zenaida doves are solitary feeders year-round. Species like snail kites have specialized, conservative diets, while others are extreme opportunistic generalists, for example, crows and gulls. Finally, birds are the most frequently and easily observed animal taxon in the wild. Their vocalizations, flight, and color make them easier to detect than many other taxa. They are also the only animal taxon for which a popular term, "birder", exists to describe the thousands of amateurs and academics who observe and report every peculiarity of their morphology, behavior, and demographics in a large array of specialized journals. The short notes from these journals are a unique data source for the study of cognition in the field (Lefebvre, 2000; Lefebvre *et al.*, 1997b, 1998; Nicolakakis and Lefebvre, 2000; Nicolakakis *et al.*, 2002; Sol, Timmermans, and Lefebvre, 2002).

The taxonomic distribution of field reports on cognitive abilities can be a powerful tool for comparative analysis. It provides a quantitative, ecologically relevant operationalization of cognition on a wide array of species; it complements the experimental method and corrects for disadvantages like the arbitrariness of many experimental tasks, the small number of species tested, and the possible confounding roles of response to captivity, stimuli associated with the task, and avoidance of human experimenters (Lefebvre, 1995a; Lefebvre, Palameta, and Hatch, 1996). Several variables are likely to bias the field reports. Up to now, phylogeny, juvenile development mode, species number per taxonomic group, research effort, interest by birders, journal source, historical period, population size, and likeliness to notice and report an innovation have all been incorporated into multivariate analyses and shown not to account for the relationship between innovations and either neural substrate size (Lefebvre *et al.*, 1998, Lefebvre, Nicolakakis, and Boire, 2001; Nicolakakis and Lefebvre, 2000) or invasion success (Sol *et al.*, 2002) in birds. Comparative analysis of field report frequencies has been applied to deception in primates (Byrne, 1993), play in mammals (Iwaniuk, Nelson, and Pellis, 2001), innovation (Lefebvre *et al.*, 1997b, 1998, 2001) and tool use (Lefebvre *et al.*, 2001) in birds, as well as social learning, innovation, and tool use in primates (Reader and Laland, 2002; see also Ch. 3).

Reports of innovative feeding techniques have always been an important part of the ornithological literature, particularly in countries of English tradition. In a 1956 article on novel feeding methods by wild birds, W. H. Thorpe encouraged both amateur and professional ornithologists to note "examples of the production of original or unusual actions by birds, however small the change". The relationship between feeding innovations and social learning has been studied in birds for many years. A decade before the studies on Japanese macaques, the first widely reported case of animal culture was the description of milk bottle opening by tits (Fisher and Hinde, 1949). The innovation was first noticed in 1921 in Swaythling, a small town in southern England. By the time Fisher and Hinde published their quantitative survey, the behavior had been reported in over 400 localities in the British Isles. Bottle opening soon became a textbook case for animal culture, although subsequent field data (Hinde and Fisher, 1951), experiments in captivity (Sherry and Galef, 1984, 1990), historical research (Ingram, 1998), and curve-fitting analyses (Lefebvre, 1995b) suggest that social contributions to learning, as implied by the common phrase "cultural transmission", may have only been one factor in the diffusion of the new behavior. Animal traditions are by no means limited to the spread of new behaviors, but innovations like bottle opening are the starting points for many studies because novelty is readily noticed in the field and, in an experiment, the introduction of a new behavior allows efficient control of alternative mechanisms. Birds are very useful for these kinds of experiment because they rely primarily on vision during feeding, and their reliance on olfactory cues that social context provides to mammals (e.g., Galef, 1996) is negligible (Campbell, Heyes, and Goldsmith, 1999). Finally, there is a very large literature on acoustic forms of social learning and traditions in birds (see Ch. 8). If some acoustic and visual forms of social learning are linked (Moore, 1992) and share a common neural substrate (Iacoboni *et al.*, 1999), the body of knowledge accumulated on birdsong could provide useful directions for the study of nonvocal traditions.

4.3 Trends in the current literature: do birds and mammals differ?

A review of the literature on avian social learning and innovation reveals three surprising trends. First, there are no avian taxa where experiments on social learning have failed, contrary to the situation in some mammals. Second, all attempts to show motor imitation in birds have been successful. Third, social learning in foraging contexts appears to be rare in birds, if one compares them with primates and takes into account the high rate of avian innovation. This last point raises an obvious caveat for the first two: if social learning reports concerning foraging in birds are rare, then any conclusions about trends in this small dataset should be tentative, all the more so if the rarity is a consequence of research biases. For the moment, we will assume that the literature on birds and mammals is an unbiased sample of the true state of affairs and examine the possible origins of the differences. We will return to the question of biases later on in the chapter.

Tables 4.1 to 4.4 list all cases we could find of socially learned feeding behaviors in birds, including anecdotal and experimental reports on free-ranging and captive species. Fisher and Hinde (1949) was used as the

Table 4.1. *Anecdotal reports (72 cases) of possible social learning in the field*

Reference	Species	What is learned?[a]	Description of behaviour
Alcock 1970	*Neophron percnopterus*	4	Throwing stones to break ostrich eggs
Barash, Donovan and Myrick 1975	*Larus glaucescens*	4	Dropping clams on hard surfaces to crack them
Beck 1982	*Larus argentatus*	4	Dropping shellfish on hard surfaces to shatter them
Bowman and Billeb 1965	*Geospiza difficilis septentrionalis*	3, 4	Puncturing the skin of seabirds and feeding on the blood
Breitwisch and Breitwisch 1991	*Passer domesticus*	2, 4	Activating sensor of automatic sliding doors to enter a cafe and scavenge crumbs
Cook, Brower and Alcock 1969	*Megarhynchus pitangua*	3	Avoiding noxious food
	Myiodynastes maculatus		
Fisher and Hinde 1949	*Parus ater*	3, 4	Removing or tearing milk bottle tops to drink milk
	Erithacus rubecula		
	Fringilla coelebs		
	Parus caeruleus		
	Parus major		
	Parus palustris		
	Passer domesticus		
	Prunella modularis		
	Sturnus vulgaris		
	Turdus merula		
	Turdus philomelos		
Fritz *et al.* 1999	*Anser anser*	3, 4	Biting and chewing stems of butterbur
Götmark 1990	*Sterna sandvicensis*	2	Locating clumped and unpredictable food sources
Greig-Smith 1978	*Nectarinia dussumieri*	2	Locating clumped and unpredictable food sources

(*cont.*)

Table 4.1. (*cont.*)

Reference	Species	What is learned?[a]	Description of behaviour
Hinde and Fisher 1951	*Corvus monedula*	3, 4	Removing or tearing milk bottle tops to drink milk
	Dendrocopos major		
	Pica pica		
Hino 1998	*Coracina cinerea*	3, 4	Changing feeding habits and diet when foraging with other species
	Dicrurus forficatus		
	Newtonia brunneicauda		
	Phyllastrephus madagascariensis		
	Tersiphone mutata		
Lawton and Guindon 1981	*Psilorhynus morio*	3, 4	Identifying and catching appropriate food items
LeCroy 1972	*Sterna dougallii*	4	Fish catching by juveniles
	Sterna hirundo		
MacDonald and Henderson 1977	*Cephalopyrus flammiceps, Certhia himalayana, Dendrocopos himalayensis, Ficedula tricolor, Muscicapa ruficauda, Muscicapa sibirica, Parus major, Parus melanolophus, Parus monticolus, Parus rubidiventris, Passer rutilans, Pericrocotus ethologus, Phylloscopus inornatus, Phylloscopus occipitalis, Phylloscopus proregulus, Phylloscopus trochiloides, Regulus regulus, Sitta europaea, Sitta leucopsis*	2	Locating clumped and unpredictable food sources

Maclean 1970	*Turdus migratorius*	2, 4	Feeding on juniper berries and toyon fruits by hovering
Meinertzhagen 1954	*Pandion haliaetus*	4	Flying and catching fish by juveniles
Murton 1970	*Columba palumbus*	3, 4	Looking at what others are eating and copying feeding actions
Murton and Isaacson 1962	*Columba palumbus*	2	Locating clumped and unpredictable food sources
Newton 1967	*Carduelis cannabina, Carduelis spinus*	3	Differences between populations in preferred seed diet
Norton-Griffiths 1967	*Haematopus ostralegus*	4	Opening mussels by fledglings
Pettersson 1956	*Chloris chloris*	3, 4	Feeding on seeds of a shrub fruit by cracking the stones
Ramsay and Cushing 1949	*Anas platyrhynchos, Anas rubripes*	3, 4	Eating dry cornmeal and washing it down with water
Rowley and Chapman 1986	*Eolophus roseicapilla*	3, 4	Choosing and obtaining food
Rubenstein *et al.* 1977	*Sporophila corvina, Sporophila torqueola, Tiaris olivacea*	2	Locating clumped and unpredictable food sources
Stenhouse 1962	*Carduelis flammea*	3	Feeding on peach and apricot blossoms
Sullivan 1984	*Picoides pubescens, Picoides villosus*	2, 4	Locating food patches and choosing most efficient feeding technique
Taylor 1972	*Carpodacus mexicanus*	3, 4	Feeding on nectar from artificial feeders by hovering
Turner 1961	*Fringilla coelebs, Passer domesticus*	3	Eating previously avoided novel food
Werner and Sherry 1987	*Pinaroloxias inornata*	3, 4	Maintaining feeding specializations (diet and foraging techniques)

[a] 1, When to eat; 2, where to eat; 3, what to eat (and not to eat); 4, How to eat.

Table 4.2. *Anecdotal reports (eight cases) of possible social learning in captivity*

Reference	Species	What is learned?[a]	Description of behavior
Cadieu and Cadieu 1996	*Serinus canaria*	3, 4	Choosing and husking seeds by juveniles
Garnetzke-Stollman and Franck 1991	*Forpus conspicillatus*	1	Synchronizing foraging among group members
Hailman 1961	*Larus atribilla*	3, 4	Pecking at sibling's bill tip to establish discriminatory feeding response
Jones and Kamil 1973	*Cyanocitta cristata*	4	Manipulating pieces of paper to reach otherwise inaccessible food
Ligon and Martin 1974	*Gymnorhinus cyanocephalus*	3, 4	Distinguishing between good and bad seeds; opening the seeds
Millikan and Bowman 1967	*Geospiza conirostris*	4	Manipulating sticks to pry food items out of narrow cracks
Stokes 1971	*Gallus gallus*	3	Chicks learning to recognize food items
Weidmann 1957	*Anas platyrhynchos*	4	Shaking reeds to obtain snails

[a] 1, When to eat; 2, where to eat; 3, what to eat (and not to eat); 4, how to eat.

Table 4.3. *Experimental reports (20 cases) of social learning in the field*

Reference	Species	What is learned?[a]	Description of behaviour
Caldwell 1981	*Casmerodius albus, Egretta caerulea, Egretta tricolor*	2	Locating clumped and unpredictable food sources
Erwin, Hafner and Dugan 1985	*Egretta garzetta*	2	Locating clumped and unpredictable food sources
Horn 1968	*Euphagus cyanocephalus*	2	Locating clumped and unpredictable food sources
Knight and Knight 1983	*Haliaeetus leucocephalus*	2	Locating clumped and unpredictable food sources
Krebs 1974	*Ardea herodias*	2	Locating clumped and unpredictable food sources
Kushlan 1977	*Ajaia ajaja, Egretta thula, Eudocimus albus, Mycteria americana, Plegadis falcinellus*	2	Locating clumped and unpredictable food sources
Langen 1996a	*Calocitta formosa*	2–4	Locating, identifying and exploiting suitable food sources
Langen 1996b	*Calocitta formosa*	4	Opening a door to gain access to food
Lefebvre 1986	*Columba livia*	4	Piercing paper cover of a box containing seeds
Midford *et al.* 2000	*Aphelocoma coerulescens*	2, 4	Digging for peanut bits buried in sand at the center of a plastic ring
Roell 1978	*Corvus monedula*	2, 4	Locating food and extracting it from a ball of clay
Sasvári and Hegyi 1998	*Parus major, Parus palustris*	2	Approaching successful foragers, whether conspecific or not
Waite 1981	*Corvus frugilegus*	2	Locating clumped and unpredictable food sources

[a] 1, When to eat; 2, where to eat; 3, what to eat (and not to eat); 4, how to eat.

Table 4.4. *Experimental reports (56 cases) of social learning in captivity*

Reference	Species	What is learned?[a]	Description of behavior
Akins and Zentall 1996	*Coturnix japonica*	4	Operating a treadle to obtain food reward using same technique as demonstrator
Alcock 1969a	*Tyrannus savana*	3	Eating previously avoided food
Alcock 1969b	*Tyrannus savana, Zonotrichia albicollis*	4	Removing cover on food tray to obtain mealworm
Altshuler and Nunn 2001	*Archilochus colubris, Selasphorus platycercus*	3, 4	Feeding from a novel nectar source (a syringe)
Avery 1996	*Carpodacus mexicanus*	3	Avoiding noxious food
Barnard and Sibly 1981	*Passer domesticus*	2	Locating clumped and unpredictable food sources
Bednekoff and Balda 1996a	*Aphelocoma ultramarina, Nucifraga columbiana*	2	Relocating caches made by conspecifics
Bednekoff and Balda 1996b	*Gymnorhinus cyanocephalus*	2	Relocating caches made by conspecifics
Cadieu and Cadieu 1998	*Serinus canaria*	3	Recognizing and eating new food by juveniles
Campbell *et al.* 1999	*Sturnus vulgaris*	4	Removing same color lid using same technique as demonstrator
Cloutier and Newberry 2001	*Gallus gallus*	3, 4	Breaking a membrane on a container and consuming blood
Dawson and Foss 1965	*Melopsittacus undulatus*	4	Opening a covered food dish to obtain seeds
De Groot 1980	*Quelea quelea*	2	Locating clumped and unpredictable food sources
Dolman *et al.* 1996	*Zenaida aurita*	3	Eating previously avoided novel food
Fritz and Kotrschal 1999	*Corvus corax*	4	Opening a covered box containing food
Fritz *et al.* 2000	*Anser anser*	4	Pushing wooden bar to open gliding door and gain access to food
Fryday and Greig-Smith 1994	*Passer domesticus*	3	Avoiding noxious food associated with visual cue
Fryday and Greig-Smith 1994	*Passer domesticus*	3	Eating novel colored food

Hatch and Lefebvre 1997	*Streptopelia roseogrisea*	3	Eating novel foods
Hatch and Lefebvre 1997	*Streptopelia roseogrisea*	4	Opening lid or pulling open drawer to access food
Klopfer 1957	*Anas platyrhynchos, Cairina moschata*	3	Avoiding food dish associated with visual cue and electrical shock
Klopfer 1959	*Chloris chloris*	3	Avoiding unpalatable food associated with a visual cue
Krebs 1973	*Parus rufescens, Parus atricapillus*	2, 4	Locating clumped food patches; choosing feeding technique
Krebs, MacRoberts and Cullen 1972	*Parus major*	2	Locating clumped and unpredictable food sources
Lefebvre *et al.* 1996	*Zenaida aurita*	4	Removing stopper from inverted test tube to release seeds
Lefebvre *et al.* 1997a	*Quiscalus lugubris*	4	Removing stopper from inverted test tube to release seeds
Mason and Reidinger 1981	*Agelaius phoeniceus*	3	Eating novel food
Mason and Reidinger 1982	*Agelaius phoeniceus*	3	Avoiding noxious food associated with visual cue
Mason, Arzt and Reidinger 1984	*Quiscalus quiscula*	3	Avoiding or prefering food associated with a specific visual cue
McQuoid and Galef 1993	*Gallus gallus*	2	Feeding in same food dish as demonstrator
Mönkkönen and Koivula 1993	*Parus montanus*	3	Feeding on novel foods
Nicol and Pope 1994	*Gallus gallus*	4	Pecking correct colored key to gain access to food
Palameta 1989	*Serinus canaria*	4	Flipping cardboard lid to gain access to a well-containing food
Rothschild and Ford 1968	*Sturnus vulgaris*	3	Avoiding noxious novel food

(*cont.*)

Table 4.4. (*cont.*)

Reference	Species	What is learned?[a]	Description of behavior
Sasvári 1979	*Parus caeruleus, Parus major, Parus palustris*	4	Lifting piece of linen to obtain hidden food
Sasvári 1985b	*Turdus merula, Turdus philomelos*	4	Pulling string out of a glass cylinder to access seeds
Schildkraut 1974	*Cyanocitta cristata*	4	Pecking a disk to obtain food
Sherry and Galef 1984	*Parus atricapillus*	3, 4	Opening and drinking from cream tubs
Templeton *et al.* 1999	*Gymnorhinus cyanocephalus, Nucifraga columbiana*	4	Removing a lid to gain access to food
Turner 1964	*Fringilla coelebs, Gallus gallus*	3	Eating previously avoided novel food
Waite and Grubb 1988	*Parus bicolor, Parus carolinensis, Picoides pubescens, Sitta carolinensis*	2	Locating food sources
Wechsler 1988	*Corvus monedula*	4	Obtaining food from dispenser using same technique as demonstrator
Zentall and Hogan 1976	*Columba livia*	4	Pecking a response key for grain

[a] 1, When to eat; 2, where to eat; 3, what to eat (and not to eat); 4, how to eat.

cut-off point for modern studies and only cases published since then have been considered. A broad perspective was adopted in defining social learning; all cases presumed by the authors to involve local or stimulus enhancement, social facilitation, observational learning, and true imitation were included in the tables. Only reports on foraging were considered, excluding vocal learning, predator avoidance, mate choice, and other nonforaging behaviors. Species with multiple reports of the same behavior were included only once in the tables; without this precaution, pigeons and chickens, for example, would have been over-represented because of their widespread use in laboratory studies. For anecdotal reports (Tables 4.1 and 4.2), cases were included if the authors mentioned social learning as a possibility, without judging whether the authors were right or wrong; this same procedure was adopted in literature surveys of innovations (see examples in Lefebvre, 2000; Lefebvre *et al.*, 1997b, 1998; Nicolakakis and Lefebvre, 2000). An independent reader, naïve to the hypotheses tested, was asked to read a random sample of the literature reviewed ($n = 50$) and to decide whether the reports should be included in the database or not. For reports that were included, the independent reader classified each report as anecdotal or experimental and noted if it took place in the field or in captivity. The principal investigator and the independent reader agreed on inclusion of reports 96% of the time and agreed on the classification of included reports 100% of the time. Contrary to innovations (Lefebvre *et al.*, 1997b, 1998) or tool use (Lefebvre *et al.*, 2001), where anecdotes simply described behaviors involving new foods or feeding implements, social learning is an inferred mechanism, not an observed fact. Experiments often show that anecdotal claims of social learning can, in part (Sherry and Galef, 1984, 1990) or in whole (Galef, 1980), be attributed to other processes. The cases presented in Tables 4.1 and 4.2 should, therefore, be treated with caution and given temporary status only, subject to confirmation by controlled experiments. The survey yielded 72 anecdotal cases from the field (Table 4.1) and eight cases from captivity (Table 4.2). Experimental work has been done on 20 cases in the field (Table 4.3) and on 56 cases in captivity (Table 4.4).

4.3.1 Social learning

The first trend in the avian literature is the absence of negative results. All species in which social learning tests have been attempted eventually yielded positive results. Negative results were reported by some researchers, for example Hitchcock and Sherry (1995) on chickadees and de Perera and Guilford (1999) on pigeons. However, these were found in species where positive results have been obtained by others (chickadees: Alcock, 1969a; Krebs, 1973; Sherry and Galef, 1984, 1990; pigeons: Alderks, 1986; Epstein, 1984; Palameta and Lefebvre, 1985). In the case of Hitchcock and Sherry (1995) and de Perera and Guilford (1999), the social learning task was applied to a specialized ability, spatial memory. The trend in birds can be contrasted with that of mammals, where some species show no sign of even the simplest form of stimulus enhancement. Cattle (Veissier, 1993) and horses (Baer *et al.*, 1983; Baker and Crawford, 1986; Clarke *et al.*, 1996) yielded negative results when a naïve observer witnessed a conspecific demonstrator eating from a feeder identified with a visual cue. In the case of horses, the negative results have been replicated in three different laboratories. The common feature of these species is that they are grazing herbivores. They are also gregarious, a variable often assumed to favor social learning (Klopfer, 1961; Reader and Lefebvre, 2001). The food they specialize on is abundant and easily accessible, however, and requires extensive digestion because of its low nutritive content, but little searching and handling.

More research is needed before negative results on two species can be generalized to an entire dietary category like herbivory. Nevertheless, if the current literature is a correct estimate of broader trends, this raises the intriguing possibility that diet is a stronger selective pressure than sociality for the evolution of socially learned foraging (Reader and Lefebvre, 2001). Up to now, only the carefully controlled study of Templeton, Kamil and Balda (1999) on pinyon jays and Clark's nutcracker has supported the idea that social learning is more efficient in more social species, once the confounding effects of other types of learning have been accounted for. In other birds, interspecific differences in social learning paralleled differences in individual learning (Sasvári, 1985a,b; reanalyzed by Lefebvre and Giraldeau, 1996), irrespective of large differences in sociality (Lefebvre *et al.*, 1996). In primates, frequency of social learning reports per species was uncorrelated with group size, once phylogenetic effects were removed (Reader, 2000; Reader and Lefebvre, 2001).

Among birds, the closest thing to a herbivorous mammal is a goose. Unlike horses and cattle, geese show social learning of new food types in the field (Fritz, Bisenberger, and Kotrschal, 1999) and of new handling techniques in experiments conducted in captivity (Fritz, Bisenberger, and Kotrschal, 2000). Granivores, another avian group whose food source is abundant (if often patchy) and easy to handle, also show social learning.

Red-winged blackbirds, for example, are agricultural pests in many parts of North America because large flocks can descend on cornfields and clean out acres of grain. The studies of Mason and Reidinger (summarized by Mason, 1988) have repeatedly demonstrated intra- and interspecific social learning in this species and have indeed been designed to find socially transmissible solutions to the pest problem posed by this species. The feral pigeon is another granivore that has often been used in social learning experiments. In the field, pigeons and other Columbiformes do not use complex searching and handling techniques for food (primarily seed and its processed derivatives like bread in cities and stored grain in ports; Lévesque and McNeil, 1985; Murton, Coombs, and Thearle, 1972). Several experiments do show, however, that pigeons are capable of social learning (Alderks, 1986; Epstein, 1984; Palameta and Lefebvre, 1985). It is possible that in pre-agricultural times, seed was a much less abundant and easily obtained food than it is today, but in the absence of at least one negative result on an avian species, we can only conclude for the moment that there is no obvious association between diet (food type abundance and complexity of searching and handling techniques) and socially learned feeding in birds.

4.3.2 Innovations

If social learning is advantageous when learning technically difficult foraging behaviors, it also constitutes an efficient method for spreading innovations. Because they are so rare, innovations have a low probability of being incorporated into an individual's repertoire unless that individual observes an innovator. Allan Wilson recognized in the early 1980s that innovation and social learning, when they co-occur in a large-brained species, provide a powerful means for new behaviors to spread rapidly through entire populations (Wilson, 1985; Wyles, Kunkel, and Wilson, 1983). If these new behaviors expose their bearers to a wider array of environmental conditions, they can increase the rate at which favorable mutations are fixed by natural selection. Wilson called this accelerating effect "behavioral drive" and was concerned about its possible effects on molecular and structural estimates of the speed of evolution (Nicolakakis *et al.*, 2002; Wilson, 1985).

Innovations have been extensively studied in birds in our laboratory for the past few years (Lefebvre 2000; Lefebvre *et al.*, 1997b, 1998; Nicolakakis and Lefebvre, 2000). The frequency of feeding innovations per taxonomic group is positively correlated with relative size of the telencephalon, in particular with that of structures that are analogous to the mammalian neocortex: the hyperstriatum ventrale and the neostriatum (Timmermans *et al.*, 2000). Reader and Laland (2002; see also Ch. 3) have found a similar relationship in primates; in this order, innovation frequency per species is correlated with relative size of the neocortex and striatum. The fact that analogous neural structures are correlated with similar cognitive traits in such distant taxa as birds and primates is powerful evidence for repeated independent evolution. We could consequently expect that other correlates of innovative behavior would be similar in birds and primates, especially if diffusion of rare, innovative behaviors is an important outcome of social learning in the two taxa. This appears not to be the case. If one looks at the relative frequencies of innovation and social learning reports in the two groups, primates and birds show different trends. In his review of the primate literature (234 species), Reader (Ch. 3) gathered a total of 558 cases of innovation and 451 cases of social learning. Of these, approximately equal numbers were field anecdotes on feeding innovations ($n = 142$) and on socially learned foraging ($n = 153$). In birds (approximately 10 000 species), innovations seem to outnumber social learning reports. Only 72 anecdotal cases of social learning in the wild are listed in Table 4.1, compared with the 1796 feeding innovation reports currently included in our database (Lefebvre, 2000; Lefebvre *et al.*, 1997b, 1998; Nicolakakis and Lefebvre, 2000; Nicolakakis *et al.*, 2002; Sol *et al.*, 2002; Timmermans *et al.*, 2000), which covers a shorter time period (1970 to the present in most zones) and a more restricted geographical area (six zones of the world) than the social learning survey (1949 to the present; worldwide). What these relative numbers seem to suggest is that a feeding innovation does not as readily spread to others in birds as it does in primates.

4.4 Why do birds and mammals seem to differ?

The differences between primates and birds could reflect real trends or they could be a result of research and publication biases. Researchers and journal editors may expect more social learning in primates because of the phyletic proximity of these species to humans, their large brains, and their extensive social relationships. However, anecdotal reports of the type used in innovation analysis are often criticized in psychology and primatology (see the open peer commentary following Byrne and Whiten, 1988). This could decrease the probability that primate innovations will be noticed, written up, and/or published. In contrast, short notes on new ranging, feeding, and nesting behaviors are encouraged in ornithology. The journal *British Birds*, for example, has an eight member "Behavior Notes Panel" specifically set up to referee these contributions. Because social learning is a technical concept that is inferred, not directly seen, and because there are many more nonacademic ornithologists than there are nonacademic primatologists, it might also be that birders notice unusual feeding behaviors more easily than they do cases of an abstract phenomenon like social learning (D. Sol, personal communication). Finally, the biases may lead to low sampling effort, which can lead to spurious trends. For example, the fact that imitation has been tested in only six avian species may have created a false positive, and negative findings will eventually emerge as more species are studied.

Other arguments, however, suggest that the differences might be real. Historically, the first widely cited modern case of social learning (Fisher and Hinde, 1949) was reported in birds by researchers from a prestigious university, Cambridge. Tool use, which is often cited as a covariate of social learning (see Chs. 3 and 11) was described (1901) and reported (1919) in Darwin's finches (see Boswall, 1977) long before it was in chimpanzees (Goodall, 1964). The discovery of a new tool-use case in birds is as newsworthy as it is in primates, as evidenced by the publication in *Nature* of Hunt's (1996) report on leaf tools in New Caledonian crows. The large number of papers on vocal imitation in birds further suggests that interest in socially learned behaviors is high in ornithology.

Finally, a rough estimate of research bias for field anecdotes can be obtained by counting experimental studies. If researchers are as interested in social learning as they are in innovations, the relative number of deliberate, organized studies involving experiments should be similar. In primates, this is the case: the number of social learning ($n = 84$) and innovation ($n = 113$) cases based on experimental work are approximately equal, and the number of cases based on anecdotes is also in the same order of magnitude ($n = 153$ for social learning and $n = 142$ for innovation). In birds, the number of social learning experiments ($n = 76$) cannot be compared with innovation experiments as this figure is not available. However, we know that the number of social learning cases based on anecdotes ($n = 72$) is similar to the number of experiments. If the number of social learning experiments in birds can be inferred from the primate pattern, we would expect no more than 70 to 150 innovation anecdotes if research effort were the sole determinant of their numbers. Instead, the sample so far includes 1796 innovation anecdotes, 30 times more than expected. We will, therefore, tentatively assume that the differences may reflect real trends and review the possible reasons for the apparent rarity of socially transmitted feeding innovations in birds.

4.4.1 Individual and social learning

In birds, many innovations are single events that surprise the ornithologist and may never be seen again in the originator or in birds that are within observational range of this individual. The innovation can reflect temporary opportunism and flexibility, but it may not be incorporated into the long-term repertoire of the animal if normal food types or handling techniques yield higher payoffs. The question, therefore, becomes one of learning in general, both individual and social. Payoffs (as they are conceptualized in behavioral ecology) and reinforcements (as they are conceived in psychology) associated with new versus old foods and techniques determine the likelihood that the innovation will be repeated. If the innovation is rare because of its difficulty, it will be unlikely that others will acquire it because observers will have a low probability of seeing innovators repeat the new behavior. Individual and social learning are thus linked. If an innovation has a higher probability of being incorporated into the long-term repertoire of the originator in primates than it does in birds, this alone could lead to differences in social learning trends. On average, most birds are more mobile than primates; this mobility in itself may decrease the probability of repeating an innovative behavior done in a particular place and context.

4.4.2 Environmental factors

Tebbich, Taborsky, Fessl and Dvorak (2002) have recently looked at ecological variation in twig tool use by Darwin's finches in the Galapagos Islands. In habitats and seasons that are extremely dry, insects withdraw into crevices to conserve water and cannot be found by gleaning on the ground. In humid habitats, gleaning is possible year-round and, in this situation, Darwin's finches do not use tools but search instead through the ground vegetation with their beaks. A similar study by Higuchi (1987) on green-backed herons documented individual differences in the use of bait-fishing in different habitats. On territories where the water is deep and herons fish from branches, lures are seldom used and individuals using them are not very successful; both lure use and success are high when water is shallow and there are many rocks and bushes for the heron to use for concealment. The studies on herons and finches suggest that birds do not use tools unless environmental conditions make alternative techniques less profitable.

Tebbich *et al.* (2001) have looked at the relative roles of social learning and individual practice in using twigs by finches caught in the more humid areas where they do not normally use twigs as tools. The striking result is that wild-caught finches spontaneously used twigs to feed on prey experimentally presented to them in cavities and that individual practice was as good as social learning at increasing the efficiency of birds over time. These results again underline the fact that the absence or low frequency of presumably cognitively demanding behaviors in many wild birds can reflect environmentally determined pay-offs rather than intrinsic abilities. The study by Tebbich *et al.* (2001) also underlines the fact that many presumably complex foraging techniques in birds may not require social learning, as Sherry and Galef (1984, 1990) have shown for bottle opening in Paridae. An obvious point for future research in the field would be to measure the relative efficiency of simple and complex, socially learned handling techniques in birds and primates. The usual foraging currency of nutrients per unit time should be used, as Tebbich *et al.* (2001) have done for twig use and gleaning. It might very well be that, in many situations, the net energetic benefit of foods obtained through complex techniques might be lower than that of foods obtained through simpler means for birds, if only because of morphological limitations. If this is so, the cognitive potential revealed in captive studies would be less relevant than the economic variables that govern foraging decisions in the wild (see the discussion of costs and benefits by Dewar in Ch. 5).

4.4.3 Tool use and morphology of food-handling organs

Many cases of socially learned foraging appear to involve food types, but van Schaik (Ch. 11) has proposed that social learning and imitation may be crucial in mastering the complex motor acts required for tool use. Goal emulation may also help observers to persist in improving the initial inefficiency that characterizes early attempts at tool use (S. M. Reader, personal communication). If van Schaik (Ch. 11) is correct, low frequencies of social learning in birds might, in part, reflect the morphological limitations that make tool use (and its accompanying social learning) relatively awkward in many birds. In a review of the avian tool-use literature, Lefebvre *et al.* (2002) found 128 cases in 108 species. This is more than some authors have expected (e.g., Thomson, 1964, who cites only one case), but the numbers are more in line with those of avian social learning (total $n = 158$) than they are with those of primate tool-use frequency. Reader (Ch. 3 and Reader and Laland, 2002) has collected 607 cases of tool use in primates, 249 of them from the wild; these numbers are in the same range as those he has collected for social learning and innovations.

Many avian species show frequent use of tools: leaf probes in New Caledonian crows, prey dropping in gulls and corvids, use of rocks as shell-smashing anvils in song thrushes, use of lures to attract fish in green-backed herons. In many other cases, however, tool use is rare and seems to be used as a last resort. A case in point is Andersson's (1989) description of "egg"-breaking attempts by a fan-tailed crow in Kenya; the "egg" was a ping-pong ball and because its "shell" could not be broken, Andersson observed the entire sequence of techniques the crow had in its repertoire. The bird first tried the easiest one in terms of cognition and motor complexity, pecking at the "shell" with its beak. It then flew up with the "egg" and dropped it. When this failed, it clumsily attempted to hammer the shell with an oversize stone, switching at last to a stone of manageable size to increase hammering efficiency. What this example illustrates is the relative inefficiency of tool use in many avian cases. Morphological constraints may limit tool-use efficiency in many birds. Birds, even flightless ones, have wings instead of arms and hands. (The same limitation may apply to whales and dolphins, which have flippers and are thus hampered in their tool-use potential.) True tools in birds, ones that are held directly by the animal, are moved with the beak, which is a better tool in itself than a primate hand but a poorer implement mover. Bird beaks have become morphologically specialized to crush hard shells (parrots), hammer nuts and trees (woodpeckers), and probe deep into flowers (hummingbirds) or tidal flats (shorebirds). Primate hands are in general less morphologically specialized for handling. What the primate hand lacks in hardness or length is made up in dexterity and in affording sight of the object during handling, two qualities that the rigid beak of birds does not have. If, as proposed by van Schaik (Ch. 11), social learning is crucial to the adoption of similar forms of tool use by members of a social group in primates, then the converse inefficiency of many tool-using birds may be one factor behind the rarity of avian social learning reports from the wild. Again, more studies of tool-using efficiency in the field are required, similar to that of Tebbich *et al.* (2002) in Darwin's finches and those of Zach (1979) and Cristol and Switzer (1999) in shell-dropping corvids.

4.4.4 Payoffs to alternative behaviors

Another factor may be differing costs and benefits of social learning and innovation across the two taxa. In behavioral ecology, the use of a behavior by an animal in a given situation is first and foremost an economic problem, and only secondarily a question of cognitive ability. Animals that are perfectly capable of using a sophisticated ability may not do so in certain circumstances because alternative behaviors pay more. In group-living animals, payoffs are often frequency dependent. A dramatic example of this is the effect of scrounging on social learning in pigeons. In this species, the average caged observer requires only a few demonstrations of a new feeding technique before it learns it (Palameta and Lefebvre, 1985). If the naïve bird is foraging with the knowledgeable one in a group, however, it can witness hundreds of demonstrations of the new technique without incorporating it in its repertoire (Giraldeau and Lefebvre, 1987; Lefebvre and Helder, 1997). This is because group feeding often allows animals to profit from the discoveries of others, a situation known in behavioral ecology as the producer–scrounger game (Barnard and Sibly, 1981; Giraldeau and Caraco, 2000). When a new technique yields a feeding payoff that can be shared, producers learn it but scroungers do not, learning instead to follow knowledgeable producers (Giraldeau and Lefebvre, 1986).

Although scrounging clearly blocks learning in pigeons, its effect may not be general enough to account for the overall difference between primates and birds. First, several birds do not show the inhibitory effect of scrounging. In the field, scrub jays (Midford, Hailman, and Woolfenden, 2000) and ravens (Fritz, Bugnyar, and Kotrschal, 1997) learn even when they scrounge, while Nicol and Pope (1999) report similar results in captive chickens. Second, inhibitory effects of scrounging have also been reported in primates (Fragaszy and Visalberghi, 1989). In Japanese macaques, adult males are also known to scrounge in situations where access to food cannot be controlled by knowledgeable individuals, for example, wheat floating on water as opposed to potatoes held in the hand; in this situation, adult males do not learn to wash wheat but instead take it from washing individuals (Kawai, 1965).

4.4.5 Group structure and attention to others

A fifth possibility is group structure and the way individuals in a group pay attention to the feeding behaviors of others. In many avian species, flocks are no more than aggregations, with individuals feeding in close proximity but showing little social interactions beyond scramble competition. In pigeon flocks, for example, juveniles forage in the company of their sibling (Cole, 1996) and adults in the company of their mate (Lefebvre and Henderson, 1986), but interactions between unmated adults and between parents and offspring do not differ from chance (L. Lefebvre and K. K. Hatch, unpublished data). In ringdoves, juveniles do not learn from their father more readily than they do from a familiar, but unrelated adult (Hatch and Lefebvre, 1997), contrary to the parent–offspring transmission that has been suggested for social learning in Japanese macaques (Kawai, 1965). Some avian species (e.g., corvids and geese) show the complex, kin-based group structure typical of primates, but many bird flocks and colonies are more similar to ungulate herds than they are to primate troops. Dunbar (1998) has proposed that the number of interactions in a group is a limiting factor for intelligence and memory and has consequently been the main selective pressure for the evolution of neocortex size in primates. The complexity of relationships in large groups is only one of the factors that are thought to select for social intelligence in primates (for reviews, see Byrne and Whiten, 1988; Whiten and Byrne, 1997). Differences in social learning between birds and primates could, in part, be the result of differences in group structure and attention to others. In the only comparative study available on primates, Cambefort (1981; Jouventin, Pasteur, and Cambefort, 1976) found that differences in social learning among vervet monkeys, mandrills, and chacma baboons were indeed in the same direction as differences in gregariousness.

4.4.6 Territoriality

A sixth factor may be territoriality. In many species of birds, individuals defend exclusive access to a feeding area, either year-round or on a seasonal basis. Mates and fledglings may share a territory, but foraging is often solitary. In many primates, whole troops defend access to feeding ranges against other troops. Defense is still present but does not entail solitary foraging. Members of the group can thus observe each other feeding, even if considerable spacing is often seen between individuals within a troop. Solitary foraging, combined with defense, may have obvious detrimental effects on social learning. Not only are others rarely present to provide new feeding information, but whenever they are, territorial individuals focus on aggression, not observation of foraging techniques. The limiting effects of territoriality on social learning have been demonstrated in at least three avian species. In Barbados, the zenaida dove aggressively defends year-round territories in most parts of the island but feeds in flocks in restricted areas like the harbor, where seed spillage is available in large, temporally unpredictable patches. At the harbor, a feeding conspecific is a source of information about ephemeral patches, which could be rapidly depleted by a hundred competitors or cleaned up by a human; in territorial zones, a feeding conspecific is instead an intruder, which is immediately chased. Experiments have shown that territorial zenaida doves do not learn from conspecifics (Dolman, Templeton, and Lefebvre, 1996; Lefebvre *et al.*, 1996), but that group-feeding harbor doves do (Carlier and Lefebvre, 1997; Dolman *et al.*, 1996).

Two other cases involve feeding innovations witnessed in the field. In England, blue tits have learned to pierce the base of flowers to drink the nectar. This innovation is extremely localized, however, and, contrary to milk bottle opening, has not spread to neighboring areas or other birds (Thompson, Ray, and Preston, 1996). The flowers bloom during a short period in the spring; foragers aggressively defend territories during this period and do not yet have fledglings to witness the behavior and assure its vertical transmission. Thompson *et al.* (1996) have suggested that the localized nature of the innovation may be a consequence of these two factors. In Barbados, territorial bullfinches have also been seen to use a localized feeding innovation. At one hotel on the Caribbean coast, bullfinches pierce small paper packets of sugar and eat the contents; sugar eating is a frequent behavior in this species, but this is usually done at open bowls. Reader, Nover, and Lefebvre (see Ch. 3) presented closed sugar packets at several sites along the Caribbean coast of Barbados but saw packet opening only at the single hotel site, suggesting a localized distribution of the innovation. Territorial exclusion is the most plausible explanation for this limited transmission, intruders being chased away by residents as soon as they approach the potential learning site. Beyond these three examples, it is impossible to tell for the moment if territoriality has a general limiting effect on avian social learning.

4.5 Conclusions

Looking at the avian and mammalian (especially primate) literature on socially transmitted feeding behaviors, we are left with a set of apparent paradoxes. Herbivorous birds like geese show social learning, contrary to herbivorous mammals like cattle and horses. Granivorous birds like budgerigars, pigeons, and quails show imitation, even if they use simple food-handling skills in the wild. Birds yield thousands of feeding innovation reports, but only a few dozen cases of social learning. Researchers seem to find avian social learning and imitation every time they look for it in captive experiments, but field reports are relatively rare. It is possible that these paradoxes result from research biases and to low interest for socially learned foraging on the part of ornithologists. More research is obviously needed to increase sample sizes on bird feeding traditions and to target specific taxa. Feeding imitation should, for example, be studied in hummingbirds, who are already known to show stimulus forms of social learning (Altshuler and Nunn, 2001). Comparative experiments on herbivorous (e.g., geese, Fritz *et al.*, 1999, 2000) versus omnivorous (e.g., ravens, Fritz and Kotrschal, 1999) species could also help us to understand the role of diet in the evolution of avian social learning. Beyond these limitations in the current dataset, however, it is possible that real differences exist between avian and primate social learning. Six potential sources for the differences have been discussed above, which could be compounded with basic differences in neural substrate size. Even if a crow has a much larger neostriatum/hyperstriatum ventrale than a quail (five times larger relative to brainstem size; Rehkämper *et al.*, 1991), a primate is still much farther from the small-brained end of its class than is a corvid. A baboon has a neocortex/brainstem ratio that is 30 times the size of that of an insectivore like the tenrec (Stephan *et al.*, 1988). The difference is even more extreme for a chimpanzee: over 50 times larger (Stephan *et al.*, 1988). There is clearly more association area in a primate brain than in even the largest-brained bird. The combination of these neural differences with differences in mobility, group structure, territoriality, payoffs to alternatives, and morphology of food-handling organs could have a multiplicative effect on many cognitive traits, offering a possible explanation for the contrasting trends in primate and avian social learning.

4.6 Acknowledgements

This research was supported by a grant from NSERC to L.L. and a graduate scholarship from NSERC to J.B.

References

Akins, C.K. and Zentall, T.R. 1996. Imitative learning in male Japanese quail (*Coturnix japonica*) using the two-action method. *Journal of Comparative Psychology*, **110**, 316–420.

Akins, C.K. and Zentall, T.R. 1998. Imitation in Japanese quail: the role of reinforcement of demonstrator responding. *Psychonomic Bulletin and Review,* **5**, 694–697.

Alcock, J. 1969a. Observational learning in three species of birds. *Ibis,* **111**, 308–321.

Alcock, J. 1969b. Observational learning by fork-tailed flycatchers (*Muscivora tyrannus*). *Animal Behaviour,* **17**, 652–657.

Alcock, J. 1970. The origin of tool-using by Egyptian vultures. *Ibis,* **112**, 542.

Alderks, C.E. 1986. Observational learning in the pigeon: effects of model's rate of response and percentage of reinforcement. *Animal Learning and Behavior,* **14**, 331–335.

Altshuler, D.L. and Nunn, A.M. 2001. Observational learning in hummingbirds. *Auk,* **118**, 795–799.

Andersson, S. 1989. Tool use by the fan-tailed raven (*Corvus rhipidurus*). *Condor,* **91**, 999.

Avery, M.L. 1996. Food avoidance by adult house finches, *Carpodacus mexicanus*, affects seed preferences of offspring. *Animal Behaviour,* **51**, 1279–1283.

Baer, K.L., Potter, G.D., Friend, T.H., and Beaver, B.V. 1983. Observational effects on learning in horses. *Applied Animal Behaviour Science,* **11**, 123–129.

Baker, A.E.M. and Crawford, B.H. 1986. Observational learning in horses. *Applied Animal Behaviour Science,* **15**, 7–13.

Barash, D.P., Donovan, P., and Myrick, R. 1975. Clam dropping behaviour of the glaucous-winged gull (*Larus glaucescens*). *Wilson Bulletin,* **87**, 60–64.

Barnard, C.J. and Sibly, R.M. 1981. Producers and scroungers, a general model and its application to captive flocks of house sparrows. *Animal Behaviour,* **29**, 543–550.

Beck, B.B. 1982. Chimpocentrism: bias in cognitive ethology. *Journal of Human Evolution,* **11**, 3–17.

Bednekoff, P.A., and Balda, R.P. 1996a. Observational spatial memory in Clark's nutcrackers and Mexican jays. *Animal Behaviour,* **52**, 833–839.

Bednekoff, P.A. and Balda, R.P. 1996b. Social caching and observational spatial memory in pinyon jays. *Behaviour,* **133**, 807–826.

Boire, D. 1989. Comparaison quantitative de l'encéphale, de ses grandes subdivisions et de relais visuels, trijumaux et acoustiques chez 28 espèces d'oiseaux. PhD Thesis, Université de Montréal, Montréal, Canada.

Boswall, J. 1977. Tool using by birds and related behaviour. *Avicultultural Magazine,* **83**, 88–97, 146–159, 220–228.

Bowman, R.I. and Billeb, S.L. 1965. Blood-eating in a Galapagos finch. *Living Bird,* **4**, 29–44.

Breitwisch, R. and Breitwisch, M. 1991. House sparrows open an automatic door. *Wilson Bulletin,* **103**, 725–726.

Byrne, R.W. 1993. Do larger brains mean greater intelligence. *Behavioral and Brain Sciences,* **16**, 696–697.

Byrne, R.W. and Whiten, A. 1988. *Machiavellian Intelligence*. Oxford: Clarendon Press.

Cadieu, J.C. and Cadieu, N. 1996. Influence of some interactions between fledglings and adults on the food choice in young canaries (*Serinus canarius*). *Journal of Ethology,* **14**, 99–109.

Cadieu, N. and Cadieu, J.C. 1998. Is food recognition in an unfamiliar environment a long-term effect of stimulus or local enhancement? A study in the juvenile canary. *Behavioural Processes,* **43**, 183–192.

Caldwell, G. S. 1981. Attraction to tropical mixed-species heron flocks: proximate mechanism and consequences. *Behavioral Ecology and Sociobiology*, **8**, 99–103.

Cambefort, J. P. 1981. Comparative study of culturally-transmitted patterns of feeding habits in the chacma baboon (*Papio ursinus*) and the vervet monkey (*Cercopithecus aethiops*). *Folia Primatologica*, **36**, 243–263.

Campbell, F. M., Heyes, C. M., and Goldsmith, A. R. 1999. Stimulus learning and response learning by observation in the European starling, in a two-object/two-action test. *Animal Behaviour*, **58**, 151–158.

Carlier, P. and Lefebvre, L. 1997. Ecological differences in social learning between adjacent, mixing populations of zenaida doves. *Ethology*, **103**, 772–784.

Clarke, J., Nicol, C., Jones, R., and McGreevy, P. 1996. Effects of observational learning on food selection in horses. *Applied Animal Behaviour Science*, **50**, 177–184.

Cloutier, S. and Newberry, R. C. 2001. Cannibalistic behaviour is influenced by social learning. *Advances in Ethology*, **36**, 138.

Cole, H. 1996. Brothers and sisters in arms: sibling coalitions in juvenile pigeons. MSc Thesis, McGill University, Montréal, Canada.

Cook, L. M., Brower, L. P., and Alcock, J. 1969. An attempt to verify mimetic advantage in a neotropical environment. *Evolution*, **23**, 339–345.

Cooper, A. and Penny, D. 1997. Mass survival of birds across the Cretaceous–Tertiary boundary: molecular evidence. *Science*, **275**, 1109–1113.

Cracraft, J. 2001. Avian evolution, Gondwana biogeography and the Cretaceous–Tertiary mass extinction event. *Proceedings of the Royal Society of London, Series B*, **268**, 459–469.

Cristol, D. A. and Switzer, P. V. 1999. Avian prey-dropping behavior. II. American crows and walnuts. *Behavioral Ecology*, **10**, 220–226.

Dawson, B. V. and Foss, B. M. 1965. Observational learning in budgerigars. *Animal Behaviour*, **13**, 470–474.

De Groot, P. 1980. Information transfer in a socially roosting weaver bird (*Quelea quelea*; Ploceinae): an experimental study. *Animal Behaviour*, **28**, 1249–1254.

de Perera, T. B. and Guilford, T. 1999. The social transmission of spatial information in homing pigeons. *Animal Behaviour*, **57**, 715–719.

Dolman, C. S., Templeton, J., and Lefebvre, L. 1996. Mode of foraging competition is related to tutor preference in *Zenaida aurita*. *Journal of Comparative Psychology*, **110**, 45–54.

Dubbledam, J. L. 1998. Birds. In *The Central Nervous System of Vertebrates*, ed. R. Nieuwenhuys, H. J. TenDonkelaar, and C. Nicholson, pp.1525–1620. Berlin: Springer Verlag.

Dunbar, R. I. M. 1998. The social brain hypothesis. *Evolutionary Anthropology*, **6**, 178–190.

Epstein, R. 1984. Spontaneous and deferred imitation in the pigeon. *Behavioural Processes*, **9**, 347–354.

Erwin, R. M., Hafner, H., and Dugan, P. 1985. Differences in the feeding behaviour of little egrets (*Egretta garzetta*) in two habitats in the Camargue, France. *Wilson Bulletin*, **97**, 534–538.

Fisher, J. and Hinde, R. A. 1949. The opening of milk bottles by birds. *British Birds*, **42**, 347–357.

Fragaszy, D. M. and Visalberghi, E. 1989. Social influences on the acquisition of tool-using behaviors in tufted capuchin monkeys (*Cebus apella*). *Journal of Comparative Psychology*, **103**, 159–170.

Fritz, J. and Kotrschal, K. 1999. Social learning in common ravens, *Corvus corax*. *Animal Behaviour,* **57**, 785–793.

Fritz, J., Bugnyar, T., and Kotrschal, K. 1997. Learning or scrounging? Implications of an experimental study with ravens (*Corvus corax*). *Advances in Ethology,* **32**, 77.

Fritz, J., Bisenberger, A., and Kotrschal, K. 1999. Social mediated learning of an operant task in greylag geese: field observation and experimental evidence. *Advances in Ethology,* **34**, 51.

Fritz, J., Bisenberger, A., and Kotrschal, K. 2000. Stimulus enhancement in greylag geese: socially mediated learning of an operant task. *Animal Behaviour,* **59**, 1119–1125.

Fryday, S. L., and Greig-Smith, P. W. 1994. The effects of social learning on the food choice of the house sparrow (*Passer domesticus*). *Behaviour,* **128**, 281–300.

Galef, B.G., Jr. 1980. Diving for food: analysis of a possible case of social learning in wild rats (*Rattus norvegicus*). *Journal of Comparative and Physiological Psychology,* **94**, 416–425.

Galef, B.G., Jr. 1996. Social enhancement of food preferences in Norway rats: a brief review. In *Social Learning in Animals: The Roots of Culture,* ed. C. M. Heyes and B. G. Galef Jr., pp.49–64. New York: Academic Press.

Garnetzke-Stollmann, K. and Franck, D. 1991. Socialisation tactics of the spectacled parrotlet (*Forpus conspicillatus*). *Behaviour,* **119**, 1–29.

Giraldeau, L. A. and Caraco, T. 2000. *Social Foraging Theory*. Princeton, NJ: Princeton University Press.

Giraldeau, L. A. and Lefebvre, L. 1986. Exchangeable producer and scrounger roles in a captive flock of feral pigeons: a case for the skill pool effect. *Animal Behaviour,* **34**, 797–803.

Giraldeau, L. A. and Lefebvre, L. 1987. Scrounging prevents cultural transmission of food-finding behaviour in pigeons. *Animal Behaviour,* **35**, 387–394.

Goodall, J. 1964. Tool-using and aimed throwing in a community of free-living chimpanzees. *Nature,* **201**, 1264–1266.

Götmark, F. 1990. A test of the information-centre hypothesis in a colony of sandwich terns *Sterna sandvicensis*. *Animal Behaviour,* **39**, 487–495.

Greig-Smith, P. W. 1978. Imitative foraging in mixed-species flocks of Seychelles birds. *Ibis,* **120**, 233–235.

Hailman, J. P. 1961. Why do gull chicks peck at visually contrasting spots? A suggestion concerning social learning of food-discrimination. *American Naturalist,* **95**, 245–247.

Harvey, P. H. and Pagel, M. 1991. *The Comparative Method in Evolutionary Biology*. Oxford: Oxford University Press.

Hatch, K. K. and Lefebvre, L. 1997. Does father know best? Social learning from kin and non-kin in juvenile ringdoves. *Behavioral Processes,* **41**, 1–10.

Hedges, S. B., Parker, P. H., Sibley, C. G., and Kumar, S. 1996. Continental breakup and the ordinal diversification of birds and mammals. *Nature,* **381**, 226–229.

Higuchi, H. 1987. Individual differences in bait-fishing by the green-backed heron *Ardeola striata* associated with territory quality. *Ibis,* **130**, 39–44.

Hinde, R. A. and Fisher, J. 1951. Further observations on the opening of milk bottles by birds. *British Birds,* **44**, 393–396.

Hino, T. 1998. Mutualistic and commensal organization of avian mixed-species foraging flocks in a forest of western Madagascar. *Journal of Avian Biology,* **29**, 17–24.

Hitchcock, C. L. and Sherry, D. F. 1995. Cache pilfering and its prevention in pairs of black-capped chickadees. *Journal of Avian Biology*, **26**, 187–192.

Horn, H. S. 1968. The adaptive significance of colonial nesting in the Brewer's blackbird (*Euphagus cyanocephalus*). *Ecology*, **49**, 682–694.

Hunt, G. R. 1996. Manufacture and use of hook-tools by New Caledonian crows. *Nature*, **379**, 249–251.

Iacoboni, M., Woods, R. P., Brass, M., Bekkering, H., Mazziotta, J. C., and Rizzolatti, G. 1999. Cortical mechanisms in human imitation. *Science*, **286**, 2526–2528.

Imanishi, K. and Altmann, S. A. (ed.) 1965. *Japanese Monkeys: A Collection of Translations*. Edmonton: University of Alberta Press.

Ingram, J. 1998. *The Barmaid's Brain*. Toronto: Viking Press.

Iwaniuk, A. N., Nelson, J. E., and Pellis, S. M. 2001. Do big-brained mammals play more? Comparative analyses of play and relative brain size in mammals. *Journal of Comparative Psychology* **115**, 29–41.

Jones, T. B. and Kamil, A. C. 1973. Tool-making and tool-using in the Northern blue jay. *Science*, **180**, 1076–1078.

Jouventin, P., Pasteur, G., and Cambefort, J. P. 1976. Observational learning of baboons and avoidance of mimics. *Evolution*, **31**, 214–218.

Karten, H. J. 1991. Homology and evolutionary origins of the "neocortex". *Brain Behavior and Evolution*, **38**, 264–272.

Kawai, M. 1965. Newly-acquired pre-cultural behavior of the natural troop of Japanese monkeys on Koshima Islet. *Primates*, **6**, 1–30.

Klopfer, P. H. 1957. An experiment on emphatic learning in ducks. *American Naturalist*, **91**, 61–63.

Klopfer, P. H. 1959. Social interactions in discrimination learning with special reference to feeding behavior in birds. *Behaviour*, **14**, 282–299.

Klopfer, P. H. 1961. Observational learning in birds: the establishment of behavioral modes. *Behaviour*, **17**, 71–80.

Knight, S. K. and Knight, R. L. 1983. Aspects of food finding by wintering bald eagles. *Auk*, **100**, 477–484.

Krebs, J. R. 1973. Social learning and the significance of mixed species flocks of chickadees (*Parus* spp.). *Canadian Journal of Zoology*, **51**, 1275–1288.

Krebs, J. R. 1974. Colonial nesting and social feeding as strategies for exploiting food resources in the great blue heron (*Ardea herodias*). *Behaviour*, **51**, 99–134.

Krebs, J. R., MacRoberts, M. H., and Cullen, J. M. 1972. Flocking and feeding in the great tit *Parus major*: an experimental study. *Ibis*, **114**, 507–530.

Kushlan, J. A. 1977. The significance of plumage colour in the formation of feeding aggregations of Ciconiiforms. *Ibis*, **119**, 361–364.

Langen, T. A. 1996a. Skill acquisition and the timing of natal dispersal in the white-throated magpie-jay, *Calocitta formosa*. *Animal Behaviour*, **51**, 575–588.

Langen, T. A. 1996b. Social learning of a novel foraging skill by white-throated magpie-jays (*Calocitta formosa*, Corvidae): a field experiment. *Ethology*, **102**, 157–166.

Lawton, M. F. and Guindon, C. F. 1981. Flock composition, breeding success, and learning in the brown jay. *Condor*, **83**, 27–33.

LeCroy, M. 1972. Young common and roseate terns learning to fish. *Wilson Bulletin*, **84**, 201–202.

Lefebvre, L. 1986. Cultural diffusion of a novel food-finding behavior in urban pigeons: an experimental field test. *Ethology*, **71**, 295–304.

Lefebvre, L. 1995a. The opening of milk bottles by birds: evidence for accelerating learning rates, but against the wave-of-advance model of cultural transmission. *Behavioural Processes*, **34**, 43–54.

Lefebvre, L. 1995b. Ecological correlates of social learning: problems and solutions for the comparative method. *Behavioural Processes*, **35**, 163–171.

Lefebvre, L. 2000. Feeding innovations and their cultural transmission in bird populations. In *The Evolution of Cognition*, ed. C. Heyes and L. Huber, pp. 311–328. Cambridge: MIT Press.

Lefebvre, L. and Giraldeau, L.-A. 1996. Is social learning an adaptive specialization? In *Social Learning in Animals: The Roots of Culture*, ed. C.M. Heyes and B.G. Galef Jr., pp. 107–128. New York: Academic Press.

Lefebvre, L. and Helder, R. 1997. Scrounger numbers and the inhibition of social learning in pigeons. *Behavioural Processes*, **40**, 201–207.

Lefebvre, L. and Henderson, D. 1986. Resource defense and priority of access to food by the mate in pigeons. *Canadian Journal of Zoology*, **64**, 1889–1992.

Lefebvre, L., Palameta, B., and Hatch, K.K. 1996. Is group-living associated with social learning? A comparative test of a gregarious and a territorial columbid. *Behaviour*, **133**, 241–261.

Lefebvre, L., Templeton, J., Brown, K., and Koelle, M. 1997a. Carib grackles imitate conspecific and zenaida dove tutors. *Behaviour*, **134**, 1003–1017.

Lefebvre, L., Whittle, P., Lascaris, E., and Finkelstein, A. 1997b. Feeding innovations and forebrain size in birds. *Animal Behaviour*, **53**, 549–560.

Lefebvre, L., Gaxiola, A., Dawson, S., Timmermans, S., Rozsa, L., and Kabai, P. 1998. Feeding innovations and forebrain size in Australasian birds. *Behaviour*, **135**, 1077–1097.

Lefebvre, L., Nicolakakis, N., and Boire, D. 2002. Tools and brains in birds. *Behaviour*, **139**, 939–973.

Lévesque, H. and McNeil, R. 1985. Abondance et activités du pigeon biset, *Columba livia*, dans le port de Montréal. *Canadian Field Naturalist*, **99**, 343–355.

Ligon, J.D. and Martin, D.J. 1974. Pinyon seed assessment by the pinyon jay *Gymnorhinus cyanocephalus*. *Animal Behaviour*, **22**, 421–429.

MacDonald, D.W. and Henderson, D.G. 1977. Aspects of the behaviour and ecology of mixed-species bird flocks in Kashmir. *Ibis*, **119**, 481–491.

Maclean, S.F. 1970. Social stimulation modifies the feeding behavior of the American robin. *Condor*, **72**, 499–500.

Mason, J.R. 1988. Direct and observational learning by redwinged blackbirds (*Agelaius phoeniceus*): the importance of complex visual stimuli. In *Social Learning: Psychological and Biological Perspectives*, ed. T.R. Zentall and B.G. Galef Jr., pp. 99–115. Hillsdale, NJ: Erlbaum.

Mason, J.R. and Reidinger, R.F. 1981. Effects of social facilitation and observational learning on feeding behavior of the red-winged blackbird (*Agelaius phoeniceus*). *Auk*, **98**, 778–784.

Mason, J.R. and Reidinger, R.F. 1982. The relative importance of reinforced versus nonreinforced stimuli in visual discrimination learning by red-winged blackbirds (*Agelaius phoeniceus*). *Journal of General Psychology*, **107**, 219–226.

Mason, J.R., Arzt, A.H., and Reidinger, R.H. 1984. Comparative assessment of food preferences and aversions acquired by blackbirds via observational learning. *Auk*, **101**, 796–803.

McQuoid, L.M. and Galef, B.G., Jr. 1993. Social stimuli influencing feeding behaviour of Burmese fowl: a video analysis. *Animal Behaviour,* **46**, 13–22.

Meinertzhagen, R. 1954. The education of young ospreys. *Ibis,* **96**, 153–155.

Midford, P.E., Hailman, J.P., and Woolfenden, G.E. 2000. Social learning of a novel foraging patch in families of free-living Florida scrub-jays. *Animal Behaviour,* **59**, 1199–1207.

Millikan, G.C. and Bowman, R.I. 1967. Observations on Galápagos tool-using finches in captivity. *Living Bird,* **6**, 23–41.

Mlíkovský, J. 1989a. Brain size in birds: 1. Tinamiformes through Ciconiiformes. *Vestnik Ceskoslovnskae Spolecnosti Zoologickae,* **53**, 33–47.

Mlíkovský, J. 1989b. Brain size in birds: 2. Falconiformes through Gaviiformes. *Vestnik Ceskoslovnskae Spolecnosti Zoologickae,* **53**, 200–213.

Mlíkovský, J. 1989c. Brain size in birds: 3. Columbiformes through Piciformes. *Vestnik Ceskoslovnskae Spolecnosti Zoologickae,* **53**, 252–264.

Mlíkovský, J. 1990. Brain size in birds: 4. Passeriformes. *Acta Societas Zoologica Bohemoslovensis,* **54**, 27–37.

Mönkkönen, M. and Koivula, K. 1993. Neophobia and social learning of foraging skills in willow tits *Parus montanus*. *Ardea,* **81**, 43–46.

Moore, B.R. 1992. Avian movement imitation and a new form of mimicry: tracing the evolution of a complex form of learning. *Behaviour,* **122**, 231–263.

Murton, R.K. 1970. Why do some bird species feed in flocks? *Ibis,* **113**, 534–536.

Murton, R.K. and Isaacson, A.J. 1962. The functional basis of some behaviour in the woodpigeon *Columbus palumbus*. *Ibis,* **104**, 503–521.

Murton, R., Coombs, C., and Thearle, R. 1972. Ecological studies of the feral pigeon *Columba livia* var. II. Flock behaviour and social organization. *Journal of Applied Ecology,* **9**, 875–889.

Newton, I. 1967. Evolution and ecology of some British finches. *Ibis,* **109**, 33–99.

Nicol, C.J., and Pope, S.J. 1994. Social learning in small flocks of laying hens. *Animal Behaviour,* **47**, 1289–1296.

Nicol, C.J. and Pope, S.J. 1999. The effects of demonstrator social status and prior foraging success on social learning in laying hens. *Animal Behaviour,* **57**, 163–171.

Nicolakakis, N. and Lefebvre, L. 2000. Forebrain size and innovation rate in European birds: Feeding, nesting and confounding variables. *Behaviour,* **137**, 1415–1427.

Nicolakakis, N., Sol, D., and Lefebvre, L. 2002. Innovation rate predicts species richness in birds, but not extinction risk. *Animal Behaviour*, in press.

Norton-Griffiths, M. 1969. The organization, control and development of parental feeding in the oyster catcher (*Haematopus ostralegus*). *Behaviour,* **34**, 55–114.

Palameta, B. 1989. The importance of socially transmitted information in the acquisition of novel foraging skills by pigeons and canaries. PhD Thesis, University of Cambridge, UK.

Palameta, B. and Lefebvre, L. 1985. The social transmission of a food-finding technique in pigeons: what is learned? *Animal Behaviour,* **33**, 892–896.

Pettersson, M. 1956. Diffusion of a new habit among greenfinches. *Nature,* **177**, 709–710.

Ramsay, A.O. and Cushing, J.E. 1949. Acquired feeding behavior in mallards. *Auk,* **66**, 80.

Reader, S.M. 2000. Social learning and innovation: individual differences, diffusion dynamics and evolutionary issues. PhD Thesis, University of Cambridge, UK.

Reader, S.M. and Laland, K. 2002. Social intelligence, innovation and enhanced brain size in primates. *Proceedings of the National Academy of Sciences, USA*, **99**, 4436–4441.

Reader, S.M. and Lefebvre, L. 2001. Social learning and sociality. *Behavioral and Brain Sciences*, **24**, 353–355.

Rehkämper, G.K. and Zilles, K. 1991. Parallel evolution in mammalian and avian brains: comparative cytoachitectonic and cytochemical analysis. *Cell and Tissue Research*, **263**, 23–28.

Rehkämper, G.K., Frahm, H.D., and Zilles, K. 1991. Quantitative development of brain structures in birds (Galliformes and Passeriformes) compared with that in mammals (Insectivores and Primates). *Brain Behavior and Evolution*, **37**, 125–143.

Roell, A. 1978. Social behaviour of the jackdaw, *Corvus monedula*, in relation to its niche. *Behaviour*, **64**, 1–124.

Rothschild, M. and Ford, R. 1968. Warning signals from a starling *Sturnus vulgaris* observing a bird rejecting unpalatable prey. *Ibis*, **110**, 104–105.

Rowley, I. and Chapman, G. 1986. Cross-fostering, imprinting and learning in two sympatric species of cockatoo. *Behaviour*, **96**, 1–16.

Rubenstein, D.I., Barnett, R.J., Ridgely, R.S., and Klopfer, P.H. 1977. Adaptive advantages of mixed-species feeding flocks among seed-eating finches in Costa Rica. *Ibis*, **119**, 10–21.

Sasvári, L. 1979. Observational learning in great, blue and marsh tits. *Animal Behaviour*, **27**, 767–771.

Sasvári, L. 1985a. Keypeck conditioning with reinforcement in two different locations in thrush, tit and sparrow species. *Behavioral Processes*, **11**, 245–252.

Sasvári, L. 1985b. Different observational learning capacity in juvenile and adult individuals of congeneric bird species. *Zeitschrift für Tierpsychologie*, **69**, 293–304.

Sasvári, L. and Hegyi, Z. 1998. How mixed-species foraging flocks develop in response to benefits from observational learning. *Animal Behaviour*, **55**, 1461–1469.

Schildkraut, D.L. 1974. Observational learning in captive blue jays. *Dissertation Abstracts International*, **35**, 6143B.

Sherry, D.F., and Galef, B.G., Jr. 1984. Cultural transmission without imitation: milk bottle opening by birds. *Animal Behaviour*, **32**, 937–938.

Sherry, D.F., and Galef, B.G., Jr. 1990. Social learning without imitation: more about milk bottle opening by birds. *Animal Behaviour*, **40**, 987–989.

Sibley, G.C. and Alquist, J.E. 1990. *Phylogeny and Classification of Birds: A Study in Molecular Evolution*. New Haven, CT: Yale University Press.

Sibley, C.G. and Monroe, B.L. 1990. *Distribution and Taxonomy of Birds of the World*. New Haven, CT: Yale University Press.

Sol, D., Timmermans, S., and Lefebvre, L. 2002. Behavioral flexibility and invasion success in birds. *Animal Behaviour*, **63**, 495–502.

Stenhouse, D. 1962. A new habit of the redpoll *Carduelis flammea* in New Zealand. *Ibis*, **104**, 250–252.

Stephan, H., Baron, G., and Frahm, H. 1988. Comparative size of brains and brain components. In *Comparative Primate Biology*, Vol. 4: *Neurosciences*, ed. H.D. Stekliss and J. Erwin, pp.1–38. New York: Alan R. Liss.

Stokes, A.W. 1971. Parental and courtship feeding in red jungle fowl. *Auk*, **88**, 21–29.

Sullivan, K.A. 1984. The advantages of social foraging in downy woodpeckers. *Animal Behaviour*, **32**, 16–22.

Taylor, P.M. 1972. Hovering behavior by house finches. *Condor,* **74**, 219–221.

Tebbich, S., Taborsky, M., Fessl, B., and Blomqvist, D. 2001. Do woodpecker finches acquire tool use by social learning? *Proceedings of the Royal Society of London, Series B,* **268**, 2189–2193.

Tebbich, S. Taborsky, M., Fessl, B., and Dvorak, M. 2002. The ecology of tool use in the woodpecker finch *Cactospiza pallida*. *Ecology Lettrs,* **5**, 656–664.

Templeton, J.J., Kamil, A.C., and Balda, R.P. 1999. Sociality and social learning in two species of corvids: the pinyon jay (*Gymnorhinus cyanocephalus*) and the Clark's nutcracker (*Nucifraga columbiana*). *Journal of Comparative Psychology,* **113**, 450–455.

Thompson, C.F., Ray, G.F., and Preston, R.L. 1996. Nectar robbing in blue tits *Parus caeruleus*: failure of a novel feeding trait to spread. *Ibis,* **138**, 552–553.

Thomson, A.L. 1964. *A New Dictionary of Birds*. London: Nelson.

Thorpe, W.H. 1956. Records of the development of original and unusual feeding methods by wild passerine birds. *British Birds,* **49**, 389–395.

Timmermans, S., Lefebvre, L., Boire, D., and Basu, P. 2000. Relative size of the hyperstriatum ventrale is the best predictor of innovation rate in birds. *Brain Behavior and Evolution,* **56**, 196–203.

Turner, E.R.A. 1961. The acquisition of new behaviour patterns: an analysis of social feeding in sparrows and chaffinches. *Animal Behaviour,* **9**, 113–114.

Turner, E.R.A. 1964. Social feeding in birds. *Behaviour,* **24**, 1–46.

Veissier, I. 1993. Observational learning in cattle. *Applied Animal Behaviour Science,* **35**, 235–243.

Waite, R.K. 1981. Local enhancement for food-finding by rooks (*Corvus frugilegus*) foraging on grassland. *Zeitschrift für Tierpsychologie,* **57**, 15–36.

Waite, T.A., and Grubb, T.C., Jr. 1988. Copying of foraging locations in mixed-species flocks of temperate-deciduous woodland birds: an experimental study. *Condor,* **90**, 132–140.

Wechsler, B. 1988. The spread of food producing techniques in a captive flock of jackdaws. *Behaviour,* **107**, 267–277.

Weidmann, U. 1957. Verhaltensstudien an der stockente. *Zeitschrift für Tierpsychologie,* **13**, 208–271.

Werner, T.K. and Sherry, T.W. 1987. Behavioral feeding specialization in *Pinaroloxias inornata*, the "Darwin's Finch" of Cocos Island, Costa Rica. *Proceedings of the National Academy of Sciences, USA,* **84**, 5506–5510.

Whiten, A. and Byrne, R.W. 1997. *Machiavellian Intelligence II: Extensions and Evaluation*. Cambridge: Cambridge University Press.

Whiten, A., Goodall, J., McGrew, W.C., Nishida, T., Reynolds, V., Sugiyama, Y., Tutin, C.E.G., Wrangham, R.W., and Boesch, C. 1999. Culture in chimpanzees. *Nature,* **399**, 682–685.

Wilson, A.C. 1985. The molecular basis of evolution. *Scientific American,* **253**, 148–157.

Wyles, J.S., Kunkel, J.G., and Wilson, A.C. 1983. Birds, behavior and anatomical evolution. *Proceedings of the National Academy of Sciences, USA,* **80**, 4394–4397.

Zach, R. 1979. Shell-dropping: decision-making and optimal foraging in northwestern crows. *Behaviour,* **68**, 106–117.

Zentall, T.R. and Hogan, D.E. 1976. Imitation and social facilitation in the pigeon. *Animal Learning and Behavior,* **4**, 427–430.

GWEN DEWAR

5

The cue reliability approach to social transmission: designing tests for adaptive traditions

5.1 Introduction

Traditions are behaviors that persist over time and are shared among group members by virtue of social learning processes (see Ch. 1). The direct observation of animals using social cues to discover or learn a behavior is perhaps the most straightforward evidence of a tradition, and numerous longitudinal, naturalistic studies and controlled laboratory experiments have yielded such evidence (see, for instance, Chs. 7, 9, 13, and 14). However, efforts to collect direct evidence are sometimes deemed impractical; consequently investigators have sought ways to infer the existence of traditions on the basis of indirect evidence. Can we identify traditions when we lack direct observations of social learning?

I present a new approach for dealing with indirect evidence. This cue reliability approach (CRA) addresses a special category of potential traditions: behaviors that (a) reflect an individual's classification of a stimulus or tactic as either safe or harmful, and (b) are costly if the individual makes classification errors. Is hemlock a safe food or a dangerous toxin? Should garter snakes be dismissed as benign trespassers or avoided as lethal predators? Animals can answer these questions by consulting local traditions. However, traditional knowledge is not necessarily the only source of information available. The CRA is designed to help us to determine if animals need social cues to classify correctly potentially dangerous stimuli or bad tactics. It begins by identifying a decision-maker's options regarding an unfamiliar stimulus or untested tactic, and the possible outcomes associated with each option. Next, the approach asks what payoffs – gains or losses in fitness – are associated with each outcome. This reveals how confident individuals need to be about the positive outcome of an option before they try it for the first time. If nonsocial sources of information are sufficient to convince individuals that the option is worth attempting, we must conclude that convergent, independent learning could explain why the behavior is widespread. Each group member might perform the behavior because he or she has discovered it through independent experience. If, however, the only way individuals can gain the required confidence is by observing conspecifics demonstrate that the behavior is safe to attempt, we can conclude that social cues are necessary for individuals to acquire the behavior. In this way, the CRA provides a stringent criterion by which to recognize traditions using indirect evidence. Social cues must do more than aid learning; they must be necessary for individuals first to attempt the behavior. Although some *bona fide* traditions may fail to meet this criterion, and thus go unrecognized, applying this stringent criterion reduces the chance of a false positive, that is, concluding that a group-wide behavior is traditional when it is not.

The CRA derives from basic principles of behavioral ecology and shares the economic orientation of well-known theoretical treatments of social learning and cultural transmission (e.g., Aoki and Feldman, 1987; Boyd and Richerson, 1985; Boyd and Richerson, 1988; Cavalli-Sforza and Feldman, 1981; Giraldeau, Caraco, and Valone, 1994; Laland, Richerson, and Boyd, 1996; Rogers, 1988). However, the CRA differs from prior theory by focusing on the economics of individual decisions to try, rather than adopt, new behaviors. The CRA does not ask why animals ultimately incorporate new strategies in their behavioral repertoires. Instead, the approach assumes that individuals might adopt new behaviors as a consequence of the positive reinforcement they receive after attempting the behaviors (Galef, 1995; Heyes, 1993). Moreover, the CRA does not address the population dynamics of transmission. Instead, the CRA is concerned with individual decision making and considers social cues solely from the standpoint of what information they contribute about the value of trying something for the first time (Box 5.1).

The CRA assumes that animals have evolved mechanisms that lead them to try something new when they perceive that doing so is worth the gamble. If true, cost–benefit analyses can help us to understand what conditions should encourage animals to attempt specific new behaviors. To illustrate, I discuss how researchers might begin to model two kinds of decision: (a) to eat or reject an unfamiliar food, and (b) to respond or fail to respond to an animal as a predator. Simple expected utility models serve as instructive starting points, since they highlight the importance of

Box 5.1. Cue reliability: an economic approach to traditions

Like the theoretical social learning literature (e.g., Aoki and Feldman 1987; Boyd and Richerson 1985; Boyd and Richerson 1988; Cavalli-Sforza and Feldman 1981; Giraldeau *et al.*, 1994; Laland *et al.*, 1996; Rogers 1988) the cue reliability approach (CRA) is economic in orientation. However, the CRA should not be considered a contribution to this literature for two reasons.

1. Unlike the theoretical social learning literature, the CRA does *not* address the *adoption* of new behaviors, but only the *first attempts* at such behaviors.
2. The CRA is concerned with individual decision making, rather than the population-level dynamics of transmission.

Therefore, the CRA does not directly address social learning *per se*, nor is it concerned with the consequences of social transmission. Instead, the CRA focuses on the role that social cues can play in alerting individual decision-makers to the existence of new behaviors that are safe enough to attempt. What is "safe enough" depends on the particular costs and benefits associated with the decision to try each behavior. To test potential traditions, the CRA uses three important concepts.

1. *Payoffs.* The payoff for a decision is the change in fitness that the decision-maker receives for making that decision.
2. *The reliability threshold.* What is the chance that a decision will yield a positive payoff? The reliability threshold specifies how probable a positive payoff must be in order to justify a decision. It is calculated by measuring the principle payoffs – positive and negative – that could result from the decision. Different kinds of decision are expected to yield different reliability thresholds.
3. *Cue reliability.* Cues are defined as "reliable" if they predict that a behavior will result in a positive payoff with a probability that exceeds the reliability threshold. Because reliability thresholds vary for different decisions, the same minimum probability that renders cues reliable in one situation may make them unreliable in another.

reliable predictive information – environmental and social cues – for the decision-maker. If animals behave as payoff maximizers, such expected utility models might be adequate to test for the existence of traditions in wild populations. But the premises and assumptions of these models must themselves be tested, and more sophisticated models may be required. The purpose of this chapter is to encourage researchers interested in nonhuman animal traditions to participate in the development and testing of CRA models.

5.2 The costs and benefits of responses to unfamiliar stimuli

5.2.1 Establishing the probability of independent discovery

For any learned, widespread behavior, there are two possible explanations: either the behavior is traditional or the behavior is widespread because each individual practicing it has acquired the behavior on his or her own. Direct approaches to identifying traditions seek to confirm the first explanation by obtaining observations of social learning between individuals. Indirect approaches to identify traditions attempt to discount the second explanation – independent acquisition by all individuals – by showing that it is very improbable.

But why should independent acquisition be improbable, particularly if the behavior in question is adaptive or profitable? Many alleged traditions may represent adaptive behaviors for coping with recurrent ecological problems such as food choice, food processing, nest site selection, or predator evasion. Such behaviors are rewarding, perhaps even self-reinforcing. Therefore, once attempted, the probability that an individual will adopt such a behavior may be very high. To be safe, then, perhaps investigators taking the indirect approach to identify traditions should exclude all adaptive or profitable behaviors.

This is the reasoning behind the regional contrast approach: an approach to indirect evidence that has been favored by many investigators (e.g., Whiten *et al.*, 1999). The regional contrast approach, also called the group contrast approach (Ch. 1) or the method of elimination (Ch. 11), examines behavioral differences between two or more groups belonging to the same species or subspecies. Because profitable behaviors are suspect, intergroup behavioral differences that can be related to local differences in profitability are eliminated from consideration. Any intergroup behavioral differences that survive this process of elimination are then deemed "traditional".

Reasonable as this sounds, the regional contrast approach is flawed because it misses a crucial point: *although profitability can explain why an individual repeats a behavior, it cannot explain why an individual attempts a behavior for the first time.* Even if positive reinforcement guarantees that individuals will adopt profitable behaviors, there remains the question of how individuals come to make the first attempt. If it is very improbable that individuals will make the first attempt without social cues, we can discount the hypothesis that a custom was acquired independently by all individuals. Consequently, taking an indirect approach to identifying traditions does not compel us to dismiss all profitable customs out of hand. Instead, we should focus on the probability of independent discovery, that is, the chances that a naïve individual will first attempt a behavior without the aid of social cues.

When is independent discovery improbable? One case is a complex behavior that yields rewards only *after* the individual has skillfully completed a lengthy manipulative and/or tool-based sequence, as described by Russon (Ch. 12). For instance, it might seem unlikely that a naïve chimpanzee will independently discover the principle of opening a nut with hammerstone and anvil, especially if he or she has no prior experience with nuts as food. And if this probability is low, the probability that all group members independently discover nut cracking is even lower – specifically, it is the chance of one independent discovery raised to the nth power, where n is the number of individuals in the group that exhibit the technique.

This example illustrates why we are intuitively persuaded that some complex food-processing techniques must be traditions. Although an individual can discover new behaviors through individual exploration, the probability of discovering some behaviors – including behaviors that require the completion of a lengthy action sequence before the individual is rewarded – is low. When we observe that such a behavior is practiced by many members of a population, the probability that *every* individual independently discovered the behavior becomes vanishingly small, and social learning is implicated as the only remaining explanation. Of course, our intuitions can fail us, so it is critical to confirm that the probability of independent discovery really is low. For instance, it once seemed plausible that social learning was the only explanation for sweet potato washing among the Japanese macaques of Koshima (Nishida, 1987). However, Visalberghi and Fragaszy (1990) demonstrated that both capuchins and macaques wash fruit spontaneously, suggesting that the probability of independent discovery of this behavior in monkeys is actually quite high. By comparison, Terkel and colleagues (Terkel, 1996) conducted extensive experiments to demonstrate that naïve black rats fail to discover pinecone stripping – a method of food processing – without the aid of social cues, thereby confirming that the probability of independent discovery is very low.

Complexity and difficulty of obtaining a reward are not the only factors that can render independent discovery improbable. Another important case is a behavior that is intrinsically dangerous to attempt. Consider, for example, a preference for a particular species of wild mushroom. Eating an unfamiliar type of mushroom does not require learning new action patterns, nor must one be highly skilled to obtain a positive payoff for eating it. But eating an unfamiliar, untested food is potentially dangerous. It might be toxic. Similarly, treating an unfamiliar creature as a benign commensal does not require learning new, complex action patterns. But relaxing vigilance is a gamble because the unfamiliar creature could be a predator. If, on their own, individuals are unlikely to try certain behaviors because the potential dangers outweigh the potential rewards, then we have a strong basis for inferring that these behaviors are traditional. These are the behaviors that the CRA is especially designed to address.

5.2.2 Modeling the decision to try something new

Over its lifetime, an individual faces many decisions that (a) require the classification of a stimulus or tactic as either safe or harmful, and (b) are costly if the individual makes classification errors. Lacking perfect knowledge of how a new stimulus or tactic ought to be classified, the naïve individual cannot be sure which of its options is the most appropriate. On the one hand, the best option might involve trying something new. On the other hand, the untested option might be inferior to a familiar alternative, in which case the experimenting individual will incur an opportunity cost and, possibly, an absolute loss in fitness. In sum, the first attempt of a new behavior is a gamble. Is it worth taking?

The answer depends on what possible outcomes are associated with the available options. To take a simple case, consider a forager who encounters an unfamiliar potential food item. He or she must decide whether to eat the item or reject it in favor of pursuing a familiar food. If we assume two possible states of the world, (a) that the net payoff for eating the item is positive and (b) that the net payoff for eating item is negative, there are four possible outcomes associated with the decision to eat or reject (Table 5.1). Each outcome entails a payoff, though in this case two outcomes result in identical payoffs, since the reward for seeking out familiar food

Table 5.1. *Payoff matrix: should a naïve forager eat or reject an unfamiliar potential food?*

	Status of unfamiliar potential food	
Decision	Safe	Bad
Eat	Positive payoff for eating safe unfamiliar food (G)	Negative payoff for eating bad unfamiliar "food" ($-B$)
Reject	Net payoff for rejecting the unfamiliar item and seeking a familiar food instead (F)	Net payoff for rejecting the unfamiliar item and seeking a familiar food instead (F)

See Box 5.2 for calculation of reliability threshold for this situation.

Table 5.2. *Payoff matrix: should a naïve individual direct an antipredator response at an unfamiliar animal?*

	Status of unfamiliar animal	
Decision	Predator	Not predator
Respond	Energetic cost of response to predator ($-R$)	Energetic cost of response to nonpredator ($-r$)
Ignore	Negative payoff for ignoring predator ($-B$)	Payoff for ignoring nonpredator (G)

See Box 5.2 for calculation of reliability threshold for this situation.

remains the same regardless of the value of the unfamiliar item. Hence, the decision is constrained by only three payoffs: the payoff for eating a "safe" unfamiliar item, the payoff for eating a "bad" unfamiliar item, and the payoff for seeking out and eating a familiar item instead.

Similarly, we can identify the possible outcomes and associated payoffs pertaining to other simple decisions, such as whether to ignore or direct an antipredator response at an unfamiliar potential predator (Table 5.2). In this case, there are again four possible outcomes, each depending on whether or not the potential predator is a real threat. If the payoff for performing an antipredator response is identical whether or not the potential predator is real, the decision-maker is again constrained by only three payoffs: the payoff for performing the antipredator response, the payoff for ignoring a true predator, and the payoff for ignoring a nonpredator.

If the decision-maker had perfect knowledge about the true state of the world, he or she would simply choose the option that yielded the highest payoff. Lacking perfect knowledge, the decision-maker can instead choose according to the economic principle of expected utility. Assuming that the decision-maker seeks the maximum *expected* payoff, the best option is calculated by weighting the payoffs according to the probabilities associated with each possible state of the world. For example, the expected payoff for eating an unfamiliar item is given by adding two sums: (a) the average payoff for eating beneficial new foods multiplied by the probability that the unfamiliar item is beneficial, and (b) the average payoff for eating harmful items multiplied by the probability that the item is harmful (Box 5.2). If this sum exceeds the expected payoff for rejecting the item and finding a familiar food, the payoff maximizer should take the gamble and eat the unfamiliar item. Similarly, the decision-maker should ignore an unfamiliar potential predator when the expected payoff for doing so exceeds that for performing the antipredator response (Box 5.2). In each case, the best option cannot be identified without knowledge of the possible outcomes and the probabilities associated with these outcomes.

Given these requirements, social cues – the behavior of conspecifics – can contribute crucial information to decision-makers about the probable state of the world. For instance, if conspecifics only very rarely eat harmful items, then an individual who observes a conspecific eating a particular food type (or otherwise detects ingestion of a food, such as from smelling the breath or excretions of a conspecific) can "conclude" that there is a very high probability that eating the item is safe. More generally, decision-makers can benefit from exploiting any cues – social or nonsocial – that provide information about the probabilities of pertinent outcomes. How high such probabilities must be to justify choices can be referred to as the "reliability thresholds". These thresholds, examples of which are presented in Box 5.2, depend on the costs of two kinds of mistake: attempting the new behavior when it is worse than the familiar alternative, and rejecting the new behavior when it is better than the familiar alternative. Specifically, a reliability threshold for attempting a new behavior is determined by the payoffs for (a) attempting the behavior, given that it yields a positive payoff; (b) attempting the behavior, given that it yields a negative payoff; (c) rejecting the behavior, given that it yields a positive payoff; and (d) rejecting the behavior, given that it yields a negative payoff.

Thus, describing a decision's payoff matrix permits us to specify how probable a positive outcome must be to make attempting a new behavior

Box 5.2. How to calculate the reliability threshold for heeding social cues

Food traditions

When does it pay to eat an unfamiliar food? The answer depends on $+G$, the average payoff for eating the item if it is safe; $-B$, the average payoff for eating the item if it is bad or harmful; and $+F$, the average net payoff for rejecting the item in favor of a familiar alternative (Table 5.1). Given the probability P that the item is safe, the expected return for eating the potential food item is

$$(P)(G) + (1 - P)(-B)$$

and eating the item is *more* profitable than seeking out the familiar alternative if

$$(P)(G) + (1 - P)(-B) > F$$

or

$$P > (F + B)/(G + B) \qquad (5.1)$$

Thus, an unfamiliar item is worth adopting if the probability that it is safe exceeds

$$(F + B)/(G + B)$$

Potential predators

Similarly, we can discover when it pays to respond with antipredator behavior to an unfamiliar animal, where $-R$ is the average payoff for responding if the unfamiliar animal is a predator; $-r$ is the average payoff for responding if the animal is not a predator; $-B$ is the average payoff for ignoring the animal given that it is a predator; and $+G$ is the average payoff for ignoring the animal given that it is not a predator (Table 5.2). Given the probability P that the animal is a predator, an antipredator response is more profitable than ignoring the animal if

$$(P)(-R) + (1 - P)(-r) > (P)(-B) + (1 - P)(G)$$

and an antipredator response is favored when

$$P > (r + G)/(B + G - R + r). \qquad (5.2)$$

Alternatively, given the probability P' that an animal is *not* a predator, it is more profitable to ignore the animal if

$$P' > (B - R)/(B + G - R + r). \qquad (5.3)$$

worthwhile. When the available information suggests that this probability exceeds the reliability threshold, the optimal decision-maker should attempt the new strategy. Otherwise, he or she should reject the new behavior and pursue the familiar alternative. If natural selection has favored individuals who maximize their expected payoffs for such decisions, we should expect animals to discriminate between reliable and unreliable cues. This should be true whether cues constitute information provided by conspecifics or by nonsocial aspects of the environment. What is important from the decision-maker's standpoint is what a cue indicates about the probability that a particular behavior will yield a positive outcome.

Note that this differs from assuming that cues advertise which behaviors are best or most profitable. To reiterate, the CRA has nothing to say about the adaptive value of *adopting* behaviors. Nor does the CRA claim that social cues are especially likely to guide animals to discover optimal behaviors. The CRA examines social and nonsocial cues from the narrow standpoint of their potential to inform individuals about the safety of *trying* new things.

Three major implications follow for the study of traditions. First, because reliability thresholds vary according to the distinctive payoffs associated with attempting or rejecting a specific behavior, no single rule about social cues (e.g., "trust all social cues" or "trust all social cues that confer at least a 75% probability that a behavior is appropriate") applies to all situations. Returning to the examples discussed above, the decision to treat an unfamiliar animal as a predator is almost certainly constrained by a different reliability threshold than the decision to eat a novel food. Moreover, reliability thresholds may vary significantly depending on the species, age, sex, rank, or condition of the decision-maker. For instance, if low-ranking individuals are denied access to high-quality familiar foods, they should have lower reliability thresholds for sampling new foods than should their better-fed, high-ranking conspecifics. Similarly, if smaller-bodied animals are more vulnerable to predation than are larger bodied animals, smaller animals should have lower reliability thresholds for responding to potential predators. These examples illustrate that the most realistic CRA models will address specific classes of behavior and, where relevant, specific types of decision-maker.

The second major implication of the approach is that opportunities for social transmission depend on the availability of reliable social cues. Similar decision-makers facing similar problems may be constrained by identical reliability thresholds yet differ in their exploitation of social cues if only some decision-makers have access to reliable cues. This may explain some of the variation between individuals, as well as between local populations and between species. To illustrate, consider the social transmission of food aversions, which has been demonstrated in blackbirds (Mason, Arzt, and Reidinger, 1984) but seems to be lacking in Norway rats (Galef, McQuoid, and Whiskin, 1990). Suppose, for the sake of argument, that this difference is related to the fact that blackbirds, but not Norway rats, can eliminate ingested noxious items by vomiting. If watching a conspecific vomit were the only social cue of food aversion reliable enough to exceed the reliability threshold, the rats' failure to learn from social cues of aversion might be attributable to a lack of reliable social cues. In order to confirm that animals should heed social cues, then, one must establish that the available social cues are reliable. Operationally, this means measuring the frequency with which a proposed social cue successfully predicts that the behavior will yield a positive outcome.

The third implication of this approach is that social cue discrimination is merely a special case of general cue discrimination. Social cues might convey useful information about the probability that a behavior is safe to attempt, but so do nonsocial cues. This means that individuals that lack reliable social cues may nonetheless attempt the appropriate behavior if reliable *non*social cues indicate that it is safe to do so. For example, a decision-maker need not wait for a demonstrator to eat a food if an observable property of the food, such as its odor, flavor, color, or texture, reliably indicates that a food is safe to eat. Because nonsocial cues may be sufficiently reliable to justify action, social cues are not always necessary to explain why a decision-maker attempts a new behavior. As a result, we cannot assume that a widespread, customary behavior is socially transmitted merely because reliable social cues are available. To rule out the possibility that the behavior is widespread because of convergent, independent discovery, it is necessary to demonstrate that the available nonsocial cues are *unreliable*. If nonsocial cues are unreliable, individuals should be discouraged from spontaneously attempting the behavior, making convergent discovery an implausible explanation for the widespread practice of behavior.

In summary, the CRA provides a theoretical framework to study possible traditions that would be potentially dangerous to discover through independent exploration. It highlights four major points:

- the payoff-maximizing decision-maker attempts a new behavior if the probability that the behavior is safe exceeds the value set by the *reliability threshold*

- cues indicating that a behavior is safe should be heeded as a function of *cue reliability,* i.e., cues should be trusted only if they indicate that the behavior is safe to attempt with a probability exceeding the reliability threshold
- the importance of reliable social cues depends on the reliability of nonsocial cues, i.e., animals do not need to attend to social cues if nonsocial cues are reliable
- we can infer that social cues have contributed to the distribution of a widespread behavior when *social* cues are *reliable* and *nonsocial* cues are *unreliable.*

Of course, the last point does not mean that nonsocial cues must be unreliable for social cues to contribute to the spread of a behavior. Rather, it is only when nonsocial cues are unreliable that researchers can *exclude* the possibility that a behavior is widespread as the result of convergent, independent learning. Assessing the reliability of both social and nonsocial cues is, therefore, essential work for researchers interested in establishing whether an apparently learned, shared behavior is a tradition. This suggests that investigators can contribute significantly to the study of traditions by collecting data that will permit estimation of reliability thresholds and the intrinsic reliability of social and nonsocial cues. Once this admittedly laborious task has been accomplished, we can distinguish whether social cues ought to be heeded and, if so, whether or not social cues are the only source of information that could prompt a decision-maker to attempt the behavior for the first time (Box 5.3).

5.2.3 Usefulness of the cue reliability approach

5.2.3.1 Limitations and practical constraints

Although the CRA offers some advantages over the regional contrast approach (see section 5.2.3.2 below), it is important to recognize that the CRA is informative only under special circumstances. First, the CRA deals only with potential traditions that have adaptive consequences for those individuals who practice them. Behaviors of unclear adaptive consequences, like stone handling among Japanese macaques (Huffman, 1996; and Ch. 10) or self-tickling with objects among chimpanzees (Whiten *et al.*, 1999), are not addressed. For such behaviors, and for apparently adaptive behaviors for which we lack the ability to estimate reliability thresholds, the CRA is uninformative. If direct evidence of learning between

Box 5.3. How to test for the existence of an adaptive tradition

Assuming that decision-makers are presented with both social and nonsocial cues about the safety of attempting a new behavior, four possible states of information exist:

1. *Both* social and nonsocial cues reliably indicate that the behavior is worth attempting
2. *Only* social cues reliably indicate that the behavior is worth attempting
3. *Only* nonsocial cues reliably indicate that the behavior is worth attempting
4. *Neither* social nor nonsocial cues reliably indicate that the behavior is worth attempting.

States 1–3 would all result in group-wide similarity of behavior as long as group members are constrained by similar reliability thresholds and encounter the same reliable cues. However, this similarity is the possible result of *social transmission* only in states 1 and 2. In state 3, group-wide behavioral similarity follows from individuals converging on the same behavior by independently attending to *nonsocial* cues. This leads to *conclusion* 1: to identify social transmission as the *potential* cause of group-wide behavioral similarity, it is necessary to establish that social cues are available and meet the requirements set by the reliability threshold.

Next, note that the mere existence of reliable social cues is an insufficient criterion of social transmission. If, as in state 1, both social and nonsocial cues are reliable, group members may converge on the same behavior by independently attending to *nonsocial* cues. Reliable social cues would be redundant and, therefore, could be potentially ignored. This leads to *conclusion* 2: to identify social transmission as the *only* cause of group-wide similarity of behaviors, it is also necessary to establish that the available *nonsocial* cues are *unreliable*.

Therefore, although social transmission might occur in any situation in which social cues are reliable, we cannot be certain that social transmission is the sole explanation for a group-wide practice unless we *rule out* cases where reliable *nonsocial* cues are available. A cue reliability appoach (CRA) seeks to determine if the reliability threshold is greater than nonsocial cue reliability and less than social cue reliability:

$$\text{nonsocial cue reliability} < \text{reliability threshold} < \text{social cue reliability}.$$

individuals is unavailable (because longitudinal observations or experimental data are lacking), the regional contrast approach represents an alternative form of analysis.

Second, the CRA requires researchers to tackle difficult new questions about what choices are available to animals and what "currency", or proxy measure of fitness, animals attempt to maximize. Researchers who apply the CRA must confront the difficulties of operationalizing the elements of the models they devise. The expected utility models described in this chapter are offered only as first approximations of what might constitute the decision-making variables important to real animals. Empirical tests will be required to determine whether or not such simple models are realistic enough to yield accurate predictions. Some studies suggest that animals do not always exploit environmental and social cues in ways that seem consistent with optimality models (e.g., Fragaszy and Visalberghi, 1996; Laland and Williams, 1998; Laland, 1999; Visalberghi and Addessi, 2000; see also Ch. 7). To evaluate such evidence, we will need to identify the costs and benefits that actually pertain in these cases and assess cue reliability. For example, nonsocial cues may be more reliable than we think, especially among captive animals that have never experienced "bad" outcomes for attempting new behaviors. Conversely, we might overestimate cue reliability if we fail to identify predation risk (see Section 5.3.1) and other factors that can raise reliability thresholds, like interference or harassment from conspecifics (Baldwin and Meese, 1979; Giraldeau and Caraco, 2000). Consequently, closer scrutiny might reveal that some cases of apparently nonoptimal cue exploitation are in fact consistent with the predictions of the CRA. Other cases, however, probably reflect oversimplifications inherent in expected utility models. For instance, the expected utility models assume that decision-makers are risk-indifferent, always seeking to maximize the expected payoff for any decision to follow or ignore a demonstrator. If some decision-makers are sensitive to risk, or probabilistic variation (Stephens and Krebs, 1986), a more sophisticated treatment is needed. In addition, the expected utility models presented here are designed only to address decisions to treat unfamiliar stimuli as safe or harmful. The models do not apply to the acquisition of complex behavioral sequences such as the hammerstone-and-anvil nutcracking discussed in Section 5.2.1. In such cases, what determines whether individuals will acquire new behaviors on their own is not exposure to a single cue, but the extensive exploration of the environment.

5.2.3.2 Advantages of the cue reliability approach: a comparison with the regional contrast approach

The CRA offers important advantages for the analysis of adaptive behaviors. For example, the CRA permits us to test potential traditions that the regional contrast would dismiss. The regional contrast approach excludes (a) behaviors that are apparently universal and (b) behaviors that are found only in communities where they are locally profitable. In both cases, the researcher reduces the chance of a false positive – the mistaken identification of behaviors as traditions. However, this increases the chance of a false negative – the rejection of a genuine tradition.

By contrast, the CRA can be extended, in principle, to any simple, adaptive behavior whether or not it is universal. This is possible because the central test of the CRA is concerned with the reliability of social and nonsocial cues. Assuming that some apparent "traditions" are really species-normal behaviors, they can be screened out as behaviors that can be attempted on the basis of heeding reliable *nonsocial* cues. For example, if a widespread preference for fruit is the result of a species-normal bias, the mediating mechanism must include some way for individuals to recognize fruit when it is encountered. Sensitivity to the observable characteristics of edible fruit (e.g., flavor, scent, texture, shape, and color) constitutes attentiveness to nonsocial cues. Assuming that the sensitivity to nonsocial cues is reliable, the CRA would rule out fruit eating as a potential tradition.

The CRA might also help us to avoid misidentifying false traditions. Consider the hypothetical case that portobella mushrooms are eaten in Corsica and ignored in Malta. Portobella mushrooms are equally profitable at both sites, but Malta has more toxic fungus species, including toadstools that resemble edible mushrooms. If true, individuals living in Malta are put in greater jeopardy for experimenting with unfamiliar mushrooms than are the inhabitants of Corsica. As a result, the same nonsocial cue ("brownish, umbrella-shaped fruiting body found in dark places") that exceeds the reliability threshold in Corsica is *unreliable* in Malta. If eaten, portobella mushrooms would be just as profitable to the Maltese forager as they are to the Corsican. However, because it is more dangerous to sample novel mushrooms in Malta, the experimentation required to discover portobella mushrooms is locally disfavored. Corsicans, by contrast, can safely experiment and discover portobella mushrooms on the basis of nonsocial cues alone. Therefore, a widespread preference for portobella mushrooms in Corsica might reflect independent, convergent learning.

This case is hypothetical, but it illustrates how researchers practicing the regional contrast approach could erroneously identify a behavioral variant as a tradition. On the basis of the regional contrast approach, it might seem that confirming the availability and similar profitability of a behavioral variant at both sites would be enough. It is not. Nonsocial cue reliability can vary between sites and thus explain why a particular behavior is prevalent at one site and absent in another.

Finally, unlike the regional contrast approach, the CRA yields itself to statistical analyses of error and variance. Because it deals with quantities that can be measured, investigators can evaluate their conclusions with all the conventional statistical tools available to scientists.

5.2.4 Summary: implications of the cue reliability approach

In conclusion, the CRA does not apply to complex behaviors that yield rewards only *after* the individual has skillfully completed a lengthy action sequence. It is also uninformative when analyzing behaviors of indeterminate adaptive value. But it is useful for analyzing adaptive behaviors that could be dangerous to discover through independent exploration. By providing an alternative test based on cue reliability, the CRA shifts focus away from the task of identifying and eliminating all possible ecological explanations for local differences in behavior, and it reveals instead the importance of predictive information about the safety of attempting the behavior. This permits us to test for traditions that are locally profitable, including those that occur universally. The CRA also helps us to avoid the mistake of ascribing intergroup behavioral variation to different traditions when in fact it can be explained by local differences in nonsocial cue reliability. The CRA highlights several important insights to the study of adaptive social transmission.

1. Because reliability thresholds and social cue reliability vary depending on the problem to be solved, *the same individual is not necessarily expected to heed social cues in all situations.*
2. Because different populations may be constrained by different reliability thresholds, *the same social cue may not equally influence all populations.*
3. Demonstrators should influence observers only insofar as demonstrator cues are reliable. *Therefore, while opportunities to learn from demonstrators may vary according to a demonstrator's age, sex, dominance rank,*

or temperament, observers should take advantage of these opportunities only when demonstrator characteristics render social cues reliable.

4. Because convergent, independent learning can be ruled out when individuals lack reliable nonsocial cues, we have evidence that a widespread behavior is traditional when *nonsocial cue reliability* < *reliability threshold* < *social cue reliability.*

The first three of these conclusions illustrate how the CRA leads to predictions that might not be otherwise obvious if we were to concern ourselves purely with questions about regional contrasts and the psychological mechanisms of social transmission. Conclusion 4 describes the logic behind a potential new test for identifying traditions in the wild. When nonsocial cues are unreliable, the probability is low that any given individual will independently discover a behavior. And when this probability is low, the probability that all group members will independently discover the behavior is lower still. As noted in Section 5.2.1, it is the chance of one independent discovery raised to the *n*th power, where *n* is the number of individuals in the group that exhibit the technique. While individuals may sometimes make the "mistake" of experimenting when it is disadvantageous to do so, the CRA assumes that *most* individuals will *not* attempt a new behavior if the net payoff for experimentation is less than the payoff for sticking with familiar alternatives. We can make a strong inference that social factors contribute to the *widespread practice* of a behavior when social cues are the *only* cues that exceed the reliability thresholds constraining *most* individuals.

Sections 5.3 and 5.4 discuss in broad terms how specific tests might be devised to understand food preferences among generalist foragers and the antipredation responses to snakes among capuchin populations. Although reference is made to the expected utility models presented in Box 5.2, I make no claims about the realism of these particular models. As noted above, these models make a number of assumptions that may require revision if they are to predict the behavior of real animals successfully. Indeed, these expected utility models – and any other, more sophisticated CRA models that might be developed in the future – will have to be tested experimentally to ascertain if animals do, in fact, recognize the reliability thresholds that the models describe. If it can be confirmed that animals discriminate between reliable and unreliable cues, and that animals act as if constrained by the relevant reliability thresholds, CRA models can offer field workers a new means of testing for traditions, including those in natural settings.

5.3 Application 1: devising a test for food traditions

5.3.1 Selecting an appropriate target population and food preference

Group-wide food preferences are promising phenomena to be investigated using the CRA. The simple decision to eat or reject an unfamiliar potential food item has obvious adaptive importance for the generalist herbivore or omnivore, which must discover new good foods while avoiding harmful or toxic substances (Freeland and Janzen, 1974; Rozin, 1976). Dietary generalists face the tasks of identifying and ranking a variety of food types and pursuing those that offer the highest energetic and/or nutritional returns (Stephens and Krebs, 1986). Therefore, even if all unfamiliar potential foods are nontoxic and yield a positive payoff, eating them may be unprofitable compared with seeking out higher-quality, familiar foods. This suggests that at least some reliability thresholds governing food choice could be high, perhaps high enough to discourage foragers from sampling unfamiliar, potential foods on the basis of nonsocial cues alone. If so, the CRA may allow us to identify some food preferences as *bona fide* traditions.

To increase the chances of identifying such a food tradition, researchers might focus on populations exhibiting group-wide preferences for a food type belonging to a category that is relatively unsafe or unprofitable. In such cases, nonsocial cues are less likely to indicate safe or good food, and foragers might need to rely on social cues to discover which foods are safe. For example, eating fungi may be more dangerous than eating fruit. If so, it will probably be easier to demonstrate that a group-wide preference for a particular fungus is traditional than to demonstrate that a group-wide preference for a particular fruit species is traditional. Identifying such categories of potential food requires that we know enough about the foragers' physiology to assess what potential foods would be more or less profitable to eat. We also need to know enough about the foragers' perceptual system to judge what nonsocial cues might be salient to them (see Section 5.3.2).

Other criteria for selecting an appropriate target population for study have to do with measuring the net payoff, F, for seeking out familiar food. The net payoff is the net change in fitness that results for choosing an option. In the case of F, this payoff might represent

$$F = f - s - h - r \qquad (5.4)$$

where all components of F have been converted to a common currency, like calories, and f is the energetic return for consuming one standard unit of the average familiar food, s is the average search cost associated with finding one standard unit of the average familiar food, h is the average handling cost associated with finding one standard unit of the average familiar food, and r is the predation risk associated with seeking out one standard unit of the average familiar food.

The ideal study population would be one for which this information is easily obtained or simplified. Consequently, desirable features to be associated with the study population include a well-documented and/or relatively small dietary repertoire (making f easier to calculate), well-understood foraging patterns and activity budgets (making the search cost easy to calculate), easily measurable average handling costs, and well-understood (preferably trivial) predation risks. Although predation risk has been quantified in common foraging currencies (e.g., Brown, 1999; Ward, Austin, and MacDonald, 2000), the problems associated with obtaining accurate measures of predation risk (e.g., Isbell, 1994) make it particularly desirable to focus on populations that lack significant risk of predation.

Taking these considerations into account, the most promising populations for study might include chimpanzees, guinea pigs, macaques, pigs, pigeons, sheep, rabbits, rats, and wild or feral dogs. Below, I discuss what kinds of information researchers would need to collect in order to test for food traditions among such well-studied generalist foragers. For illustration purposes, this discussion assumes that the expected utility model in Box 5.2 adequately describes the decisions of these foragers.

5.3.2 How to calculate the reliability threshold for food choice

To estimate the reliability threshold using the expected utility model in Box 5.2, one must know the average fitness payoffs for (a) eating safe novel food, (b) eating unsafe novel "food", and (c) rejecting novel food in favor of seeking out a familiar food for the population under study. It is convenient to begin with the last of these, because the payoff for seeking out familiar food (F) is the yardstick by which the other payoffs can be measured. To calculate this payoff, one must decide how to measure f, the reward for eating the average familiar food. One possibility is to estimate the average caloric density (kilocalories per gram) of familiar food, i.e., to calculate what percentage, by weight, each familiar food type makes up of the overall diet, multiply this percentage by the caloric density of the familiar food type, and sum for all food types. However, it is important to recognize that the payoff for familiar food changes over time, rising and falling as the availability of high-quality familiar foods waxes and wanes. Animals should be more likely to experiment with new potential foods during food shortages when the payoff for familiar food has dropped, and, thus, the reliability threshold has dropped as well. The pertinent estimate of the payoff for familiar food is thus the *lowest* one to have influenced novel food sampling during the lifetimes of the youngest individuals that exhibit the group-wide food preference to be tested. This is the threshold most likely to have favored experimental food sampling by all group members.

The payoff for familiar food may also vary across environments and across age–sex classes and ranks. The payoff is higher for individuals in food-rich environments and for individuals with special access to high-quality foods. For instance, some high-quality fruits are encased in shells that only the strongest capuchin monkeys – usually adult males – can open (see Ch. 13, for example). Hence, compared with females and immature males, adult males may receive higher payoffs for eating familiar foods. In such cases, it may be important to estimate distinct familiar food payoffs and, thus, distinct reliability thresholds for different age–sex classes.

To estimate the payoff for eating unfamiliar potential foods, researchers must identify how foragers categorize foods. What cues do foragers use to distinguish food categories? For instance, do foragers discriminate between sweet fruits and bitter fruits? Answering these questions is important because foragers may be constrained by different reliability thresholds depending on the categories they recognize. For example, if sweet fruits are more profitable than bitter fruits, sweet fruits are probably associated with a lower reliability threshold. To determine what cues and food categories are salient to foragers, researchers can consult the experimental literature on taste thresholds, olfactory thresholds, and visual perception. This literature addresses an array of taxa, including nonhuman primates (e.g., Dominy *et al.*, 2001; Hladik and Simmen, 1996), birds (e.g., Rowe and Guilford, 1999; Schuler and Roper, 1992) ungulates (e.g., Forbes, 1998; Goatcher and Church, 1970), rodents (e.g., Glendinning, 1993; Sclafani, 1991), bees (e.g., Dukas and Waser, 1994), and butterflies (e.g., Weiss, 1997).

Once the pertinent food category has been selected, the target population's habitat can be surveyed for members of that category that the target population ignores. For example, if the category of interest were mushrooms, researchers would collect samples of all available mushroom types currently rejected by the target population. Chemical analysis of each potential food type would permit the identification of types containing secondary compounds of potential harm to consumers (e.g., Huffman, 1997; Wink *et al.*, 1993). Those potential food types lacking such harmful compounds could be labeled as "safe" and their caloric value calculated. Their average value, as measured in standard units of the average familiar food, yields an estimate of the payoff for eating safe novel food. Potential food types containing harmful compounds in high enough concentration to impair activity or health could be identified as "bad". The average loss they cause, as measured in standard units, would yield the payoff for eating bad potential "food". Note that, on the one hand, the payoffs for novel food – safe or bad – do not incorporate search costs because novel potential food is not sought out but encountered spontaneously. On the other hand, researchers may need to incorporate food-handling or food-processing costs into their calculations.

As was the case for the familiar food payoff, the payoffs for safe and bad unfamiliar potential foods may vary for different habitats and foragers. For example, because soil condition or plant defenses may render toxins more prevalent or potent at some locations (e.g., McKey, 1978), the expected payoff for eating a bad potential "food" may vary from site to site. Likewise, local conditions might cause the value of safe potential foods to vary. Finally, if the capacity to detoxify and/or tolerate toxins varies with age–sex class membership (Freeland and Janzen, 1974), different foragers may face different payoffs for eating the same unfamiliar potential foods.

In summary, researchers should keep sources of variation in mind when attempting to derive an estimate of the reliability threshold. In particular, to obtain the most precise and accurate threshold estimates, it may be important to model *different* reliability thresholds for distinct categories of potential foods, insofar as foragers recognize such categories and distinguish the potential payoffs associated with each. It is possible that some categories of potential foods are safe and profitable enough when eaten in small quantities, resulting in a low threshold for experimentation. Other categories of potential foods may be associated with such high thresholds that only virtual certainty can justify experimentation. It is also possible that individual characteristics, such as sex, age, body size, and reproductive status, could significantly affect reliability thresholds. Habitat differences might influence thresholds as well. Therefore, when estimating a threshold, it is important to specify what kind of food, forager, and habitat are used for reference, and to recognize that the resulting estimate might apply only in these circumstances.

5.3.3 How reliable are local cues?

Compared with estimating reliability thresholds, assessing cue reliability is relatively straightforward. Nonsocial cue reliability is estimated by calculating the frequency with which observable characteristics predict that food is safe for a given category of potential food types in the habitat. These observable characteristics may be visual, olfactory, tactile, or taste stimuli and are the same ones that define the threshold. Social cue reliability can be estimated by observing the frequency with which demonstrators eat food that is also "safe" from the observer's perspective. Given the possibility that individuals reap different payoffs for pursuing familiar foods and may also exhibit different tolerances to toxins depending on their condition, rank, and age–sex class membership, social cue reliability might vary depending on who "demonstrates" and who observes. However, it seems likely that most social animals have access to at least some reliable social cues.

5.3.4 Shortcuts

When special conditions apply, it may be possible to avoid some of the work described above and derive estimates of the threshold with less information. In particular, we can eliminate the need to estimate the values of the payoffs for both familiar and unfamiliar foods if we are willing to assume (a) the payoff for eating bad unfamiliar potential foods is zero (rather than negative) and (b) the payoff for eating safe, unfamiliar potential food does not exceed the caloric reward for eating the average familiar food.

The first assumption is uncontroversial because it can only result in an estimate of the reliability threshold that errs on the side of being too low. Such an underestimate might increase the chance of a false negative: the rejection of a *bona fide* food tradition. This is especially true if a suspected food tradition is associated with a potential food category that is dangerous overall, such that the actual expected cost of eating a bad food is high. However, the assumption that the payoff for bad unfamiliar food is zero reduces the chance of a false positive: the misidentification of a groupwide food preference as traditional when it is not. Consequently, the assumption cannot jeopardize the stringency of the test, which is our primary concern.

By contrast, the second assumption can harm the stringency of the test if it is unjustified. I suggest that this assumption is reasonable, however, if we have reason to believe that the target population is well adapted to its environment. If the foraging population has had sufficient time to adapt to local conditions, it seems likely that the average profitability of foods in the dietary repertoire is *at least* as high as the average profitability of safe foods that are not a part of the repertoire. This may not be true during times of extreme or unprecedented food shortage, but it should hold true during shortages that are a regular feature of the environment, for example annual periods of food scarcity. For well-adapted foragers that do not experience extreme or unprecedented food shortages, the formula for the reliability threshold can be reduced to

$$(1 - s - h - r)/1 \tag{5.5}$$

or

$$1 - s - h - r \tag{5.6}$$

where the expected payoff for eating safe novel food is equal to the expected caloric gain for eating familiar food, 1 standard unit, and s is the cost of searching for a familiar alternative to the novel food, h is the average cost of handling or processing familiar food, and r is the predation risk associated with seeking out familiar food as measured in these standard units. If the target population also meets the ideal criteria of negligible handling costs and predation risk, these variables drop out as well and the cost of searching for familiar food alone can provide us with an estimate of the reliability threshold for this special case.

In conclusion, for well-adapted foragers that have *not* lived through an extreme or unprecedented food shortage, the CRA to testing whether or not a shared food preference is traditional can be reduced to the following steps four steps.

1. Decide what category the proposed food type belongs to, for example mature leaves, bark or mushrooms. The reliability threshold to be estimated will apply only to this category, and the observable characteristics associated with this category will define the nonsocial cues.
2. Measure the search cost for seeking out familiar food and, if applicable, the average cost of handling familiar food and the average predation risk associated with seeking out familiar food. Calculate the reliability threshold.

3. Calculate the reliability of the relevant nonsocial cue(s); that is, the probability that a potential food type exhibiting the observable characteristics selected in step (1) is safe. If the reliability threshold is meant to apply to leafy foods, for instance, the reliability of nonsocial cues is the percentage of leafy foods in the local habitat that are safe.
4. Compare nonsocial cue reliability calculated in step (3) with the estimate of the reliability threshold (2).

If nonsocial cue reliability does not exceed the threshold estimate, there is strong evidence that group-wide preferences for food types of the specified category are maintained by social transmission. Only social cues are reliable enough to justify sampling the food if it is unfamiliar. If, however, the nonsocial cue reliability exceeds the estimated threshold, independent, convergent learning could explain the group-wide agreement of food choice. There is insufficient evidence to conclude that the group-wide preference is traditional.

5.4 Application 2: devising a test for responses to potential predators

5.4.1 Selecting an appropriate target population

Like food choice, the identification of predators is a problem of obvious ecological importance. Alarm calls, flight, and mobbing are typical predator responses in many taxa, and a large body of research indicates that predator recognition may be learned socially (e.g., Curio 1988; Mathis, Chivers, and Smith, 1996; Mineka and Cook,1988). However, some aspects of predator recognition may develop invariably in almost all individuals, regardless of experience (e.g., Coss, 1991), and it seems likely that animals come equipped or "prepared" (Seligman, 1971) with an evolved bias to recognize some stimuli as predators more readily than others (e.g., Curio, 1988; Mineka and Cook, 1988).

When might we be able to conclude that a particular example of predator recognition in a wild population is a *bona fide* tradition? The crucial test would compare nonsocial cue reliability with the reliability threshold for responding to potential predators (Box 5.2). If this reliability threshold is very low (such that an animal is better off assuming that an unfamiliar stimulus is a predator, even on the basis of very weak cues), social cues may be unnecessary to explain why individuals alarm call, flee, or mob particular types of animal. Nonsocial cues may provide ample evidence to justify the antipredator response.

For this reason, it may be more useful to look for cases in which the members of a group consistently *fail* to recognize a given species as a potential predator. When should an individual treat an unfamiliar animal as a harmless nonpredator? In this case, the reliability threshold is the complement of the threshold for antipredator response. It is high when the threshold for antipredator response is low. The reliability threshold for *ignoring* a potential predator is given in Equation 5.2, in Box 5.2. Note that the threshold will be high when B, the penalty for ignoring a true predator, is high relative to G, the payoff for ignoring a nonpredator. Therefore, to maximize our chances of detecting a traditional policy of "relaxed indifference" towards a potential predator species, we want to identify situations where the potential prey population has much to lose by making a mistake yet consistently ignores a species belonging to a category associated with dangerous predators. Once such a situation has been identified, the appropriate predator response should be specified and the relevant payoffs estimated.

The response of capuchin monkeys to snakes may be a promising area of study. Capuchins at Palo Verde appear to ignore indigo snakes as a potential predator (see Ch. 13). By contrast, capuchins at Santa Rosa and Lomas Barbudal have been observed to alarm call and/or mob indigo snakes. Assuming these observations reflect a real difference between groups, what do we need to know to determine if the difference is traditional?

5.4.2 How to calculate the reliability threshold for decisions to ignore potential predators

For the purposes of illustrating how the CRA might work, I assumed that the expected utility model adequately describes the decisions that capuchins make about snakes. I also assumed that two choices are available to a capuchin that encounters a snake at striking range: (a) to ignore it, which means going about business as usual; and (b) to retreat to a safe distance and engage in alarm calling and mobbing behavior for a specified length of time. This simplification overlooks a third option, which is merely to retreat without engaging in any other antipredator behavior. However, assuming that the combination of retreat and mobbing behavior is more costly than mere retreat, the simplification will not threaten the stringency of the CRA test. This is because the payoff for responding to false predators ($-r$) occurs in the denominator of the formula for the reliability threshold (Equation 5.2). Therefore, the more costly the antipredator behavior, the lower the reliability threshold for ignoring potential predators. And the lower the estimate of the reliability threshold, the less likely it is that false-positive errors will occur.

I also assumed that snake mobbing and alarm call behaviors do not pose a significant danger to capuchin mobbers because they harass snakes from a safe distance (S. Perry, personal communication). Consequently, the primary expense of harassment is the energy expended by mobbers. Furthermore, suppose that the (negative) payoffs for mobbing are similar whether or not the snake is a true predator; that is, it takes as much energy to mob a dangerous snake as it does a harmless one (this ignores the possible inclusive fitness benefits for mobbing a true predator as opposed to a false predator). If these assumptions are true, then $-r = -R$, and Equation 5.2 above can be reduced to

$$P > (B - R)/(B + G) \quad (5.7)$$

where P is the probability that the animal is *not* a predator, B is the absolute value of the negative payoff for ignoring a predatory snake, $-R$ is the negative payoff for mobbing a predatory snake, and G is the positive payoff for ignoring a harmless, non-predatory snake. What are these payoffs in operational terms? If we use calories as a proxy for fitness, one way to answer this question is to determine how much time is involved in the average snake-mobbing episode. With this information, and information about the rate at which calories are expended during mobbing activities, $-R$ can be estimated. Similarly, assuming that snake mobbings occur during periods that would otherwise be spent foraging, G could be estimated by obtaining the average rate of caloric return for foraging over the equivalent length of time it takes to mob. The most abstruse payoff, $-B$, would then reflect the average loss in fitness (as measured in calories) resulting from permitting a predatory snake to strike from a close distance. This average would reflect the outcomes of both successful and unsuccessful strikes, weighted by the probabilities of their occurrence.

Obtaining an estimate of $|B|$ might seem prohibitively difficult, but even a rough estimate may be useful, as long as it is an overestimate that will not threaten the stringency of the test. Moreover, what is crucial when estimating the reliability threshold is not the absolute measures of the payoffs, but the values of these payoffs *relative* to each other. For instance, define G, the payoff for ignoring the snake given that it is harmless, as the arbitrary unit of measurement. Given that G is the net caloric return for foraging over the average length of time it takes to mob a snake, this unit could be called the "*snack*". The researcher's task is to reckon an upper boundary estimate for $|B|$ as measured in *snack* units. Moreover, to the degree that snake mobbing by capuchins consists primarily of watching the snake from a safe distance and barking at it, it seems very likely that R, the absolute value of the cost of mobbing, is no more than one *snack* unit. That is to say, capuchins probably do not spend more energy mobbing than the net energy they could have earned if they spent the time foraging. When this assumption holds true, $G = R$ and we can replace the previous formula with

$$P > (B - 1)/(B + 1). \tag{5.8}$$

Therefore, at minimum, fieldworkers need to be able to estimate how high B is relative to 1 *snack*, the net return for foraging over the average length of time it takes to mob a snake.

5.4.3 How reliable are local nonsocial cues?

As was the case for food choice, the crucial measurement to obtain for nonsocial cue reliability about snakes is relatively straightforward. What percentage of snakes at Palo Verde are harmless to capuchins? Assuming that capuchins encounter both harmless and predatory snakes at rates reflecting the snakes' representation in the habitat, the answer to this question provides us with a good estimate of the probability that any encountered snake is harmless. If this probability fails to exceed the estimated reliability threshold, we have a strong case for a tradition of ignoring indigo snakes.

5.5 Conclusions

The CRA may represent a valuable new tool to study potential traditions among wild populations. By identifying cue reliability as a crucial factor influencing an individual's first attempt at a behavior, the approach suggests a new way to test whether or not a widespread behavior is a tradition. This new way is effective even when (a) the intergroup distribution of a learned behavior is influenced by ecological factors and (b) a learned behavior is habitual or customary at all known sites.

The CRA also provides a theoretical basis for generating new, testable questions about the occurrence of traditions, and it highlights what data need to be collected to exclude the possibility that behaviors are widespread by virtue of convergent, independent learning. Although it is unlikely that CRA tests will always be feasible, the CRA can clarify what researchers need to know to demonstrate that social cues are necessary to explain an adaptive custom in a wild population. For example, observations made at Mahale suggest that chimpanzees of all ages were encouraged to sample guavas, mangoes, and lemons by having seen a variety of demonstrators doing so (Takahata *et al.*, 1986). However, since we lack information about cue reliability and the reliability threshold, we cannot say whether chimpanzees *needed* social demonstrators to prompt their (presumed) first attempts at eating the fruits. Given that the fruits were human cultigens, artificially selected to exhibit exaggerated cues of edibility (i.e., sweeter taste and reduced bitterness), it seems very likely that they could be judged safe to eat on the basis of nonsocial cues alone. This seems especially likely in the light of evidence that chimpanzees have a higher tolerance for bitter flavors than do human beings (Nishida, Ohigashi, and Koshimizu, 2000). If we are interested in identifying simple, adaptive customs that require the influence of social cues to become widespread, the CRA suggests that the best places to look are those where reliability thresholds are high and nonsocial cue reliability is low.

Similarly, intuition may argue that medicinal plant use (Huffman and Wrangham, 1994) must be traditional because it would be very difficult to learn through trial-and-error. However, once we focus on cue reliability, we may find that medicinal plant use is easier to discover independently than we thought. For instance, if bitter taste is a reliable cue of antihelminthic properties (Huffman, 1997; Johns, 1994), animals may be able to discover medicinal plants without the help of demonstrators. The problem can be characterized by a food choice reliability threshold where the question is not "is this novel item worth eating compared with seeking out familiar food?" but, rather, "is this novel item worth eating compared with doing nothing and remaining ill or parasitized?" If the payoff for doing nothing is low, the reliability threshold for eating potentially medicinal plants is reduced. Moreover, the CRA reminds us that social cue reliability should not be taken for granted. Assuming that medicinal plant use is socially transmitted, what behaviors or signals from conspecifics could be reliable indicators that a plant is "good medicine?" Reliable social cues must be identified before the traditional status of medicinal plant use can be tested.

Finally, the CRA suggests that the regional contrast argument for a tradition – that a strategy is traditional if it is equally advantageous at two sites but practiced only at one – is flawed. It is not the profitability of practicing the strategy that is crucial but instead the risk entailed by attempting something new. The CRA helps to clarify that the true alternative explanation for an apparent tradition is not local profitability or universality, but convergent, independent learning encouraged by reliable nonsocial cues.

5.6 Acknowledgements

I am indebted to John Mitani and Bobbi Low for their help in launching my research in behavioral ecology and social transmission. I am also grateful to Dorothy Fragaszy and Susan Perry for giving me the opportunity to participate in the conference *Traditions in Nonhuman Primates*, where this paper was presented. I thank conference participants for their thoughtful criticisms and E. Addessi, M. Flower, D. Fragaszy, K. Hunley, S. Jesseau, J. Kendal, K. Laland, J. Mitani, S. Perry, L. Rose, B. Stallman, D. Unger, E. Visalberghi, J. Westin and an anonymous reviewer for their comments on the manuscript.

References

Aoki, K. and Feldman, M. W. 1987. Toward a theory for the evolution of cultural communication: Coevolution of signal transmission and reception. *Proceeding of the National Academy of Science, USA*, **84**, 7164–7168.

Baldwin, B. A. and Meese, G. B. 1979. Social behaviour in pigs studied by means of operant conditioning. *Animal Behaviour*, **27**, 947–957.

Boyd, R. and Richerson, P. J. 1985. *Culture and the Evolutionary Process*. Chicago, IL: University of Chicago Press.

Boyd, R. and Richerson, P. J. 1988. An evolutionary model of social learning: the effects of spatial and temporal variation. In *Social Learning: Psychological and Biological Perspectives*, ed. T. R. Zentall and B. G. Galef Jr., pp. 29–48. Hillsdale, NJ: Erlbaum.

Brown, J. 1999. Vigilance, patch use and habitat selection: foraging under predation risk. *Evolutionary Ecology Research*, **1**, 49–71.

Cavalli-Sforza, L. L. and Feldman, M. W. 1981. *Cultural Transmission and Evolution: A Quantitative Approach*. Princeton: Princeton University Press.

Coss R. G. 1991. Context and animal behavior III. The relationship between early development and evolutionary persistence of ground squirrel antisnake behavior. *Ecological Psychology*, **3**, 277–315.

Curio, E. 1988. Cultural transmission of enemy recognition in birds. In *Social Learning: Psychological and Biological Perspectives*, ed. T. R. Zentall and B. G. Galef Jr., pp. 75–98. Hillsdale, NJ: Erlbaum.

Dominy, N. J., Lucas, P. W., Osorio, D., and Yamashita, N. 2001. The sensory ecology of primate food perception. *Evolutionary Anthropology*, **10**, 171–186.

Dukas, R. and Waser, N. 1994. Categorization of food types enhances the foraging efficiency of bumblebees. *Animal Behaviour,* **48**, 1001–1006.

Forbes, J. M. 1998. Dietary awareness. *Applied Animal Behaviour Science,* **57**, 287–297.

Fragaszy, D. M. and Visalberghi, E. 1996. Social learning in monkeys: primate primacy reconsidered. In *Social Learning and Imitation: The Roots of Culture,* ed. C. M. Heyes and B. G. Galef Jr., pp. 49–64. New York: Academic Press.

Freeland, W. J. and Janzen, D. H. 1974. Strategies in herbivory by mammals: the role of plant secondary compounds. *American Naturalist,* **108**, 269–289.

Galef, B. G., Jr. 1995. Why behaviour patterns that animals learn socially are locally adaptive. *Animal Behaviour,* **49**, 1325–1334.

Galef, B. G., Jr., McQuoid, L. M., and Whiskin, E. E. 1990. Further evidence that Norway rats do not socially transmit learned aversions to toxic baits. *Animal Learning and Behavior,* **18**, 199–205.

Giraldeau, L. A. and Caraco, T. 2000. *Social Foraging Theory.* Princeton, NJ: Princeton University Press.

Giraldeau, L. A., Caraco, T., and Valone, T. J. 1994. Social foraging: individual learning and cultural transmission of innovations. *Behavioral Ecology,* **5**, 35–43.

Glendinning, J. I. 1994. Is the bitter rejection response always adaptive? *Physiological Behavior,* **56**, 1217–1227.

Goatcher, W. D. and Church, D. C. 1970. Taste responses in ruminants. IV. Reactions of pygmy goats, normal goats, sheep and cattle to acetic acid and quinine hydrochloride. *Journal of Animal Science,* **31**, 373–382.

Heyes, C. M. 1993. Imitation, culture and cognition. *Animal Behaviour,* **46**, 999–1010.

Hladik, C. M. and Simmen, B. 1996. Taste perception and feeding behavior in nonhuman primates and human populations. *Evolutionary Anthropology,* **5**, 58–71.

Huffman, M. 1996. Acquisition of innovative cultural behaviors in nonhuman primates: a case study of stone-handling, a socially transmitted behavior in Japanese macaques. In *Social Learning in Animals: The Roots of Culture,* ed. C. M. Heyes and B. G. Galef Jr., pp. 267–286. San Diego, CA: Academic Press.

Huffman, M. 1997. Current evidence for self-medication in primates: a multidisciplinary perspective. *Yearbook of Physical Anthropology,* **40**, 171–200.

Huffman, M. and Wrangham, R. W. 1994. Diversity of medicinal plant use by chimpanzees in the wild. In *Chimpanzee Cultures,* ed. R. W. Wrangham, W. C. McGrew, F. B. M. de Waal, and P. G. Heltne, pp. 129–148. Cambridge, MA: Harvard University Press.

Isbell, L. A. 1994. Predation on primates: ecological patterns and evolutionary consequences. *Evolutionary Anthropology,* **3**, 61–71.

Johns, T. 1994. Ambivalence to the palatability factors in wild plant foods. In *Eating on the Wild Side: The Pharmacologic, Ecologic, and Social Implications of Using Noncultigens,* ed. N. L. Etkin, pp. 46–61. Tuscon, AZ: University of Arizona Press.

Laland, K. N. 1999. Exploring the dynamics of social transmission with rats. In *Mammalian Social Learning: Comparative and Ecological Perspectives,* ed. H. O. Box and K. R. Gibson, pp. 174–187. New York: Cambridge University Press.

Laland, K. N. and Williams, K. 1998. Social transmission of maladaptive information in the guppy. *Behavioral Ecology,* **9**, 493–499.

Laland, K. N., Richerson, P. J., and Boyd, R. 1996. Developing a theory of animal social learning. In *Social Learning in Animals: The Roots of Culture*, ed. C. M. Heyes and B. G. Galef Jr., pp. 129–154. San Diego, CA: Academic Press.

Mason, J. R., Arzt, A. H., and Reidinger, R. F. 1984. Comparative assessment of food preferences and aversions acquired by blackbirds via observational conditioning. *Auk*, **101**, 796–803.

Mathis, A., Chivers, D. P., and Smith, R. J. F. 1996. Cultural transmission of predator recognition in fishes: intraspecific and interspecific learning. *Animal Behaviour*, **51**, 185–201.

McKey, D. 1978. Soil, vegetation and seed-eating by black colobus monkeys. In *The Ecology of Arboreal Folivores*, ed. G. G. Montgomery, pp. 423–237. Washington, DC: Smithsonian Institute Press.

Mineka, S. and Cook, M. 1988. Social learning and the acquisition of snake fear in monkeys. In *Social Learning: Psychological and Biological Perspectives*, ed. T. R. Zentall and B. G. Galef Jr., pp. 51–73. Hillsdale, NJ: Erlbaum.

Nishida, T. 1987. Local traditions and cultural transmission. In *Primate Societies*, ed. B. B. Smuts, D. L. Cheney, R. M. Seyfarth, R. W. Wrangham, and T. T. Struhsaker, pp. 462–474. Chicago, IL: University of Chicago Press.

Nishida, T., Ohigashi, H., and Koshimizu, K. 2000. Tastes of chimpanzee plant foods. *Current Anthropology*, **41**, 431–438.

Rogers, A. 1988. Does biology constrain culture? *American Anthropologist*, **90**, 819– 831.

Rowe, C. and Guilford, T. 1999. Novelty effects in a multimodal warning signal. *Animal Behaviour* **57**, 341–346.

Rozin, P. 1976. Specific aversions as a function of specific hungers. *Journal of Comparative Physiological Psychology*, **64**, 237–242.

Schuler, W. and Roper, T. J. 1992. Responses to warning coloration in avian predators. *Advances in the Study of Behavior*, **21**, 111–146.

Sclafani, A. 1991. Starch and sugar tastes in rodents: an update. *Brain Research Bulletin*, **27**, 383–386.

Seligman, M. 1971. Phobias and preparedness. *Behavior Therapy*, **2**, 307–320.

Stephens, D. W. and Krebs, J. R. 1986. *Foraging Theory*. Princeton, NJ: Princeton University Press.

Takahata, Y., Hiraiwa-Hasegawa, M., Takasaki, H., and Nyundo, R. 1986. Newly acquired feeding habits among the chimpanzees of the Mahale Mountains National Park, Tanzania. *Human Evolution*, **1**, 277–284.

Terkel, J. 1996. Cultural transmission of feeding behavior in the black rat (*Rattus rattus*). In *Social Learning in Animals: The Roots of Culture*, ed. C. M. Heyes and B. G. Galef Jr., pp. 267–286. San Diego, CA: Academic Press.

Visalberghi, E. and Addessi, E. 2000. Seeing group members eating familiar food enhances the acceptance of novel foods in capuchin monkeys. *Animal Behaviour*, **59**, 1–8.

Visalberghi, E. and Fragaszy, D. 1990. Food-washing behaviour in tufted capuchin monkeys, *Cebus apella*, and crabeating macaques, *Macaca fascicularis*. *Animal Behaviour*, **40**, 829–836.

Ward, J., Austin, R., and MacDonald, D. 2000. A simulation model of foraging behaviour and the effect of predation risk. *Journal of Animal Ecology*, **69**, 16–30.

Weiss, M. R. 1997. Innate colour preferences and flexible colour learning in the pipevine swallowtail. *Animal Behaviour*, **53**, 1043–1052.

Whiten, A., Goodall, J., McGrew, W. C., Nishida, T., Reynolds, V., Sugiyama, Y., Tutin, C. E. G., Wrangham, R. W., and Boesch, C. 1999. Cultures in chimpanzees. *Nature,* **39**, 682–685.

Wink, M., Hofer, A., Bilfinger, M., Englert, E., Martin, M., and Schneider, D. 1993. Geese and dietary allelochemicals – food palatability and geophagy. *Chemoecology*, **4**, 93–107.

BENNETT G. GALEF JR.

6

"Traditional" foraging behaviors of brown and black rats (*Rattus norvegicus* and *Rattus rattus*)

> The brown rat, in particular, appears especially able to develop local traditions, more so perhaps than other more-closely examined mammals, possibly including the anthropoids.
>
> STEINIGER, 1950, P. 368

6.1 Introduction

Imagine, if you will, an energetic, young graduate student who has established a study site near Para, Brazil, where she spends 3 years observing a geographically isolated population of capuchin monkeys that no other primatologist has looked at. Imagine further that our graduate student soon finds, to her great surprise and pleasure, that all of the members of one troop of capuchins at Para, unlike any previously studied capuchins, regularly hunt and eat small lizards. Many months of demanding field work show that the lizards are the source of more than 20% of the calories and 36% of the protein ingested by troop members.

Discovering a complex, biologically meaningful pattern of behavior that is unique to a particular population of monkeys would be a significant event in the career of any behavioral scientist. Surely, before very long, our imaginary graduate student is going to want to tell her colleagues, and quite possibly members of the media as well, about her discovery. To do so, she is going to have to decide how to refer to the unusual behavior that her field studies have documented.

If our imaginary graduate student were to make the conventional choice, and there is little reason to doubt that she would, she would soon be referring to the lizard hunting she has observed as "cultural", as a "tradition" of the capuchins at Para. Her decision may seem a trivial one, but dozens of similar decisions made over decades have had unintended effects leading the unwary to conclude that intellectual problems have been solved that have not even been addressed.

6.1.1 Defining tradition

The English word "tradition" derives from the Latin *traditio* meaning either the action of handing something over to another or of delivering up a possession (Lewis and Short, 1969). In ordinary speech, a behavior described as "traditional" is one that has been learned in some way from others and is passed on to naïve individuals (Gove, 1971). Consequently, calling a pattern of behavior "traditional" implies (or, at the least, will surely lead a listener to infer) that social learning of some kind has played a role in its development. Those unfamiliar with the literature on traditions of animals may even infer that the behavior described as traditional or cultural is actively transmitted by the knowledgeable to the naïve by teaching, imitation, or some other complex process, as are most elements of human culture (Galef, 1992).

Of course, the word used to describe a phenomenon is of little importance so long as the label does not interfere with understanding, as describing population-specific behaviors as traditional seems sometimes to do (Whiten and Ham, 1992). What is important is that we not allow the use of words from the common language as technical terms to cloud our thinking about behavioral phenomena.

Why field workers have until fairly recently labeled as "traditional" essentially any pattern of behavior common in one population of a species and rare or absent in others is not obvious. Whatever the origins of the practice, it is problematic for those interested in the processes responsible for the development of specific patterns of behavior. Behavioral differences among groups can often be explained as the result of asocial developmental processes (see Ch. 11). Consequently, referring to any population-specific behavior as traditional before it has been established that it is transmitted from individual to individual by social learning conceals the need for developmental analysis.

Tradition, like adaptation (Williams, 1966), is an onerous concept that should be employed only when there is evidence that social learning of some kind actually plays a role in dissemination of the supposedly traditional behavior. Otherwise, description of a behavior as "traditional" serves only to camouflage ignorance of the developmental processes responsible for the spread of behaviors so labeled.

Calling a population-specific behavior traditional before the causes of its development have been identified has a further unfortunate consequence. Those with a primary interest in areas other than behavioral development may assume that once it has been established that a behavior is, in fact, traditional in an animal population (i.e., that it is learned in some way by the naïve as a result of interaction with knowledgeable others), the causes of its diffusion are known.

6.1.2 Tradition and social learning

Gaulin and Kurland (1976, p. 374) may have overstated the case in asserting that "Unless the spread of a behavioral trait is attributable to a particular diffusion mechanism, the concept of tradition is completely uninformative". Surely, the concept of tradition differentiates those instances of behavioral variance resulting from social transmission from those resulting either from genetic processes or from behavioral differences reflecting response to variation in the asocial environment. Still, Gaulin and Kurland (1976) focused attention on an important issue. Social learning processes, from "teaching" (Caro and Hauser, 1992) to "local enhancement" to "true imitation" (Thorpe, 1963), can result in transmission of behavior from one individual to another. Consequently, for those interested in understanding either behavioral development or social learning processes, calling a population-specific behavior "traditional" answers relatively few questions and raises many.

6.2 Alternative explanations of behavioral variation

Variance among individuals in behavioral development can be conceived of as caused by interaction of three types of information: (a) genetically transmitted information received from parents, (b) information acquired individually as a result of direct transactions with the asocial environment, and (c) information acquired by individuals as a consequence of interactions with conspecifics (Galef, 1976). Obviously, simply discovering a difference in the behavior of two populations does not demonstrate that social learning produced that difference. Less widely appreciated is the converse proposition. Discovery of singular properties of either the gene pool or ecology of a population that exhibits a unique pattern of behavior does not mean that social learning is excluded as a cause of diffusion of that behavior.

The relationship among findings in genetics, ecology, and the study of social learning has produced sufficient misunderstanding (see, for example, the exchange in *Science* between Strum (1975, 1976) and Gaulin and Kurland (1976)) that discussion of a concrete example may prove useful.

6.3 An example: vampire finches of Wolf and Darwin Islands

Measurement of body parts of adult male, sharp-beaked ground finches (*Geospiza difficilis*) on Wolf (Wenman) and Darwin (Culpepper) Islands (40 km apart and 100 km from the closest other island) in the Galapagos Archipelago has resulted in classification of *G. difficilis* on these two islands as a distinct subspecies (*septentrionalis*) (Lack, 1947, 1969; Schluter and Grant, 1982, 1984). Such classification may lead to the inference that the unique morphology of *G. difficilis* on Wolf and Darwin Islands reflects differences between the genotypes of *G. difficilis septentrionalis* and those of *G. difficilis* found elsewhere in the Galapagos. Indeed, DNA analyses in progress at the time this manuscript was in preparation are providing direct evidence that *G. difficilis* found on Wolf and Darwin Islands is genetically distinct from other population of the species (P. Grant, personal communication, September 8, 1999).

Sharp-beaked ground finches found on Wolf and Darwin Islands differ from those found elsewhere in the Galapagos not only in heritable morphological characters but also in their environment and behavior. For example, Wolf and Darwin Islands are not inhabited by the predatory owls and hawks that are found elsewhere in the Galapagos Archipelago. Possibly as a consequence, *G. difficilis septentrionalis* exhibits "a tameness that is most striking" (Bowman and Billeb, 1965, p. 41).

Wolf Island is also the only place in the Galapagos where *Opuntia* (prickly-pear) cacti are found that do not also support species of ground finches (*G. scandens* and *G. conirostis*) that are specialized feeders on *Opuntia*. Perhaps because of the absence of efficient competitors on Wolf Island, *G. difficilis* birds found there, unlike conspecifics elsewhere in the Galapagos, probe *Opuntia* flowers for nectar and pollen.

More startling, *G. difficilis* subspecies on Darwin and Wolf Islands, but not others of their species, perch on the tails of masked and red-footed boobies (large, white-bodied seabirds of the genus *Sula*), draw blood by pecking at the base of boobies' feathers, and feed on blood flowing from the wounds thus created. Also on Wolf and Darwin Islands, but not elsewhere, *G. difficilis* uses its relatively long bill to pierce seabird eggs and eat their contents (Bowman and Billeb, 1965; Koster and Koster, 1983; Schluter and Grant, 1982, 1984).

In sum, the *G. difficilis septentrionalis* subspecies exhibits four population-specific behaviors: unusual tameness, feeding on cactus flowers, feeding on birds' eggs, and feeding on blood. The last of these four population-specific behaviors is the one most frequently referred to in the literature as a "tradition" of finches on Wolf and Darwin Islands, so I shall focus discussion on it. The question, of course, is whether the wealth of available information regarding the taxonomy, ecology, and natural history of sharp-beaked ground finches is sufficient to determine whether the unique patterns of behavior exhibited by *G. difficilis* on Wolf and Darwin Islands are "traditional" in the strict sense of the term.

6.3.1 Is blood feeding an animal tradition?

To test the hypothesis that the unusual behaviors exhibited by *G. difficilis septentrionalis* are traditional, information is needed about social interactions that might increase the probability that an individual born on Wolf or Darwin Island would exhibit behaviors typical of the *G. difficilis* found there. Although hypotheses relating to the development of such unique behaviors will surely incorporate information about ecology and genetics, their test requires study of behavioral development in individuals. Analyses at population, ecological, or genetic levels are simply not sufficient.

For example, Bowman and Billeb (1965) have suggested, regarding the habit of blood feeding, that (a) during the dry season, when insects (the typical fare of *G. difficilis*) are reduced in numbers, boobies are frequently infested with black hippoboscid flies that are, at least to a human observer, very conspicuous against the boobies' white plumage, and (b) finches might pursue flies on boobies and develop the blood-feeding habit as a result of accidentally puncturing a booby's skin while attempting to capture a fly.

Although such an account fails to address directly the question of why *G. difficilis* on Wolf and Darwin Islands feeds on the blood of boobies, whereas *G. difficilis* found elsewhere does not, the explanation is at a level of analysis appropriate to that issue. To understand the origins of blood feeding we need information about how the behavior develops in individuals.

Heritable differences in tameness might permit *G. difficilis septentrionalis* to approach boobies when other subspecies of *G. difficilis* would not. Heritable differences in beak shape might increase the ease with which this subspecies gains access to blood. There might also be heritable differences among subspecies of *G. difficilis* in the tendency to attack seabirds. However, ecological differences among islands of the Galapagos Archipelago might make blood feeding particularly valuable to finches on Darwin and Wolf Islands, maintaining a behavior in which all *G. difficilis* subspecies would engage, if they were exposed to similar ecological conditions.

Last, it is also possible that a very rare incident allowed one *G. difficilis septentrionalis* bird living on Wolf or Darwin island to learn to attack boobies and feed on their blood, and that the habit of blood feeding developed in others as a result of learning from this innovator. Indeed, blood feeding may have developed or be maintained in response to all five of the factors mentioned above interacting in complex ways in the unique situation, environmental, genetic as well as social, in which all the birds of this subspecies live. Determining causes of the unusual behaviors of sharp-billed ground finches on Wolf and Darwin Islands would require experiments, in addition to observation and correlational analyses. Such experiments have not been and, given the protected state of the genus, may never be conducted with Darwin's finches. However, behaviors that are engaged in by members of some populations of a species but not others have been found in species less fragile than the ground finches of the Galapagos Archipelago.

6.3.2 Primate traditions

In the anthropological or psychological literatures, particular attention has been given to evidence consistent with the view that at least some of the unusual behaviors observed in only one or a few chimpanzee, capuchin, or dolphin troops may be behavioral traditions (for review see Whiten *et al.*, 1999 and Chs. 13 and 14). However, in apes and in capuchin monkeys, as in the Galapagos finches discussed above, the hypothesis that population-specific patterns of behavior observed in free-living populations are traditional does not rest on experimental evidence. Rather, the conclusion that such species exhibit true behavioral traditions depends largely on exclusion of alternative explanations of the origins of population-specific behaviors (for an exception, see Ch. 5).

6.4 Traditions of rats

It is seldom mentioned in discussions of the possibility of population-specific patterns of behavior in primates that the most convincing evidence of behavioral traditions in free-living, nonhuman animals is to be found not in the geographical distribution of patterns of tool use by our great ape cousins or of social behaviors in our more distant primate relatives but in the singing of passerine birds and the feeding habits of Norway and black rats (for a refreshing exception, see McGrew, 1998).

The fact that evidence of behavioral traditions is not restricted to our close phylogenetic relatives is important because it serves as a reminder that evidence of traditional patterns of behavior in animals, no matter how convincing, is not evidence of mental processes in animals similar to those supporting traditions in humans (Galef, 1992). Indeed, analyses of traditions in nonprimates, particularly in Norway and black rats, have demonstrated repeatedly that animal traditions can rest on rather simple behavioral substrates.

6.4.1 Field evidence of traditions in Norway and black rats

Norway rats (*Rattus norvegicus*) are the most successful, nonhuman mammals on the planet and are found breeding from Nome, Alaska (64° 32′ N), where they live on human garbage (Kenyon, 1961), to South Georgia Island (54° 90′ S), where they subsist on a diet of tussock grass, beetles, and ground-nesting birds (Pye and Bonner, 1980). Much of the biological success that rats enjoy results from their ability to adapt their foraging to an extraordinary range of ecological conditions.

Not surprisingly, given the plasticity of the foraging behavior of Norway rats, most population-specific behaviors in the species involve foraging of one sort or another. Norway rats living on the banks of ponds in a hatchery in West Virginia catch fingerling fish and eat them (Cottam, 1948). Many members of some colonies of Norway rats living on the banks of the Po River in Northern Italy dive for and feed on mollusks inhabiting the river bottom, whereas no members of nearby colonies with equal access to mollusks prey upon them (Gandolfi and Parisi, 1972, 1973; Parisi and Gandolfi, 1974). On the island of Norderoog in the North Sea, Norway rats frequently stalk and kill sparrows and ducks (Steiniger, 1950), though they have not been reported to do so elsewhere. Colonies of black rats (*Rattus rattus*) thrive in the pine forests of Israel by removing scales from pinecones and eating the seeds that the scales conceal, a behavior not reported in other populations of black rats (Terkel, 1996), and so on.

6.4.2 Laboratory studies of "traditions" in free-living rats

Numerous instances of possible socially transmitted behavior have been analyzed in laboratory studies of rat behavior: everything from movement in a T-maze to predation on house mice and avoidance of candle flames (e.g., Church, 1957; Flandera and Novakova, 1974; Lore, Blanc and Suedfeld, 1971). However, most systematic, experimental investigations of traditions in rats have involved analyses of instances of population-specific patterns of behavior that, like those mentioned in the preceding section, were first described by those studying free-living rats.

Fortunately, population-specific behavior observed in rats can often be reproduced in the laboratory. Consequently, development of such behaviors can be examined experimentally, and assertions that population-specific behaviors seen in free-living animals are, in fact, traditional can be critically evaluated.

6.4.3 Learning what to eat

6.4.3.1 Field observations

Fritz Steiniger (1950), an applied ecologist who spent many years studying ways to improve methods of rodent control, discovered that it was particularly difficult to exterminate rat colonies by repeatedly placing the same poison bait in a rat-infested area. When Steiniger used the same bait a number of times, despite initial success in reducing pest numbers, later bait acceptance was very poor, and colonies soon returned to their initial sizes (Steiniger, 1950). Young rats that were born into colonies that contained animals that had survived their first ingestion of a poison bait, and had consequently learned not to eat it, avoided the bait without ever even tasting it for themselves. Steiniger (1950) believed (incorrectly, as it turned out) that inexperienced rats were dissuaded by experienced individuals from ingesting potential foods by those that had learned that the bait was toxic.

6.4.3.2 A laboratory analogue

Young wild rats' total avoidance of diets that adults of their colony have learned to avoid ingesting is a robust phenomenon that can be brought into the laboratory with little difficulty (Galef and Clark, 1971a). We captured adult wild rats on a garbage dump in southern Ontario and placed them in groups of five or six in 2 m^2 laboratory cages. For 3 hours each day, each experimental colony was provided with two easily distinguished, equally nutritious foods.

To begin a typical experiment, we introduced sublethal doses of a toxin into one of the two foods placed in a colony's cage each day. Under such conditions, colony members rapidly learned to avoid ingesting the poisoned food, and continued to do so even when subsequently offered uncontaminated samples of the previously toxic bait.

After a colony had been trained, we had to wait until a female colony member gave birth and her young grew to weaning age. Then, we could use closed-circuit television to observe adults and pups throughout daily feeding sessions and record the number of times that pups ate each of the two uncontaminated foods in their cage: one of which adult colony members were eating and the other they were avoiding.

We found repeatedly that weaning young ate only the food that the adults of their colony were eating and totally avoided the alternative (Galef and Clark, 1971a). Even when we removed pups from their natal enclosures and offered them the same two foods that had previously been available to them, the pups continued to eat only the food that adults of their colony had eaten (Galef and Clark, 1971a). Clearly, we had a laboratory situation in which young rats showed a population-specific pattern of food choice similar to that shown by the free-living wild rats Steiniger (1950) had studied in Germany two decades earlier.

6.4.3.3 Analysis of the phenomenon

My students and I have spent much of the last 30 years determining how feeding patterns of adult rats influence food choices of the young that interact with them (for reviews see Galef, 1977, 1988, 1996a,b). We have not been working painfully slowly. Rather, we have discovered that there are many ways in which social interactions affect rats' selection of foods and feeding sites, and years of investigation, both in our laboratory and elsewhere, have been required to begin to unravel the complexities involved. Below, I explore briefly some of the processes occurring throughout life that result in rats tending to select the same foods to eat as their fellows.

6.4.3.4 Prenatal effects

A rat fetus exposed to a flavor while still in its mother's womb (as a result, for example, of injection of that flavor into its dam's amniotic fluid) will, when grown, drink more of a solution containing that flavor than will control rats lacking such prenatal experience (Smotherman, 1982). More realistically, feeding garlic to a pregnant rat enhances the postnatal preference of her young for the odor of garlic (Hepper, 1988).

6.4.3.5 Effects while suckling

Evidence from several laboratories has indicated that flavors of foods that a rat dam eats while lactating affect the flavor of her milk, and exposure to such flavored milk affects the food preferences of weaning pups (Galef and Henderson, 1972; see also, Bronstein, Levine, and Marcus, 1975; Galef and Sherry, 1973; Martin and Alberts, 1979).

Clearly, a process is at work during the nursing period that can increase the probability that successive generations of rats will choose to eat the same foods. As weaning proceeds, both the number of such processes and the magnitude of their impact of food choice increases.

6.4.3.6 Effects while weaning

Galef and Clark (1971b) used time-lapse videography to observe each of nine wild rat pups take their very first meals of solid food. All nine pups ate for the first time under exactly the same circumstances. Each took its first meal at the same time that an adult member of its colony was eating and each ate at the same place that the adult was feeding, not at an alternative feeding site a short distance away.

Further studies revealed that weaning rat pups do not follow adults as they move to feeding sites but instead use visual cues to detect and approach feeding adults from a distance (Galef and Clark, 1971b). In fact, anesthetizing an adult rat and placing it near one of two otherwise identical feeding sites makes the site occupied by the anesthetized adult far more attractive to pups than the unoccupied site, and young pups both visit and eat more at the occupied site than at the unoccupied one (Galef, 1981).

6.4.4 Residual olfactory cues

6.4.4.1 Feeding site selection

Adult rats do not need to be physically present at a feeding site to cause conspecific young to prefer to feed there. As rats leave a feeding site, they deposit scent trails that direct young rats seeking food to locations where food was ingested (Galef and Buckley, 1996). Also, feeding adult rats deposit residual olfactory cues both in the vicinity of a food source

(Galef and Heiber, 1976; Laland and Plotkin, 1991) and on foods they are eating (Galef and Beck, 1985). These odors are attractive to pups and, like the presence of an adult rat at a feeding site, cause young rats to prefer marked sites to unmarked ones.

Normal response to residual cues found around a feeding site depends on preweaning experience of pups with their dam and siblings. Pups reared without contact with conspecifics (Hall, 1975) do not find feeding sites marked with feces of adult rats attractive; pups reared in social isolation and given a few days to interact with a lactating female and pups are subsequently attracted to a feeding site by the presence there of fecal material (Galef, 1981).

6.4.4.2 Feeding site selection and food choice

Although both adult rats and residual olfactory cues present at a feeding site increase a site's attractiveness to weaning rats, such effects are, obviously, not in themselves sufficient to produce socially transmitted food preferences. However, if feeding sites that are used and marked by adult rats contain foods different from those found at sites that adults are not exploiting and marking, then socially learned food preferences can result from socially learned feeding site preferences (Galef and Clark, 1971a).

Wild Norway rats are extremely hesitant to ingest any potential food that they have not previously eaten (Barnett, 1958; Galef, 1970), and young wild rats socially induced to eat their first meals at a site containing a food become familiar with that food and are very reluctant to eat anything else (Galef and Clark, 1971a). Consequently, social influences on feeding site selection may act indirectly (Galef, 1985) to produce traditions of food preference and avoidance in rats of the kind Steiniger (1950) described.

6.4.4.3 Direct transmission of flavor preferences

After a naïve "observer" rat interacts with a recently fed conspecific "demonstrator", the observer exhibits substantial enhancement of its preference for whatever food its demonstrator ate (Galef and Wigmore, 1983; Posadas-Andrews and Roper, 1983; Strupp and Levitsky, 1984). Both food-related odors escaping from the digestive tract of a demonstrator and the scent of bits of food clinging to its fur and vibrissae allow conspecifics to identify foods others have eaten (Galef, Attenborough and Whiskin, 1990; Galef, Kennett and Stein, 1985; Galef and Whiskin, 1992). However, socially enhanced food preferences depend on rats experiencing food odors together with other stimuli emitted by live conspecifics (Galef *et al.*,

1985, 1988; Galef and Stein, 1985; Heyes and Durlach, 1990). For example, rats exposed to pieces of cotton batting dusted with a food and moistened with distilled water do not develop a preference for the food. However, rats exposed to the same food either dusted on the head of an anesthetized conspecific or on a piece of cotton batting that has been moistened with a dilute carbon disulfide solution (carbon disulfide is a constituent of rat breath) exhibit strong preferences for the food to which they were exposed (Galef *et al.*, 1988; Galef and Stein, 1985).

Such effects of exposure to a recently fed rat on the food choices of its fellows are surprisingly powerful (Galef, Kennett, and Wigmore, 1984; Richard, Grover, and Davis, 1987). If observer rats first taught to avoid totally ingesting a diet by following its ingestion with an injection of toxin are then placed with a conspecific demonstrator that has eaten the diet to which an aversion has been learned, these observers frequently totally abandon their aversion to the diet associated with illness. Further, most rats that interact with conspecifics fed a diet adulterated with cayenne pepper, which is inherently unpalatable to rats, subsequently prefer peppered diet to unadulterated diet (Galef, 1986a). However, as the degree of aversiveness of a food increases, the impact of social influences on its acceptance decreases (Galef and Whiskin, 1998a).

6.4.5 Multigenerational traditions

Evidence that rats can influence one another's choice of foods is overwhelming. However, for a "tradition" to become established in a population, at least some individuals who acquire the traditional pattern of behavior must engage in it long enough to induce others to behave similarly.

As Heyes (1993) has pointed out, socially learned behaviors are not insulated from modification by individual learning during the time between their acquisition and transmission. Consequently, demonstrations that socially transmitted behaviors are sufficiently stable to permit repeated retransmission, and consequent diffusion through a population, are necessary to establish the sufficiency of social learning to support behavioral traditions (Laland, Richerson, and Boyd, 1993). In part because of the expense of maintaining large numbers of animals in the laboratory, such demonstrations are few in number.

6.4.5.1 Digging for food

Laland and Plotkin (1990, 1992) employed a procedure in which a rat that had learned socially to dig for buried food served as a model for a naïve rat, which, after learning socially to dig for buried food, became a model for another naïve rat etc. Such chaining captures some features of diffusion of socially learned behaviors through free-living populations of animals. However, Laland and Plotkin's (1990, 1992) procedures involved simple iteration of a basic social learning situation (in which a naïve individual learns by interaction with a trained model) and failed to capture many features of life outside the laboratory that might interfere with propagation of behavior. In particular, there was no opportunity for individual learning about alternative behaviors in the interval between social acquisition and transmission of digging behavior, and the presence of alternatives is of considerable possible importance in determining the fidelity of transmission of a socially learned behavior (Galef and Whiskin, 1997, 1998b).

6.4.5.2 Food preferences

Galef and Allen (1995) established small colonies of rats and trained all members of each colony to eat only one of two equipalatable foods available *ad libitum*. After training, one member of the trained colony was removed every 24 hours and replaced with a naïve individual. The process was continued long after all original colony members had been removed, with replacement each day of the colony member that had been in the colony longest. Colonies maintained the food preferences taught to their founders for weeks after all the founders had been replaced (Fig. 6.1).

The longevity of such traditions of food choice was affected by a number of factors, including colony size, rate of replacement of colony members, and number of hours each day that colony members had access to foods (Galef and Allen, 1995; Galef and Whiskin, 1997).

6.5 Summary

Results of more than a quarter century of research demonstrate unequivocally that, under laboratory conditions, rat colonies can maintain stable traditions of food preference. Consequently, we know that at least some of the many mechanisms for social learning about foods uncovered in laboratory studies of social influences on food choice have the potential to support traditions of food preference of the sort Steiniger (1950) described in free-living rats. Although we do not yet know which processes demonstrated in the laboratory to support social learning of food preferences are actually responsible for feeding traditions in free-living populations of rats (Galef, 1984), we do know that social learning can lead to traditions

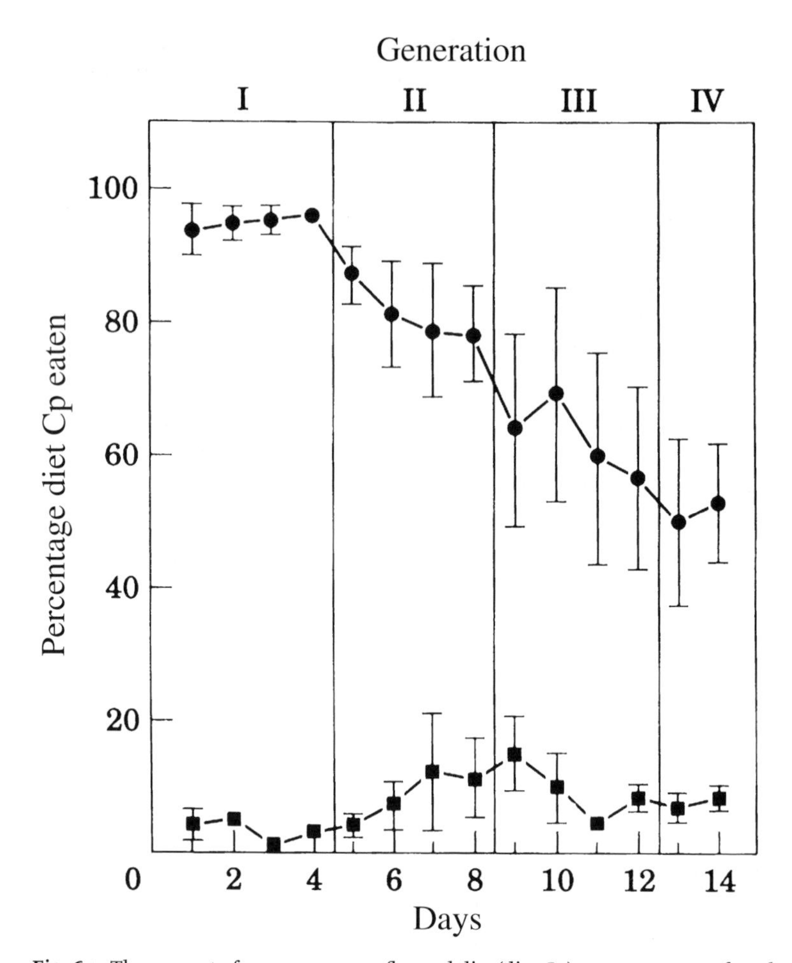

Fig. 6.1. The amount of cayenne-pepper-flavored diet (diet Cp) as a percentage of total amount (mean ± SEM) eaten by subjects offered both diet Cp and wasabi-flavored diet in colonies whose founding members ate only diet Cp or wasabi-flavored diet. On day 1, enclosures contained only founding members, on days 2–4 both founding colony members and replacement subjects, and on days 5–14 successive generations of replacement subjects (Galef and Allen, 1995; by permission of Academic Press).

in these animals, something that has not been demonstrated in any other genus of nonhuman mammal.

6.6 Learning how to eat

Field observations suggest that social influences can affect not only what rats eat, but how they eat as well.

6.6.1 Diving for food in nature

Gandolfi and Parisi (1972) reported that most members of some colonies of Norway rats living along the banks of the Po River in Italy would dive in the river and feed on mollusks on the river bottom, whereas no members of nearby colonies with equal access to mollusks did so. Gandolfi and Parisi (1972) interpreted their findings as consistent with the hypothesis that predation on submerged mollusks spreads through colonies by social learning. If discovery of mollusks on the riverbed were a rare event, and colony members could learn to dive for mollusks by observing other rats doing so, then the reported bimodality in frequency of diving in rat colonies along the Po could be explained.

Although the hypothesis that the habit of diving in shallow water for food spreads through colonies of wild rats as a result of social learning is attractive, confirmatory evidence has proved difficult to collect even in seminatural settings. Nieder, Cagnin, and Parisi (1982) observed mollusk predation by rats in a large (22 m × 10 m) outdoor enclosure built over a branch of the Po River. The enclosure provided opportunity for unobtrusive observation of mollusk predation in a small population of rats. Unfortunately, although the data collected in the enclosure did suggest that social learning of some sort might have been involved in diffusion of the habit of mollusk predation through a rat population, they are ambiguous.

6.6.2 Diving for food in the laboratory

The potential contribution of social processes to development of the habit of diving for food in shallow water has also been examined under controlled conditions (Galef, 1980). The experiments involved simplified laboratory analogues of the natural situation and, therefore, cannot be extrapolated uncritically to the more complex uncontrolled environment. However, results of the experiments did provide evidence bearing on the issue of whether it is necessary to invoke social learning to explain the distribution of the habit of diving for food reported by Gandolfi and Parisi (1972, 1973).

Second- and third-generation laboratory-bred female wild rats captured on garbage dumps in southern Ontario were placed together with their offspring in enclosures with separate nesting and diving areas connected by meter-long tunnels. In the diving area, subjects could retrieve pieces of chocolate from beneath 15 cm of water.

Adult rats that had not been explicitly trained to dive for food never dived, even if housed with a rat that had been trained to dive for food by placing chocolate squares in an empty tank and, over a period of weeks, increasing the water level to 15 cm. However, approximately 20% of juvenile wild rats reared in the enclosures came to dive for food. Juveniles were as likely to learn to dive whether their dam regularly dived and retrieved chocolates from under water or never did so. Such results suggest that observation of a diving conspecific does not, in itself, induce rats to dive.

6.6.2.1 Social learning of swimming?

In a subsequent study, young wild rats were trained to swim across the surface of a small body of water to reach food. When introduced into enclosures connected to a diving area, where food was available below 15 cm of water, more than 90% of subjects trained to swim spontaneously dived for food.

The finding that swimming rats are effectively diving rats limits the potential role of social learning in the spread of diving behavior through a population. If rats learn to swim independently, and if swimming rats dive, then social learning could serve only to direct rats to dive in one area rather than another. However, development of swimming might itself be socially influenced. If so, then social learning might indirectly potentiate propagation of diving behavior by facilitating propagation of swimming behavior.

An experiment in which wild rat pups were reared by dams that either swam or did not swim to food in an apparatus where highly palatable food could be reached by swimming 1.7 m down an alley revealed no difference in the age of initiation of swimming by pups as a function of whether their dam swam. All pups began to swim before they reached 40 days of age.

6.6.3 Relating laboratory to field studies

The findings that, at least in the laboratory, wild rat pups readily learn independently to swim and that almost all swimming wild rat pups spontaneously dive for food in shallow water suggest that absence of diving by members of some colonies that live along the Po River may be in greater need of explanation than the diving exhibited by members of other colonies.

Conceivably, all rats living along the Po River know how to dive for mollusks, but they do not dive when sufficient nutriment is available ashore. If so, one might expect rats that had been trained to dive for food to cease diving if adequate rations were made available to them on land. In fact, rats that reliably dived for food while food was available on land for only 3 in 24 hours stopped diving when given *ad libitum* access to food ashore, even if the food available on land was considerably less palatable than that available under water (Galef, 1980).

Taken together, the laboratory results offer little support for the hypothesis that the distribution of the habit of diving for food observed among colonies of rats living along the Po River results from social leaning of the habit in some colonies, but not others. To the contrary, the laboratory data suggest that all rats may know how to dive for food, but they do so only when adequate food is not available on dry land.

In retrospect, some observations made in the field are consistent with the notion that availability of food on land may be the major determinant of whether members of rat colonies living along the banks of the Po River feed on submerged mollusks. For example, Gandolfi and Parisi (1973, p. 69) reported that in those locations where mollusk predation occurs, mollusks "represent one of the main sources, if not the main source of food for rats" and Parisi and Gandolfi (1974, p. 102) suggested that "the time dedicated by rats to mollusk capture depends greatly on the availability of other foods".

Our laboratory findings suggest that these informal field observations may be more informative than those who made them realized. Possibly, members of colonies that regularly dive for mollusks would stop diving for food if palatable food were available in their territories, and removal of food from the territories of colonies whose members do not normally dive might cause them to start diving. The relevant field experiments have not been carried out, but obviously could be, and might, at least in principle, exclude social learning as an explanation of the distribution of diving behavior along the Po.

6.7 Stripping pinecones for seeds

6.7.1 Field observations

Some years ago, Aisner and Terkel (1992) discovered that black rats (*R. rattus*) living in the pine forests of Israel subsist on a diet of pine seeds in an otherwise sterile habitat. Extraction of pine seeds by stripping pinecones of their scales and eating the seeds the scales concealed allows black rats in Israel to fill a niche occupied elsewhere in the world by tree squirrels, which are not present in the Middle East.

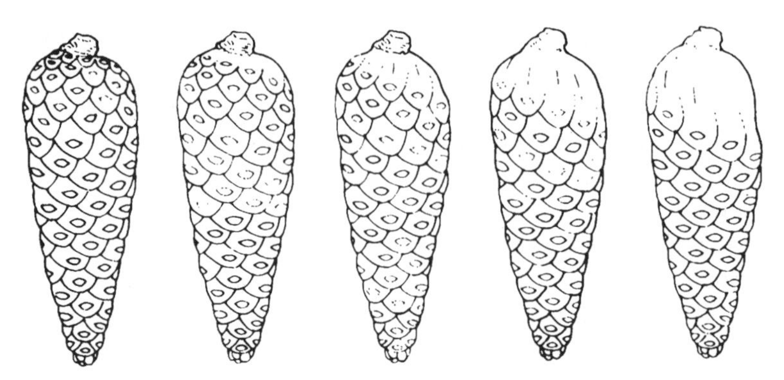

Fig. 6.2. Pinecones in different stages of opening with the number of rows of previously stripped scales increasing from left to right (Terkel, 1996; by permission of Academic Press).

6.7.2 Laboratory analyses

Laboratory studies of pinecone stripping by wild-caught rats revealed that it is difficult for rats to remove the tough scales from pinecones and gain access to the energy-rich seeds they protect and, in doing so, gain more energy from eating the seeds than is expended in acquiring them. To harvest pine seeds efficiently, rats must take advantage of the physical structure of pinecones, first stripping scales from the base of a cone and then removing the scales spiraling around the cone's shaft to its apex one after another (Terkel, 1996; Fig. 6.2).

Laboratory studies of development of the energetically efficient pattern of stripping pinecones found that only 6 of 222 hungry laboratory-reared wild rats given access to a surplus of pinecones for several weeks learned the efficient pattern of cone stripping for themselves (Zohar and Terkel, 1995). The remaining 216 animals either ignored the pinecones altogether or gnawed at them in ways that did not lead to a net energy gain from eating pine seeds. Similarly, pups gestated by dams that efficiently stripped pinecones of their seeds, but reared by foster mothers that did not strip cones, failed to learn to strip pinecones (Aisner and Terkel, 1992). However, more than 90% of pups came to open cones for themselves when reared by a foster mother that stripped pinecones efficiently in the presence of her foster young. Clearly, some aspect of the postnatal interaction between a dam stripping pinecones and the young she rears suffices for transmission of the efficient means of pinecone stripping from one generation to the next (Aisner and Terkel, 1992; Zohar and Terkel, 1992).

Additional experiments led to the conclusion that experience of young rats in completing the stripping of pinecones started appropriately by either an experienced adult rat or a human experimenter, who used a pair of pliers to imitate the pattern of scale removal by a rat, starting to strip a cone efficiently, enabling 70% of young rats to become efficient strippers (Terkel, 1996).

Terkel's (1996) observations indicated that when a black rat mother opens pinecones by stripping scales and eating exposed seeds her young gather around her and attempt to obtain seeds. Once the young are old enough, they snatch partially opened cones from their mother and continue the stripping process by themselves. Thus, activities of feeding mother rats appear to facilitate acquisition of pinecone stripping by their offspring, first, by focusing attention of juveniles on pinecones as potential sources of food and, later, by providing their young with partially opened pinecones that the young can learn to exploit as sources of food (Terkel, 1996).

As McGrew (1998), a leading proponent of the view that apes exhibit culture, pointed out, no study of a traditional behavior exhibited by the free-living members of any species, even humans, has been carried out with the rigor or elegance of Terkel's analysis of the social transmission of pinecone stripping by black rats. Consequently, today, as 50 years ago when Steiniger made the statement that serves as an epigram for this chapter, we have a better understanding of the origins of behavioral traditions in free-living rats than in any other nonhuman, mammalian species.

6.8 Conclusions

Analyses of behavioral processes resulting in population-specific patterns of behavior often require numerous experiments each involving dozens of animals with similar prior life histories (e.g., Galef, 1980, 1996b; Terkel, 1996). It is relatively easy to procure the numbers of experimentally naïve rodents, birds, fishes, or insects needed for such studies. Almost always, it is impossible to procure similar numbers of primates (for example, see Ch. 7) or cetaceans (see Ch. 9). Consequently, analyses of behavioral processes supporting traditions in relatively simple systems are likely to be more complete than analyses of traditions in species with large cortices, and lessons learned from analyses of such simple systems will have to inform our understanding of more complex, but less-available species.

Some of these lessons are now discussed and their possible relationship to analyses of some primate "traditions".

6.8.1 Simple mechanisms, complex outcomes

Terkel's studies of rats stripping pinecones of seeds show that social transmission of complex motor patterns can rest on simple social learning processes. Adult rats direct attention of conspecific young to pinecones and provide young with partially opened cones to exploit. Young take advantage of these affordances and learn for themselves the sequence of motor acts needed to strip cones efficiently. No imitation, no teaching, no emulation, and no observational conditioning (Galef, 1976; Whiten and Ham, 1992) are involved in transmission from one generation to the next of a motor skill possibly as complex as that exhibited by apes ingesting difficult vegetable foods (Byrne and Byrne, 1994).

Mediation of social learning by environmental affordances is not unique to pinecone opening by black rats, although the complexity of the motor patterns involved in consumption of pine seeds makes the example a particularly striking one. For example, Norway rats create trails as they move through underbrush on trips to and from foraging sites. These trails lead to traditional patterns of space utilization within colonies (Calhoun, 1962; Telle, 1966).

6.8.2 Environment determination of expression of behavior

6.8.2.1 Pinecone stripping by rats

A few rats in every hundred given pinecones learn independently to strip cones efficiently (Zohar and Terkel, 1995). However, the behavior is common, so far as is known (Smith and Balda, 1979), only in areas where rats do not have to compete with squirrels for pine seeds. Even though pinecone-stripping behavior is clearly socially learned by the majority of rats that eat pine seeds, environmental influences suffice to explain why Israeli black rats strip pine cones and black rats living elsewhere do not (Galef, 1995). There is no need to imagine a "genius" Israeli rat that discovered the proper method for opening pinecones and whose remarkable innovation is the origin of pinecone stripping by Israeli rats.

6.8.2.2 Diving rats

The effect of food distribution on expression of diving behavior in Norway rats in the laboratory is direct and obvious. Rats that can find adequate food ashore refuse to dive even when they have experience of more palatable food under water. Whether a similar process is responsible for the observed distribution of diving behavior in natural circumstances remains to be determined.

6.8.2.3 Chimpanzee culture

In a publication likely to have significant impact on future discussions of traditions in animals, Whiten *et al.* (1999) provided a list of 65 behaviors that vary in frequency of occurrence in seven geographically distinct, free-living populations of chimpanzees, each studied for many years. The authors subdivide their list of candidate "traditions" into four categories: (a) patterns absent at no site, (b) patterns not achieving habitual frequency at any site, (c) patterns for which absence can be explained by local ecological factors, and (d) patterns customary or habitual at some sites yet absent at others with no ecological explanation.

The 39 behaviors listed in category (d) are discussed as "cultural", the implication being that the distributions of these 39 behaviors across populations, unlike behaviors listed in categories (a–c), result from social learning, rather than from environmental causes. Whiten *et al.* (1999) did not discuss the implications of the fact that 22 of these 39 behaviors are "common" or "habitual" in one or several populations but only "present" in others. For example, "ant fish" (a probe used to extract ants) is "common" in two populations, "present" in two populations, and "absent" in three.

If ant fishing is "cultural", then explanation is required for why ant fishing is common in only two of the four populations where it has been observed. Such explanation is likely to be ecological, as is the case in pinecone-stripping and diving rats. Perhaps all chimpanzees learn to fish for ants, but the probability that members of different populations ant fish varies from 0 to 100 depending on local environmental conditions, which determine the relative efficiency of ant fishing as a means of obtaining nutrients (Galef, 1992).

The data from chimpanzees are quite different from those emerging from van Schaik's studies of orangutan use of tools to secure food (Ch. 11). Here there is quite convincing evidence that social learning rather than ecology is responsible for population differences in behavior. All members of each of van Schaik's two study populations either do or do not exhibit tool use in exploiting a resource that is exploited by members of both populations.

6.8.3 Information transmission or information acquisition

Norway rats exploit their fellows as sources of information about where, when, and what to eat in several different ways (Galef, 1996a). This redundancy suggests that in Norway rats, as in honeybees (von Frisch, 1967), socially acquired information is important in development of adaptive behavioral repertoires. There is, however, no evidence that any route to social learning about foods or feeding sites involves knowledgeable animals modifying their foraging behavior so as to provide information to others. Indeed, anesthetized rats could provide information about where and what to eat, and they are at least as effective as conscious providers of similar information (Galef, 1981; Galef and Stein, 1985).

Social learning of food preferences in rats appears to be an extractive process in which naïve individuals appropriate information from their social environment, not a process involving active transmission of information by knowledgeable individuals. King (1994a,b) reached much the same conclusion from her field studies of social foraging in baboons, as did Fragaszy and Visalberghi (1996) in their laboratory work with capuchin monkeys.

6.8.4 Traditions or animal traditions

As discussed in the introduction to this chapter, the Latin root of the word tradition implies an active role for the source of information in generation of population-specific behaviors. Consequently, "traditional" may be an inappropriate adjective to use to describe socially learned, population-specific behaviors of animals. The life-sustaining behavior of one animal can provide unintended signals or cues to others, enabling them to increase the efficiency with which they interact with the physical environment (Galef, 1986b). As indicated in the preceding section, animal traditions, unlike traditions of humans, seem to involve extraction rather than active transmission of information. It might, therefore, be salutary to refer to traditions in animals as "animal traditions" to remind ourselves (and others) that active transmission of information, as is implicit in use of the word "tradition", may not be involved in socially learned, population-specific patterns of behavior in nonhuman species.

6.8.5 Social and individual learning

Consideration of studies of social foraging by Norway rats also suggests that even the term social learning may, in some ways, be misleading. In the short term, exposure to conspecifics that have eaten a food can increase a rat's attention to that food (Galef and Clark, 1971a,b) and increase a rat's liking of it (Galef, Whiskin, and Bielavska, 1997). However, in the long term, individual rats choose substances to ingest as a result of feedback they receive subsequent to ingesting the foods that they sample (Galef, 1995; Galef and Whiskin, 2001). In the study of animal traditions, it is the long term that is of greatest interest (but see Perry *et al.*, Ch. 14, for another view).

Information extracted from others can bias an individual to sample one food rather than another, but the long-term selection of food is not "socially learned" in the literal sense of that term. Strictly speaking, social learning about foods by Norway rats is not "social learning" but "socially biased individual learning" (Galef, 1995; Heyes, 1993; Fragaszy and Visalberghi, 2001). Reader's (Ch. 3) failure to find evidence of independent social- and individual-learning processes in his analysis of the relationship between brain structure and frequency of innovation in primates is consistent with this view.

6.8.6 Summary

Information about factors affecting the frequency of expression of "socially learned" behaviors is far easier to collect in animals that, like rats, are easily procured in large numbers and are relatively inexpensive to maintain. Results of laboratory studies of causes of population-specific behaviors of rats suggest that simple observation cannot determine whether a population-specific behavior reflects socially biased learning or differences in the nonsocial environment in which different populations live. Such laboratory studies suggest that even apparently complex behavioral traditions can rest on simple social learning processes. Complex socially learned behaviors need not involve active transmission of information or sophisticated social learning processes such as teaching, emulation, or imitation.

Laboratory studies indicate further that the frequency of expression of both socially and individually learned behaviors can be markedly affected by subtle environmental factors. Caution must, therefore, be exercised when trying to deduce causes of development of behavior from relative frequencies of expression of that behavior in free-living, allopatric populations.

There is, of course, potential for error in extrapolation from behavior of laboratory rodents to that of free-living members of other species (Galef, 1996c). Still, judicious use of studies of the development and maintenance of population-specific patterns of behavior in relatively simple systems (e.g., rodents, honeybees, passerine birds, etc.) should continue to inform our understanding of the development of population-specific behaviors more generally.

6.9 Acknowledgements

This chapter is a synthesis (with considerable revision and additional material) of several earlier articles (Galef, 1976, 1977, 1980, 1988, 1992, 1996a–c) cited in full in the references. Preparation of the manuscript was facilitated by funds from the Natural Sciences and Engineering Research Council of Canada.

I thank the editors of the present volume, and M. Panger, L. Rose and E. Visalberghi for their thoughtful comments on earlier drafts.

References

Aisner, R. and Terkel, J. 1992. Ontogeny of pine-cone opening behaviour in the black rat (*Rattus rattus*). *Animal Behaviour*, **44**, 327–336.

Barnett, S. A. 1958. Experiments on "neophobia" in wild and laboratory rats. *British Journal of Psychology*, **49**, 195–201.

Bowman, R. I. and Billeb, S. I. 1965. Blood-eating in a Galapagos finch. *Living Bird*, **4**, 29–44.

Bronstein, P. M., Levine, M. J., and Marcus, M. 1975. A rat's first bite: the non-genetic, cross-generational transfer of information. *Journal of Comparative and Physiological Psychology*, **89**, 295–298.

Byrne, R. W. and Byrne, J. M. E. 1994. The complex leaf-gathering skills of mountain gorillas (*Gorilla g. beringei*): variability and standardization. *American Journal of Primatology*, **31**, 241–261.

Calhoun, J. B. 1962. *The Ecology and Sociology of the Norway Rat*. Bethesda, MD: US Department of Health, Education, and Welfare.

Caro, T. M. and Hauser, M. D. 1992. Is there teaching in nonhuman animals. *Quarterly Review of Biology*, **67**, 151–174.

Church, R. M. 1957. Two procedures for the establishment of "imitative behavior". *Journal of Comparative and Physiological Psychology*, **50**, 315–318.

Cottam, C. 1948. Aquatic habits of the Norway rat. *Journal of Mammalogy*, **29**, 299.

Flandera, V. and Novakova, V. 1974. Effect of the mother on the development of aggressive behavior in rats. *Developmental Psychobiology*, **8**, 49–54.

Fragaszy, D. M. and Visalberghi, E. 1996. Social learning in monkeys: primate primacy reconsidered. In *Social Learning and Imitation: The Roots of Culture*, ed. C. M. Heyes and B. G. Galef Jr., pp. 65–84. New York: Academic Press.

Fragaszy, D. M. and Visalberghi, E. 2001. Recognizing a swan: socially-biased learning. *Psychologia*, **44**, 82–98.

Galef, B. G., Jr. 1970. Aggression and timidity: responses to novelty in feral Norway rats. *Journal of Comparative and Physiological Psychology*, **75**, 358–362.

Galef, B. G., Jr. 1976. Social transmission of acquired behavior: a discussion of tradition and social learning in vertebrates. *Advances in the Study of Behavior,* **6**, 77–100.
Galef, B. G., Jr. 1977. Mechanisms for the social transmission of food preferences from adult to weanling rats. In *Learning Mechanisms in Food Selection*, ed. L. M. Barker, M. Best and M. Domjan, pp. 123–150. Waco, TX: Baylor University Press.
Galef, B. G., Jr. 1980. Diving for food: analysis of a possible case of social learning in wild rats (*Rattus norvegicus*). *Journal of Comparative and Physiological Psychology,* **94**, 416–425.
Galef, B. G., Jr. 1981. Development of olfactory control of feeding-site selection in rat pups. *Journal of Comparative and Physiological Psychology,* **95**, 615–622.
Galef, B. G., Jr. 1984. Reciprocal heuristics: a discussion of the study of learned behavior in laboratory and field. *Learning and Motivation,* **15**, 479–493.
Galef, B. G., Jr. 1985. Direct and indirect behavioral pathways to the social transmission of food avoidance. In *Experimental Assessments and Clinical Applications of Conditioned Food Aversions*, ed. P. Bronstein and N. S. Braveman, pp. 203–215. New York: New York Academy of Sciences.
Galef, B. G., Jr. 1986a. Social interaction modifies learned aversions, sodium appetite, and both palatability and handling-time induced dietary preference in rats (*Rattus norvegicus*). *Journal of Comparative Psychology,* **100**, 432–439.
Galef, B. G., Jr. 1986b. Olfactory communication among rats of information concerning distant diets. In *Chemical Signals in Vertebrates, Vol. IV: Ecology, Evolution, and Comparative Biology*, ed. D. Duvall, D. Muller-Schwarze, and R. M. Silverstein, pp. 487–505. New York: Plenum Press.
Galef, B. G., Jr. 1988. Communication of information concerning distant diets in a social, central-place foraging species (*Rattus norvegicus*). In *Social Learning: Psychological and Biological Perspectives*, ed. T. R. Zentall and B. G. Galef Jr., pp. 119–140. Hillsdale, NJ: Erlbaum.
Galef, B. G., Jr. 1992. The question of animal culture. *Human Nature,* **3**, 157–178.
Galef, B. G., Jr. 1995. Why behaviour patterns that animals learn socially are locally adaptive. *Animal Behaviour,* **49**, 1325–1334.
Galef, B. G., Jr. 1996a. Social influences on food preferences and feeding behaviors of vertebrates. In *Why We Eat What We Eat*, ed. E. Capaldi, pp. 207–232. Washington, DC: American Psychological Association.
Galef, B. G., Jr. 1996b. Social enhancement of food preferences in Norway rats. In *Social Learning and Imitation: The Roots of Culture*, ed. C. M. Heyes and B. G. Galef, Jr., pp. 49–64. New York: Academic Press.
Galef, B. G., Jr. 1996c. Tradition in animals: field observations and laboratory analyses. In *Readings in Animal Cognition*, ed. M. Bekoff and D. Jamieson, pp. 91–106. Cambridge: MIT Press.
Galef, B. G., Jr. and Allen, C. 1995. A new model system for studying animal traditions. *Animal Behaviour,* **50**, 705–717.
Galef, B. G., Jr. and Beck, M. 1985. Aversive and attractive marking of toxic and safe foods by Norway rats. *Behavioral and Neural Biology,* **43**, 298–310.
Galef, B. G., Jr. and Buckley, L. L. 1996. Use of foraging trails by Norway rats. *Animal Behaviour,* **51**, 765–771.
Galef, B. G., Jr. and Clark, M. M. 1971a. Parent–offspring interactions determine time and place of first ingestion of solid food by wild rat pups. *Psychonomic Science,* **25**, 5–16.

Galef, B. G., Jr. and Clark, M. M. 1971b. Social factors in the poison avoidance and feeding behavior of wild and domesticated rat pups. *Journal of Comparative and Physiological Psychology*, **25**, 341–357.

Galef, B. G., Jr. and Heiber, L. 1976. The role of residual olfactory cues in the determination of feeding site selection and exploration patterns of domestic rats. *Journal of Comparative and Physiological Psychology*, **90**, 727–739.

Galef, B. G., Jr., and Henderson, P. W. 1972. Mother's milk: a determinant of food preferences of weaning rats. *Journal of Comparative and Physiological Psychology*, **78**, 213–219.

Galef, B. G., Jr. and Sherry, D. F. 1973. Mother's milk: a medium for transmission of information about mother's diet. *Journal of Comparative and Physiological Psychology*, **83**, 374–378.

Galef, B. G., Jr. and Stein, M. 1985. Demonstrator influence on observer diet preference: analyses of critical social interactions and olfactory signals. *Animal Learning and Behavior*, **13**, 131–138.

Galef, B. G., Jr. and Whiskin, E. E. 1992. Social transmission of information about multiflavored foods. *Animal Behaviour*, **20**, 56–62.

Galef, B. G., Jr. and Whiskin, E. E. 1997. Effects of asocial learning on longevity of food-preference traditions. *Animal Behaviour*, **53**, 1313–1322.

Galef, B. G., Jr. and Whiskin, E. E. 1998a. Limits on social influence on food choices of Norway rats. *Animal Behaviour*, **56**, 1015–1020.

Galef, B. G., Jr. and Whiskin, E. E. 1998b. Determinants of the longevity of socially learned food preferences of Norway rats. *Animal Behaviour*, **55**, 967–975.

Galef, B. G., Jr. and Whiskin, E. E. 2001. Interaction of social and asocial learning in food preferences of Norway rats. *Animal Behaviour*, **62**, 41–46.

Galef, B. G., Jr. and Wigmore, S. W. 1983. Transfer of information concerning distant foods: a laboratory investigation of the "information-centre" hypothesis. *Animal Behaviour*, **31**, 748–758.

Galef, B. G., Jr., Kennett, D. J., and Wigmore, S. W. 1984. Transfer of information concerning distant foods in rats: a robust phenomenon. *Animal Learning and Behavior*, **12**, 292–296.

Galef, B. G., Jr., Kennett, D. J., and Stein, M. 1985. Demonstrator influence on observer diet preference: effects of simple exposure and presence of a demonstrator. *Animal Learning and Behavior*, **13**, 25–30.

Galef, B. G., Jr., Mason, J. R., Pretti, G., and Bean, N. J. 1988. Carbon disulfide: a semiochemical mediating socially-induced diet choice in rats. *Physiology and Behaviour*, **42**, 119–124.

Galef, B. G., Jr., Attenborough, K. S., and Whiskin, E. E. 1990. Responses of observer rats to complex, diet-related signals emitted by demonstrator rats. *Journal of Comparative Psychology*, **104**, 11–19.

Galef, B. G., Jr., Whiskin, E. E., and Bielavska, E. 1997. Interaction with demonstrator rats changes their observers' affective responses to flavors. *Journal of Comparative Psychology*, **111**, 393–398.

Gandolfi, G. and Parisi, V. 1972. Predazione su *Unio pictorum* L. da parte del ratto, *Rattus norvegicus* (Berkenhout). *Acta Naturalia*, **8**, 1–27.

Gandolfi, G. and Parisi, V. 1973. Ethological aspects of predation by rats, *Rattus norvegicus* (Berkenhout) on bivalves, *Unio pictorum*, L. and *Cerastoderma lamarcki* (Reeve). *Bullettino di Zoologia*, **40**, 69–74.

Gaulin, S. J. C. and Kurland, J. A. 1976. Primate predation and bioenergetics. *Science*, **191**, 314–315.

Gove, P. B. 1971. *Webster's Third New International Dictionary of the English Language Unabridged*. Springfield, MA: G. and C. Merriam.

Hall, W. G. 1975. Weaning and growth of artificially reared rats. *Science*, **190**, 1313–1315.

Hepper, P. G. 1988. Adaptive fetal learning: Prenatal exposure to garlic affects postnatal preference. *Animal Behaviour*, **36**, 935–936.

Heyes, C. M. 1993. Imitation, culture, and cognition. *Animal Behaviour*, **46**, 999–1010.

Heyes, C. M. and Durlach, P. J. 1990. "Social blockade" of taste-aversion learning in Norway rats (*R. norvegicus*): is it a social phenomenon? *Journal of Comparative Psychology*, **104**, 82–87.

Kenyon, K. W. 1961. Birds of Amchitka, Alaska. *Auk*, **78**, 305–326.

King, B. J. 1994a. Primate infants as skilled information gatherers. *Pre- and Perinatal Psychology Journal*, **8**, 287–307.

King, B. J. 1994b. *The Information Continuum*. Santa Fe, NM: SAR Press.

Koster, F. and Koster, M. 1983. Twelve days among the "vampire finches" of Wolf Island. *Noticias de Galapagos*, **38**, 4–10.

Lack, D. 1947. *Darwin's Finches*. Cambridge: Cambridge University Press.

Lack, D. 1969. Subspecies and sympatry in Darwin's finches. *Evolution*, **23**, 252–263.

Laland, K. N., and Plotkin, H. C. 1990. Social learning and social transmission of foraging information in Norway rats (*Rattus norvegicus*). *Animal Learning and Behavior*, **18**, 246–251.

Laland, K. N. and Plotkin, H. C. 1991. Excretory deposits surrounding food sites facilitate social learning of food preferences in Norway rats. *Animal Behaviour*, **41**, 997–1005.

Laland, K. N. and Plotkin, H. C. 1992. Further experimental analysis of the social leaning and transmission of foraging information amongst Norway rats. *Behavioural Processes*, **27**, 53–64.

Laland, K. N., Richerson, P. J., and Boyd, R. 1993. Animal social learning: toward a new theoretical approach. *Perspectives in Ethology*, **10**, 249–277.

Lewis, C. T. and Short, C. 1969. *A Latin Dictionary*. Oxford: Clarendon Press.

Lore, R., Blanc, A., and Suedfeld, P. (1971). Empathic learning of a passive-avoidance response in domesticated *Rattus norvegicus*. *Animal Behaviour*, **19**, 112–114.

Martin, L. T. and Alberts, J. R. 1979. Taste aversions to mother's milk: the age-related role of nursing in the acquisition and expression of a learned association. *Journal of Comparative and Physiological Psychology*, **93**, 430–445.

McGrew, W. C. 1998. Culture in nonhuman primates. *Annual Review of Anthropology*, **27**, 301–308.

Morgan, C. L. (1894). *Introduction to Comparative Psychology*. London: Scott.

Nieder, L., Cagnin, M., and Parisi, V. 1982. Burrowing and feeding behaviour in the rat. *Animal Behaviour*, **30**, 837–844.

Parisi, V. and Gandolfi, G. 1974. Further aspects of the predation by rats on various mollusc species. *Bollettino di Zoologia*, **41**, 87–106.

Posadas-Andrews, A. and Roper, T. J. 1983. Social transmission of food preferences in adult rats. *Animal Behaviour*, **31**, 265–271.

Pye, T. and Bonner, W. N. 1980. Feral brown rats, *Rattus norvegicus*, in South Georgia (South Atlantic Ocean). *Journal of the Zoological Society of London*, **192**, 237–235.

Richard, M. M., Grover, C. A., and Davis, S. F. 1987. Galef's transfer-of-information effect occurs in a free -foraging situation. *Psychological Record*, **37**, 79–87.

Schluter, D. and Grant, P. R. 1982. The distribution of *Geospiza difficilis* in relation to *G. fuliginosa* in the Galapagos Islands: tests of three hypotheses. *Evolution*, **36**, 1213–1226.

Schluter, D. and Grant, P. R. 1984. Ecological correlates of morphological evolution in a Darwin's finch species. *Evolution*, **38**, 856–869.

Smith, C. C. and Balda, R. P. 1979. Competition among insects, birds and mammals for conifer seeds. *American Zoologist*, **19**, 1065–1083.

Smotherman, W. P. 1982. Odor aversion learning by the rat fetus. *Physiology and Behavior*, **29**, 769–771.

Stanford, C. B. 1998. *Chimpanzee and Red Colobus: The Ecology of Predator and Prey*. Cambridge, MA: Harvard University Press.

Steiniger, F. 1950. Beitrage zur Soziologie und sonstigen Biologie der Wanderratte. *Zeitschrift für Tierpsychologie*, **7**, 356–379.

Strum, S. C. 1975. Primate predation: interim report on the development of a tradition in a troop of olive baboons. *Science*, **187**, 755–757.

Strum, S. C. 1976. Untitled. *Science*, **191**, 314–317.

Strupp, B. J. and Levitsky, D. E. 1984. Social transmission of food preferences in adult hooded rats (*Rattus norvegicus*). *Journal of Comparative Psychology*, **98**, 257–266.

Telle, H. J. 1966. Beitrag zur Kenntnis der Verhaltensweise von Ratten, vergleichend dargestellt bei *Rattus norvegicus* und *Rattus rattus*. *Zeitschrift für Angewandte Zoologie*, **53**, 129–196.

Terkel, J. 1996. Cultural transmission of feeding behavior in the black rat (*Rattus rattus*). In *Social Learning in Animals: The Roots of Culture*, ed. C. M. Heyes and B. G. Galef Jr., pp. 17–49. San Diego, CA: Academic Press.

Thorpe, W. H. 1963. *Learning and Instinct in Animals*, 2nd edn. London: Methuen.

von Frisch, K. 1967. *The Dance Language and Orientation of Bees*. Cambridge, UK: Belknap Press.

Whiten, A. and Ham, R. 1992. On the nature and evolution of imitation in the animal kingdom: reappraisal of a century of research. *Advances in the Study of Behaviour*, **22**, 239–283.

Whitten, A., Goodall, J., McGrew, W. C., Nishida, T., Reynolds, V., Sugiyama, Y., Tutin, C. E. G., Wrangham, R. W., and Boesch, C. 1999, Cultures in chimpanzees. *Nature*, **399**, 682–685.

Williams, G. C. 1966. *Adaptation and Natural Selection: A Critique of Some Current Evolutionary Thought*. Princeton, NJ: Princeton University Press.

Zohar, O. and Terkel, J. 1992. Acquisition of pinecone stripping behaviour in black rats (*Rattus rattus*). *International Journal of Comparative Psychology*, **5**, 1–6.

Zohar, O. and Terkel, J. 1995. Spontaneous learning of pinecone opening behavior by black rats. *Mammalia*, **59**, 481–487.

ELISABETTA VISALBERGHI AND ELSA ADDESSI

7

Food for thought: social learning about food in capuchin monkeys

7.1 Introduction

It used to be thought that shared behaviors are learned from others and that this was especially true of infants and their mothers. In recent years, many scientists have advocated parsimony in interpreting the diffusion of innovative behaviors in primates (Galef, 1991; Heyes and Galef, 1996; Lefebvre, 1995; Miklósi, 1999; Tomasello and Call, 1997; Visalberghi and Fragaszy, 1990a). This view has prompted systematic investigations of the learning processes involved in the spread of innovations and fueled debates on the nature of cultural traditions (Boesch and Tomasello, 1998; Whiten *et al.,* 1999). Capuchin monkeys are among the few primate species in which systematic research has been carried out on the acquisition and social learning of tool-using skills (Anderson, 2000; Fragaszy and Visalberghi, 1989; Visalberghi, 1993), on the patterns of object-related and goal-directed behaviors (Custance, Whiten, and Fredman, 1999; Fragaszy, Vitale, and Ritchie, 1994), and on the patterns of food-processing behaviors (e.g., "food washing") (Visalberghi and Fragaszy, 1990b; for an extensive review see Visalberghi and Fragaszy, 2002). Overall, these studies have demonstrated that social influences such as stimulus enhancement, local enhancement, and object reenactment are indeed present, whereas imitative learning (defined as learning a novel behavior by observing it performed by a demonstrator) is not (Visalberghi, 2000; Visalberghi and Fragaszy, 2002). Therefore, although the species name *Cebus imitator* assigned to capuchin monkeys by the prominent taxonomist Thomas (1903) seems unwarranted, we have begun to realize that other social learning processes seem to influence capuchins' behavior.

Feeding is *a condicio sine qua non* for an animal's survival, and food selection and processing are behavioral domains in which social learning is likely to occur. Living in social groups, as most primate species do, has costs and benefits (Dunbar, 1988; Wrangham, 1980). Among the benefits, Lee (1994 p.270) has listed opportunities for exchange of information among individuals: "close proximity increases the number of opportunities for observation and the rapidity of information procurement"; consequently, group living can be of great advantage in learning when, how, and what to feed upon (e.g., Giraldeau, 1997).

In theory, primates can learn to identify safe foods individually by trial and error or they can learn what to eat from conspecifics. As Dewar (Ch. 5) argues, social learning is rather unimportant if individuals can assess whether a food is "good" or "bad" from reliable nonsocial cues. Conversely, social learning is particularly needed when nonsocial cues are lacking or unreliable, and especially when eating a "bad" food can be fatal. Unfortunately, we know very little about which of the many food items present in nature are toxic for a given species. However, since toxicity is not always advertised by a bitter taste (Hladik and Simmen, 1996), nonsocial cues can be unreliable.

Primatologists have often assumed that primates learn to identify foods they eat from conspecifics and that dietary convergence or diffusion of new feeding habits in wild groups results from social learning (e.g., Kummer, 1971; Nishida, 1987; for a critical review see Visalberghi, 1994). Direct observation of knowledgeable conspecifics has been considered an important factor in improving juveniles' foraging skills (Janson and van Schaik, 1993) and the facts that infants have their first experiences with food in the social milieu and that they "are intensely curious about what their mothers eat" (Janson and van Schaik, 1993, p. 64) were considered to indicate a formative role on infants' later feeding behavior (Box, 1984; Fedigan, 1982; Goodall, 1986; Kummer, 1971; Watts, 1985). King (1999, p. 21) even argued that "primate infants seem to have been selected to be information extractors" (see also King 1994a,b).

Learning that a food is toxic from watching behavior of others (instead of by trial and error) could be an effective way of reducing the risk of ingesting poison. According to Janson and van Schaik (1993), juvenile primates may learn to avoid a food from the behavior of conspecifics. However, there is very little or no experimental evidence that observing a conspecific avoiding a food decreases an observer's consumption of the same food; and this is also the case when the aversively conditioned models are "relevant" individuals, such as the dominant male for group members, or a mother for her offspring (Hikami, 1991; Hikami, Hasegawa, and Matsuzawa, 1990). Also for rats, the species for which food avoidance behavior has been most thoroughly investigated, there is little or no evidence that food avoidance is learned observationally (see Ch. 6).

The few informal observations suggesting that monkeys learn food avoidance by observing conspecifics or that experienced individuals warn naïve ones about food harmfulness (Cambefort, 1981; Fletemeyer, 1978; Jouventin, Pasteur, and Cambefort, 1976) are not supported by strong evidence (see Visalberghi, 1994 for a critical review). For example, Watts (1985) described an instance in which a mother gorilla pushed away the stem of an unidentified plant (not eaten by other gorillas) that her young daughter had pulled towards herself. This anecdote is open to alternative interpretations. It could be either a case of maternal intervention to prevent food ingestion, as suggested by Watts, or more parsimoniously one of the many instances in which a mother takes food away from its offspring. It is possible that this event captured the scientist's attention because it occurred with an item not included in the gorillas' diet and which the mother discarded after having taken it from her infant. Similar episodes of parental discouragement have been reported for wild apes (e.g., Fossey, 1979; Goodall, 1973; Hiraiwa-Hasegawa, 1990; Nishida *et al.*, 1983).

It is indeed possible that "learning which plants are edible is aided by the fact that infants are showered with remains of food plants from the first day of life, and their feeding is usually synchronized with the mother's" (Byrne, 1999, p. 339). However, it would be of interest to evaluate experimentally the extent to which infants and youngsters assess the palatability (agreeable to the palate or taste) and toxicity (causing impairment, injuries, or death) of a food by extracting information from what the others do rather than through trial and error themselves, to qualify the type of information (visual, olfactory, gustatory, etc.) infants use to guide their food choices and to assess whether the mother *actively* provides information or sets situations for her infant to learn about foods.

Capuchins are omnivorous and feed on a wide variety of food sources, most of which are seasonal (Brown and Zunino, 1990; Kinzey, 1997; Sussman, 2000; Terborgh, 1983), and some of which (e.g., insects, leaves, etc.) may contain toxic substances or need specific processing techniques (Izawa, 1978; Izawa and Mizuno, 1977; Robinson 1986). The acquisition of information about food from group members seems particularly relevant in either generalist species, whose diets include food sources that might contain toxic substances (e.g., rats: Galef, 1993; capuchin monkeys: Visalberghi and Fragaszy, 1995), or species whose diets include plants producing defensive chemicals (e.g., ruminants: Provenza, 1995; folivorous howler monkeys: Whitehead, 1986; see also Freeland and Janzen, 1974). Tolerance among group members is necessary for transfer of information, since it allows for proximity between individuals. Tufted capuchins exhibit a high degree of interindividual tolerance, especially towards infant and juveniles, in the wild (Izawa, 1980; Janson, 1996; Perry and Rose, 1994) as well as in captivity, where food is sometimes transferred from one individual to another (de Waal, Luttrell, and Canfield, 1993; Fragaszy, Feuerstein, and Mitra, 1997a; Thierry, Wunderlich, and Gueth, 1989).

Therefore, a tolerant and omnivorous species such as tufted capuchins could be expected to learn about food from group members. Unfortunately, field data do not make it possible to determine the contribution of social context to learning, nor to assess what kind of impact social companionships have on the diffusion of traditions (see Ch. 1). Such goals are better pursued using controlled experimental approaches that deal with proximate mechanisms underlying behavior. This chapter tries to shed light on these issues by reporting the results of systematic studies in tufted capuchin monkeys (*Cebus apella*) carried out in Rome (Italy) and Athens (Georgia, USA) in collaboration with Dorothy Fragaszy. To assess social bias on individual learning, our experimental designs usually included an individual condition (in which subjects were tested alone and social influences were removed) and a social condition (in which subjects were tested together with their group members). Whenever possible, we will refer to capuchins and especially to the studies aimed at examining the individual's response to food when alone or with group members.

Sections 7.2 and 7.3 describe our experimental studies on capuchins and provide a theoretical background and brief overview of the relevant literature. Section 7.2 describes how an individual behaves towards food, how it responds towards novel foods, and how it responds to familiar foods whose palatability has changed, in the absence of group members. Section 7.3 deals with the behavior of individuals socially tested in the same experimental paradigms and discusses what the social context adds to the individual's experience with food, comparing findings obtained in the social and the individual conditions. Section 7.4 describes our own recent experiments focused on the factors affecting the individual's response to novel foods when the individual is socially tested;

Section 7.5 describes feeding traditions of wild capuchins and provides some preliminary data indicate a convergence of food preferences in our groups of captive capuchins. Finally, Section 7.6 discusses the present lack of experimental evidence that capuchins learn about foods by observing what others eat or discard, or by being prevented from ingesting unpalatable or toxic foods by the intervention of more knowledgeable individuals. Future research approaches to these questions are also discussed.

7.2 What to eat: the responses of an individual to novel foods and foods that change in palatability

7.2.1 The individual's response to foods

An individual's physiology, anatomy, behavior, biological environment, and social environment all play substantial roles in food choice (Galef, 1996). Like most animals, capuchins and other primates like sweet foods and dislike bitter substances. Though primate species differ in their thresholds for sugars, such as fructose and sucrose, and for quinine (Simmen and Hladik, 1998), interindividual variability within the same species is remarkably low (e.g., squirrel monkeys: Laska, 1997; spider monkeys: Laska, Carrera Sanchez, and Rodriguez Luna, 1998). Though data on nonhuman primates are lacking, it is likely that, similarly to humans, other primate species learn to like a certain amount of salt in their food (Beauchamp, Cowart, and Moran, 1986) and to prefer foods that are higher in calories (Galef, 1996; Laska, Hernandez Salazar, and Rodriguez Luna, 2000). Therefore, when individuals belonging to the same species and with comparable dietary energetic requirements (e.g., same age and sex class, similar physiological states, etc.) encounter the same food sources they are likely to develop comparable food acceptance profiles.

7.2.2 How an individual discovers if a novel food is edible: the neophobic response

A monkey can encounter a food that it likes (for example because it is sweet), that it is neutral about, or that it dislikes (for example because it is bitter). If the food is not familiar, the monkey is basically neophobic and eats only a small amount of it. If the food is bitter, it will avoid it in the future. If the food is neutral or good, the monkey will first eat a small amount, and if ingestion does not produce illness (vomiting, gastrointestinal distress, nausea), it will eat it when encountering it again. Decades ago, neophobic reaction towards new foods was thoroughly investigated and discovered to be present in a variety of animal species (e.g., rats, Barnett, 1958). Primates are also neophobic, though some species are more neophobic than others (Johnson, 2000; Menzel, 1997; Yamamoto *et al.*, 2000).

To face seasonal changes in food availability, capuchins need to find and exploit new food sources. Captive tufted capuchins are neophobic towards novel foods (Visalberghi and Fragaszy, 1995), whereas group differences in neophobia are present both in captivity in the different social settings (A. T. Galloway, D. Fragaszy, A. M. McCabe, and E. Visalberghi, unpublished data) and in the wild. Wild capuchins in the Iguaçu National Park in Argentina are extremely neophobic towards novel foods and objects (Agostini, 2001), whereas capuchins living in a 43 000 ha ecological reserve (Parque Nacional de Brasilia, Brazil) and accustomed to visitors are not. The latter have probably learned that humans leave behind foods that are safe to eat, and they are willing to exploit food sources left by humans in the area (Siemers, 2000; E. Visalberghi, personal observation). However, information is too fragmentary to understand the relative contributions of experiential, social, and environmental factors to neophobia.

According to Freeland and Janzen (1974), since the ingestion of plant items is likely to lead to drug interactions or impairment of the function of gut flora, once herbivorous mammals have established a range of food species and items that they can consume safely, they continue to eat them. However, generalist species often include new foods in their diet to overcome shortages of staple foods and seasonal changes in food availability. Generalist species "preferentially feed on the foods with which they are familiar, and continue to feed on them for as long as possible" and "simultaneously indulge in a continuous food sampling program" (Freeland and Janzen, 1974, p. 281). Visalberghi (1994) suggests a link between neophobia and having a well-established diet. Staple foods serve as a secure base from which to venture out cautiously to learn the possible consequences of ingestion of new foods. Conversely, lack of staple foods may lead to a general reduction in neophobia and an increase in the risk of being poisoned. This is exactly what happened to a troop of Japanese macaques moved from Japan to Texas. These monkeys, released in a completely new environment, lacked familiar foods and "proved themselves to be fearless, innovative, and eclectic in what they would eat, sampling all the local flora soon after arrival" (Fedigan, 1991, p. 60). Afterwards, the macaques soon began to narrow the range of the foods they ate (L. M. Fedigan, personal communication).

Neophobia would reflect the payoff for investigating new things compared with the payoff for avoiding new things (see Ch. 5). For the Japanese macaques released in Texas, the absence of staple foods and the risk of starvation might initially have made the payoff for being conservative and neophobic very low and the payoff for consuming novel foods very high; however, as soon as specific plants became their staple food, the payoff of being neophobic increased (see Ch. 5).

If ingestion of a novel food has noxious consequences, an individual will associate it with its consumption and that food will not be eaten anymore. This phenomenon of food-aversion learning (or Garcia effect: Garcia, Kimeldorf, and Koelling, 1955; Garcia and Koelling, 1966), first documented in rats, is widespread among animal species (e.g., hamsters: Zahorik and Johnston, 1976; herbivorous mammals: Zahorik and Houpt, 1981). Food-aversion learning is a robust learning mechanism that operates at the individual level and primarily concerns novel foods. In food-aversion learning, the ingestion of a food that is associated with strong negative experience(s), such as gastrointestinal illness, leads to complete avoidance of the noxious food; avoidance persists when the food is no longer noxious, and the avoidance learning process is quicker when the food is novel than when the food is familiar. Food-aversion learning has been reported in the several primate species so far tested (e.g. Japanese macaques, squirrel monkeys, vervet monkeys, etc.; for a review see Visalberghi, 1994). Since some of these species (e.g., squirrel monkeys) have a feeding ecology similar to that of capuchins, we can assume that food-aversion learning is likely to occur in capuchins as well.

If negative postingestive consequences do not occur, a novel food gradually becomes familiar. A powerful factor influencing consumption of a novel food is how often it is encountered. For captive capuchins, a food remains unfamiliar (i.e., they respond to it neophobically) only for the first few encounters. The temporal course of capuchins' neophobic responses towards eight different novel foods that were repeatedly presented to individuals when alone was such that these foods were eaten to the same extent as familiar foods after just five presentations of 5 minutes each (Visalberghi, Valente, and Fragaszy, 1998).

Children are also cautious about novel foods and, similarly to capuchins, their neophobic response decreases with exposure to them (Birch and Marlin, 1982). Food acceptance increases if young children taste the novel foods during repeated encounters, whereas just smelling or looking at them is not enough to decrease neophobia (Birch *et al.*, 1987). Children familiarize not only with the novel food's characteristics (i.e., its visual appearance, texture, flavor, odor) but also learn about the negative (see above) and positive consequences of ingesting it. Positive consequences can be associated to the food and increase its consumption. Birch *et al.* (1990) demonstrated that children aged three to four years learn to prefer food with a high caloric content over one with low caloric content and use different flavors as immediate cues to distinguish foods.

From the foregoing, it appears that, when social influences are lacking, individuals are likely to develop similar food preferences as a consequence of caloric and sugar content.

7.2.3 The response of an individual to the change in palatability of a familiar food

In the wild, palatability and/or toxicity of plant foods can change over time in response to the concentration of secondary metabolites. Some of these substances, such as glucosides and alkaloids, have a bitter taste (Garcia and Hankins, 1975) that facilitates their detection at low concentration. The success of a generalist species, which is likely to encounter familiar foods that change in flavor seasonally (Jones, Keymer, and Ellis, 1978) and to face the problem of avoiding those which are potentially toxic, depends on the species' flexible exploitation of food resources. We expect that capuchins will explore and taste foods that change in palatability and that they will consume them when palatable and not consume them when unpalatable.

Capuchins were tested in a paradigm aimed at investigating their behavior when encountering a familiar food of which the taste has been experimentally changed (Visalberghi and Addessi, 2000a). The monkeys were presented with a familiar palatable food (cheese curd, oats, and bran mixed together) the palatability of which changed according to the experimental phase. In phase 1, capuchins were individually presented with this familiar food; in phase 2 they received the same familiar food with pepper added to it, making it unpalatable, and in phase 3 they received the same familiar palatable food of phase 1. Five sessions were carried out in each phase. The capuchins adapted immediately to the change in food palatability by reducing (phase 2) and increasing (phase 3) the amount of food eaten. During phase 2, most of the individuals kept tasting the peppery food and its unpalatability prompted an increase in olfactory exploration and food processing (rubbing the cheese curds and extracting them from the rest of the food). These findings show that capuchins readily adjust to changes in palatability of a familiar food, and that encounters with the food when unpalatable do not affect its consumption when palatable once again.

Two points deserve attention. First, capuchins do not behave towards a familiar food whose palatability has changed as if it were a food that has caused negative postingestive consequences. In this experiment, ingestion of the peppery food was necessary to prevent further consumption. The food was eaten again when the pepper was removed, and the food's familiarity did not prevent an immediate drastic reduction in its consumption when pepper was added. Conversely, ingestion of a food that is matched with strong negative experience(s), such as gastrointestinal illness, leads to the complete avoidance of the noxious food; this avoidance persists when the food is no longer noxious, and the avoidance-learning process is quicker when the food is novel (Garcia *et al.*, 1955; Garcia and Koelling, 1966; for a review of primate studies see Visalberghi, 1994). Second, the monkeys' response to nontoxic unpalatability (like the peppery food) also differs from their response to novel food. When they repeatedly encounter an unpalatable food they keep tasting it, but do not consume much, whereas when they repeatedly encounter a novel food they increase consumption of it over time.

7.3 What to eat: social influences on an individual's responses to novel foods and foods that change in palatability

7.3.1 Social influences on the individual's response towards food

For primates, feeding is undoubtedly a social affair. Although the presence of group members is not likely to influence taste perception, it is likely to affect behavioral responses towards food. In most species, individuals often feed close to one another and have a chance to see and smell another's food (or mouth). King (1999) has argued that when infant baboons (*Papio cynocephalus*) are in close proximity to foraging adults, the infants seek information about food by approaching adults and sniffing their muzzles, apparently to receive sensory cues about foods being eaten.

King (1999) distinguished between social information acquisition by the infant and social information donation by adults. She argued that, in the latter case, adults direct some action or behavior at immature individuals, enabling them potentially to receive more information than they would otherwise.

In a tolerant species such as capuchins (de Waal, 2000; Fragaszy *et al.*, 1997a; Fragaszy, Visalberghi, and Galloway, 1997b), when an individual A holds a piece of food it often tolerates the physical proximity of individual B (especially, but not exclusively, if B is a juvenile or infant). Individuals without food (but sometimes also with food) approach individual A and show interest in A's food and feeding activity by looking and sniffing A's food close up. Moreover, the same food freely available elsewhere does not prompt the same interest as A's food; therefore, we believe that the combination food–individual A, and possibly A's activity with the food, increases B's interest towards the food. In other words, the salience of the food is increased simply by the fact that A has it and not because B wants to get A's food (see also Thierry *et al.*, 1989).

It is plausible that interest fosters learning about food and that individual B, who shows interest in A's food, may, by doing so, acquire information about food from individual A. A may be aware of its role as information provider, or it may not, and may actively or passively provide information (all combinations being possible). A systematic comparison of A's and B's interactions in feeding contexts, differing in the extent to which individual B benefits in learning about food from individual A, and in the extent to which individuals compete for food (not otherwise available), may tease apart what role social influences have in feeding.

7.3.2 Social influences on food learning and on the neophobic response

In humans, a key factor inducing acceptance of novel foods is the social setting in which the food is presented. Birch, Zimmerman, and Hind (1980) presented preschool children with novel snacks in four different conditions. The snacks were found by children in their locker without apparent reason (nonsocial condition), on the table during snack time, given by the teacher during play time without apparent reason (noncontingent attention condition), or as a reward for having done something during play time (reward condition). Children's preferences for the novel snacks were strongly influenced by condition, and these differences lasted over time. In particular, the last two conditions, in which the teacher played an active role in giving the snack to the child, induced significantly higher preferences for snacks. In rats also, a food gains "value" if another rat is near it, and even an anesthetized rat may function as a social bias increasing the salience of a feeding site (Galef, 1981; Galef *et al.*, 2001; see also Ch. 6).

Our goal is to assess how encountering food in the presence of group members biases individual's response to foods. Section 7.2 described responses to novel foods when individuals were tested alone; this section describes responses observed in the social condition of those experiments. Capuchins are neophobic towards food, and Visalberghi and Fragaszy (1995) found that capuchins eat more of a novel food if group members are also eating nearby; this social facilitation[1] of eating occurs with novel foods but not with familiar foods. Social facilitation lasts for the first few encounters with a novel food, as though capuchins consider a food to be "novel" for a short time only (Visalberghi *et al.*, 1998). When encountering this same food later, consumption was not affected by having previously encountered it alone or with group members. In short, after a few encounters with unfamiliar food, feeding behavior of solitary and social feeders was indistinguishable, indicating that when individuals are observed eating the same foods it is not possible to establish whether the underlying processes of acquisition occurred in solitary or social contexts. Similarities in acceptance of novel foods may arise from an individual's experiences alone, and social learning is not necessarily the underlying process.

In the experiments we described below (Fragaszy *et al.*, 1997a,b; Visalberghi and Fragaszy, 1995), interest was defined as B's mouth coming within a distance of 12 cm or less of A's food, when A was eating, holding, or closely exploring the food. One of the aims of these experiments was to evaluate whether interest occurred more often when the food was novel, that is, when there was something to learn from others about it. In Visalberghi and Fragaszy (1995), since capuchins ate more familiar than novel foods, the number of scans in which an individual ate, held, or explored food was higher in familiar food condition than in novel food condition (7.4 and 1.5 scans out of 30, respectively). However, very interestingly, though opportunities for interest were more frequent in the familiar food condition, capuchins showed interest in someone else's

[1]Clayton (1978) defined social facilitation as an increase in the frequency of a behavior pattern in the presence of others displaying the same behavior pattern at the same time.

food almost exclusively in sessions in which novel foods were given.[2] Also, young capuchins showed more interest in somebody else's food if it was novel (Fragaszy *et al.*, 1997b; A. T. Galloway, D. Fragaszy, A. M. McCabe and E. Visalberghi, unpublished data).

In another study focused on food-related social interactions and food transfer, Fragaszy *et al.* (1997a) presented capuchin groups with monkey chow and pecan nuts and scored the behavior of young individuals towards other group members. The behaviors scored were interest (here defined as B's mouth coming within a distance of 12 cm or less of A's food), a variety of tolerated interactions between infants and peers and between infants and adult group members (attempting to take A's food, taking A's food, eating from A's hand, collecting pieces of food lying within approximately 12 cm of another feeding individual) and the response of A to B (A tolerates, avoids, or opposes B's behaviors).

Although both foods were abundant, infants' interactions were significantly more frequent towards nuts (a preferred food) than towards monkey chow when these foods were held or eaten by group members. Infants interacted more with adults than with peers, and adults were equally tolerant towards infant capuchins that could open nuts and those that could not (over the course of the study, out of 11 youngsters, only six were observed opening nuts). It is possible, as suggested by Fragaszy *et al.* (1997a), that the fact that adults are more likely to possess open nuts than peers, providing more frequent opportunities, may account for the difference in interest towards peers and adults.

It was also noted that when interest was directed to peers, it was preferentially directed to peers able to open nuts. Therefore, the interest towards individuals holding/cracking/eating a nut can be motivated by the desire for that food as well as by the purpose of monitoring the individual's behavior in order to learn how to crack open the nut. The experimental design does not allow these two possibilities to be distinguished. Moreover, equivalent tolerance (allowing interest/taking/collecting from all infants, not just those unable to open nuts) by adults indicates that adults are not active in providing information and do not take youngsters' skills into account. In a similar experiment, in which the focus was on the infants' behavior towards novel food, infants did not show more interest towards

[2] We statistically analyzed these data and found that the median number of times in which an individual showed interest in the novel food was significantly higher than in the familiar food condition (novel food: 1 (0–10); familiar food: 0 (0–0), Wilcoxon $t = 0$, $p < 0.02$, $n = 11$). Moreover, interest was mainly directed towards the dominant male (73% of the time).

somebody else's food before ingesting the same food than after having ingested it (Fragaszy *et al.*, 1997b).

We can conclude by saying that capuchins are very interested in the feeding activities of their group members, and especially so when a food is novel or difficult to process. Despite this, they do not seem to regulate their behavior on the basis of the information they might have acquired in this way.

7.3.3 Social influences on the response to the change of palatability of a familiar food

Section 7.2.3 described an experiment in which pepper was added to a familiar food (Visalberghi and Addessi, 2000a) when individuals were tested alone. These results can be compared with those obtained in the social condition of the same experiment: food consumption and overall response to the changes in food palatability were not affected by testing condition (social versus individual). For example, seeing a sudden change (increase or decrease) in the amount of food eaten by group members did not influence a subject's behavior. In addition, although capuchins responded quickly to changes in flavor and were often in proximity to one another, proximity and interest towards somebody else's food did not occur preferentially when the behavior of conspecifics might provide useful indirect information about palatability, that is, in the sessions in which the food was presented for the first time with or without pepper. Nor has prevention ever been observed, not even in the mother–infant pair where you would expect intervention to be more likely to occur. For example, the eight-month old infant was never prevented from taking the peppery food by her mother. In phases 1 and 3, the mother attempted to prevent her infant from taking the good food from her mouth or hand. Nevertheless, on two occasions the infant succeeded in taking food from her. No systematic pattern (phases 1 and 3 versus phase 2) was found in the number of times in which the mother or the infant ate first when the food was palatable (phase 1 and 3) and unpalatable (phase 2).

Finally, during the many years in which we have observed capuchins' spontaneous behavior, as well as their responses to foods in experimental settings (including those mentioned above), we have never witnessed a single episode in which an individual prevented another from eating something unpalatable or toxic. In addition, we have no evidence that an individual learns that a food is unpalatable or toxic by observing group members not eating, avoiding, or discarding it. On the contrary, we once witnessed an adult male vomiting a novel food after which two adult females, who also observed the episode, approached and ate his vomit.

7.4 New insights on social influences on the acceptance of novel foods: sorting out the factors

Social facilitation of eating novel foods has adaptive values as a quicker way of overcoming neophobia and/or learning about a safe diet (Galef, 1993). Social facilitation of eating novel foods by capuchins (Visalberghi and Fragaszy, 1995; and see above for details) can be interpreted in either or both these ways. To shed light on which of these adaptive hypotheses is correct, Visalberghi and Addessi (2000b) investigated the factors promoting social facilitation of eating novel foods.

7.4.1 Social facilitation of eating increases acceptance of novel foods

Visalberghi and Addessi (2000b) assessed whether an individual's consumption of novel foods is different when (a) the individual is alone (alone condition), (b) group members are visible through a transparent Plexiglas panel in the nearby cage with no food (group present condition), and (c) group members are present and eating a familiar food from a box attached to the Plexiglas panel (group plus food condition). In all three conditions the novel food is presented to the subject in a box attached to the panel .

Fifteen subjects were tested with three novel foods, each food assigned to one of the three conditions. The subject's food consumption was scored by the subject's eating behavior every 10 seconds and by measuring the grams of food ingested (i.e., eaten) by the subject (the grams of food left by the subject at the end of the trial was subtracted from the initial weight to give the total weight of the food provided to the subject). Results showed that capuchins performed significantly more eating behavior (measured as number of 10 second sample points) and ingested significantly higher amounts (measured in grams) in the group plus food condition than in the alone condition (see Fig. 7.1). The values of these two measures in the group present condition did not differ from those obtained in the alone and the group plus food conditions. In the group plus food condition, the number of group members eating near the panel and the average number of eating sample points of the subjects were significantly correlated. For further details, see Visalberghi and Addessi (2000b) and Addessi and Visalberghi (2001).

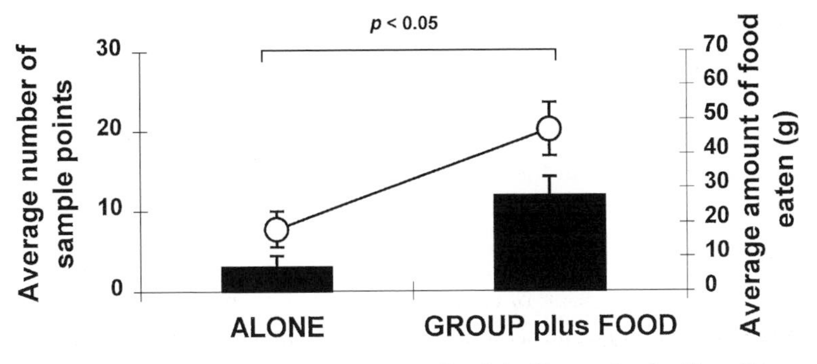

Fig. 7.1. Food ingested and eaten in the "alone" and the "group plus food" conditions (Wilcoxon test). ■, Eating behavior; □, food eaten. (For further details see Visalberghi and Addessi, 2000b.)

7.4.2 Social facilitation of eating does not foster learning about a safe diet

It is important to stress that, in this experiment, social facilitation of eating (which led to the increased consumption of a novel food) occurred even if group members on the other side of the panel were eating a strikingly different food from that available to the subject. A further experiment (Visalberghi and Addessi, 2001) investigated whether social facilitation of eating was more pronounced when group members were eating the same food (same color condition) than when group members were eating a different food (different color condition). It should be clarified that here "same" and "different" refer to food color since the experimental subject could not have direct access to nor taste or smell the group members' food (which was in a transparent box attached to the side of the Plexiglas panel opposite the subject's box with novel food). Results showed that whereas the number of eating sample points was significantly higher in the same color condition, the amount of food ingested was not. Therefore, the match in the color of the novel food and the food provided to the group members affected eating behavior but not ingestion. It has to be stressed that the time spent eating a given amount of food is influenced by the pace and/or speed of eating; it follows that, given the same amount of food, the number of eating samples of an individual eating slowly is going to be higher than that of an individual eating quickly.

To learn about a safe diet, a capuchin should have paid attention to *what* others were eating, and should have eaten more of a novel food *only if* its own food matched the food that group members were eating. Since our results showed that social facilitation of eating novel foods occurs regardless of what is eaten, we can discard the hypothesis that it fosters learning about a safe diet. However, it can be argued that wild capuchins are never faced with just one novel food. On the contrary, capuchins encounter more than one food at a time from which they have to choose. It is possible that, given a choice between two novel foods, only one of which matches in color (or some other property) the food that group members are eating, a capuchin would direct its preference towards the food that matches. An experiment designed to explore this possibility is underway in our laboratory.

7.5 Feeding habits and traditions in capuchins

Years ago, Chapman and Fedigan (1990) studied the diets of three neighboring groups of *Cebus capucinus* living in Santa Rosa National Park, Costa Rica (the home ranges of two groups partly overlap). They found a considerable variability among groups in relative amounts of fruits, plants, and insects eaten. According to Chapman and Fedigan (1990), dietary differences among populations are affected by (a) presence or absence of a food; (b) food profitability in terms of nutrients, energetics, toxins, and availability; and/or (c) learned group traditions. By measuring the densities of all major plant foods ($n = 16$) in the three home ranges, it was found that two were plant foods eaten by all groups and the magnitude of use was in accord with availability, and four were plant foods eaten by all groups, but the magnitude of use did not correspond to availability. Of the remaining 10 plants not eaten by at least one group, seven were available to all groups. Therefore, since dietary differences could not be attributed to simple measures of food abundance, Chapman and Fedigan (1990) considered both the food profitability hypothesis and the learned group traditions hypothesis as likely but could not distinguish between them. In fact, reliable claims about social learning are particularly difficult for field workers to substantiate.

Panger and co-workers (2002; see also Ch. 14) have identified several differences in the foraging behavior of *C. capucinus* populations living in three tropical dry forest sites in Costa Rica. Their analyses show that population-specific behaviors are not the result of obvious genetic or ecological differences and, therefore, are likely to represent traditions.

Partly prompted by the findings in wild capuchins, we recently determined the preferences towards seven familiar foods (pear, banana, tangerine, pellet, bread, romana lettuce, and boiled potatoes) of 26 captive capuchins living in four groups. The aim was to assess whether food preferences were more similar among individuals living in the same social group than among individuals living in different groups (G. Sabbatini, M. Stammati, and E. Visalberghi, unpublished data). We determined food preferences by presenting each capuchin with a choice between two foods. Each subject was presented with all the possible pairs of foods three times (21×3 choices $= 63$ choice tests). Results showed that the overall order of preference was tangerine, banana, potatoes, pear, bread, pellet and romana lettuce (from most to least). Capuchins differed significantly in their preferences, and food preferences for tangerine, pellet, and potatoes differed significantly among groups.

Do our findings suggest the existence of feeding traditions in our group? Given that all our foods were equally available and similarly profitable for all subjects, our findings support the idea that the convergence among the preferences of the individuals living in the same group might be a result of social learning. However, in our groups, as well as groups in the wild, individuals are more genetically related than individuals belonging to different groups. Therefore, before accepting the hypothesis that convergences in food preferences are socially learned, we should exclude the influence of genetic factors.

These findings on food choices, although preliminary, suggest that there are also surprising behavioral convergences among individuals living together in captivity. For both wild capuchins and captive ones, differences among groups and convergences within groups were unnoticed until researchers specifically looked for them. In the scenario described in Sections 7.3 and 7.4, in which we argue that learning from others about food plays a minor role, the possible traditions reported for groups of wild capuchins as well as our preliminary data on behavioral convergences in food choices and processing are indeed a puzzle that we are not yet able to solve.

7.6 Discussion and suggestions for future research

Can laboratory experiments such as those described in the previous sections shed light on the processes that lead individuals to learn about foods? Probably yes, at least to some extent. The results described above do not support a few common assumptions about how diets are refined and suggest new hypotheses to be validated through future experiments and field observations.

A common assumption has been that naive individuals gain information about food by observing what other group members do. In fact, vision is important for food discrimination in primates (Jacobs, 1995). Moreover, food unpalatability can readily be associated with visual cues and odor (Laska and Metzker, 1998), while sight and taste are important in the acquisition of food aversion (Matsuzawa *et al.*, 1983). However, experimental data show that tufted capuchins do not seem to use visual information from the behavior of other group members to decide what to eat. On the one hand, capuchins eat more of a novel food if other group members are eating, regardless of whether their food matches that eaten by group members (Visalberghi and Addessi, 2000b); on the other hand, infant capuchins were just as likely to show interest in a group member *after* eating a novel food as *before* (Fragaszy *et al.*, 1997b).

It is possible, however, that naive individuals (and especially infants) are able to learn from the feeding activities of group members through other cues. In particular, it can be argued that sniffing and/or tasting foods that other capuchins are eating are good candidates for producing social influences, for at least two reasons. First, transfer of food from one individual to another occurs commonly in capuchins; in particular, young individuals are allowed to smell and taste foods that other group members are eating. Second, in several mammal species, individuals are better at associating the consequences of food ingestion to food smell and taste than to its visual appearance (e.g., rats, Galef, 1993; Schafe and Bernstein, 1996). Therefore, future research should investigate whether acceptance of novel foods and learning whether a food is safe are affected by sniffing and tasting another individual's food.

In contrast with what is often assumed, we did not find evidence that capuchins learn to avoid a food by watching conspecifics avoid that food (Visalberghi and Addessi, 2000a) or that experienced individuals warn naive ones about unpalatable or toxic food (see Visalberghi and Fragaszy, 1996). In humans, spitting out food, vomiting, or making facial expressions indicating disgust at the taste of food are reliable cues that a food is unpalatable and/or toxic; could nonhuman primates also learn about food from these cues? In nonhuman primates, spitting out food, vomiting, or making facial expressions of disgust are either absent

(Preuschoft, 2000) or rarely performed and not salient for capuchin group members witnessing them (e.g., vomiting, E. Visalberghi and M. Valente, unpublished results). It is possible that the lack of these responses, or their lack of salience and/or reliability as cues, makes them unsuited for social learning.

In any case, it is important to stress that, from a cognitive point of view, learning to avoid a food by watching the behavior of group members avoiding it or warning another individual not to eat that food are rather demanding tasks (Visalberghi and Fragaszy, 1996). Hearst (1991) provided many examples drawn from animal and human experimental studies showing that it is much easier for an individual to detect presence than absence, and easier for an individual to learn something from the arrival of a stimulus than from the removal of a stimulus. In other words, it is easier to learn what to eat by observing what another individual eats than to learn what to avoid by observing what another individual avoids. Similarly in a social context, it is easier for an individual to direct the observer's attention to what it is eating, than to what it is not eating (see also Ch. 6). Only sophisticated communication and comprehension enable one individual to draw another's attention to the food the individual is avoiding, or to behavior that is not occurring (see Visalberghi and Fragaszy, 1996).

Future studies of social influences on learning about foods should be accompanied by evaluation of their adaptive significance. We could measure whether social influences of group members promote faster and/or better exploitation of new foods than occurs when an individual is exposed to the same environmental conditions without social partners. We also could measure whether social influences from more knowledgeable group members contribute to the adoption of a nutritionally adequate diet by other members (see Ch. 5).

Finally, we believe that the extent to which monkeys are influenced by social partners depends on what they are learning about. Fear seems to be a salient and reliable cue and social learning of fear has been documented when monkeys observed others reacting strongly towards a novel stimulus (for example, a snake) by giving alarm vocalizations and expressing fear (Cheney and Seyfarth, 1990; Mineka and Cook, 1993; Srivastava, 1991; see also Ch. 14). We expect future research to confirm that primates rely heavily on social learning when the identification of dangerous predator is involved (as demonstrated by Mineka and Cook (1993); see also Ch. 2) and that primates learn individually about toxic/poisonous foods through the amazingly powerful food-aversion learning process, which is far safer than relying on conspecifics avoiding that food. Moreover, we expect social influences will be shown to bias individuals' interest and eating activities towards foods that are eaten by others. This bias, far from being specifically directed towards a target food, nevertheless channels activities in such a way as to increase the chances of "getting it right". Possible "errors" can be corrected by an individual's physiological response and feedback, or they may lead to an increase in overall variability in the population.

7.7 Conclusions

In the past, feeding behavior in nonhuman primates was looked at with eyes biased by the strong lens of the tremendous impact that social learning has on the ways humans behave towards food (Rozin, 1996). On the one hand, our experiments have shown that capuchins are very interested in others' foods but this interest is not restricted to situations in which it may lead to acquisition of useful information. In particular, capuchins do not look for information selectively when needed; they do not carefully scrutinize the appearance of another's food (and this should caution us in attributing to monkeys the ability to learn what to feed upon by observing what group members feed upon), and they do not guide other's feeding behavior. On the other hand, the individual's cautious approach to novel foods, the ingestion of very limited amounts of novel food, the innate preference for certain substances, and the capacity for food-aversion learning are all factors that allow the individual to learn about foods and reduce the risk of getting poisoned. Therefore, learning individually (rather than socially) seems a viable option.

However, there are other ways in which the social context provides opportunities to learn about foods. The presence of others and what they do channel the individual's interest, attention, and activities. Capuchins have high levels of interindividual tolerance and naïve individuals may accidentally taste the food eaten by other group members and, by doing so, learn to eat it. The tendencies to coordinate activities in space and time (Coussi-Korbel and Fragaszy, 1995) are a simple and powerful social bias on individual learning; coordination allows individuals to do similar things in similar places at a similar time and thus increases an individual's chance to do what others do (and to eat what others eat). We may conclude by saying that capuchins do not seem to learn *from* others but *with* others.

7.8 Acknowledgements

We thank all the students, researchers, keepers, and capuchins, which, with different roles, participated in the experiments we discuss in this chapter. Furthermore, we thank G. Dewar, D. Fragaszy, J. Galef, S. Perry, and an anonymous referee for their comments on a previous version of the manuscript. We also thank Bioparco SPA for hosting our laboratory.

References

Addessi, E. and Visalberghi, E. 2001. Social facilitation of eating novel foods in tufted capuchin monkeys (*Cebus apella*): input provided, responses affected and cognitive implications. *Animal Cognition* **4**, 297–303.

Agostini, I. 2001. Neofobia in *Cebus apella nigritus*: uno studio sperimentale nel Parco Nazionale di Iguazù, Argentina. Università degli Studi di Roma "La Sapienza". Facoltà di Scienze matematiche, fisiche e naturali.

Anderson, J. R. 2000. Tool-use, manipulation and cognition in capuchin monkeys (*Cebus*). In *New Perspectives in Primate Evolution and Behaviour*, ed. C. Harcourt, pp. 91–110. Otley, UK: Westbury Publishing.

Barnett, S. A. 1958. Experiments on "neophobia" in wild and laboratory rats. *British Journal of Psychology*, **49**, 195–201.

Beauchamp, G. K., Cowart, B. J., and Moran, M. 1986. Developmental changes in salt acceptability in human infants. *Developmental Psychobiology*, **27**, 353–365.

Birch, L. L. and Marlin, D. W. 1982. I don't like it; I never tried it: effects of exposure on two-year old children's food preferences. *Appetite*, **3**, 353–360.

Birch, L. L., Zimmerman, S., and Hind, H. 1980. The influence of social-affective context on preschool children's food preferences. *Child Development*, **51**, 856–861.

Birch, L. L., McPhee, L., Shoba, B. C., Pirok, E., and Steinberg, L. 1987. What kind of exposure reduces children's food neophobia? Looking vs. tasting. *Appetite*, **9**, 171–178.

Birch, L. L., McPhee, L., Steinberg, L., and Sullivan, S. 1990. Conditioned flavour preferences in young children. *Physiology and Behaviour*, **47**, 501–505.

Boesch, C. and Tomasello, M. 1998. Chimpanzee and human cultures. *Current Anthropology*, **39**, 591–614.

Box, H. O. 1984. *Primate Behaviour and Social Ecology*. London: Chapman & Hall.

Brown, A. D. and Zunino, G. E. 1990. Dietary variability in *Cebus apella* in extreme habitats: evidence for adaptability. *Folia Primatologica*, **54**, 187–195.

Byrne, R. W. 1999. Cognition in great ape ecology. In *Mammalian Social Learning: Comparative and Ecological Perspectives*, ed. H. O. Box and K. R. Gibson, pp. 332–350. Cambridge: Cambridge University Press.

Cambefort, J. P. 1981. A comparative study of culturally transmitted patterns of feeding habits in the chacma baboon *Papio ursinus* and the vervet monkey *Cercopithecus aethiops*. *Folia Primatologica*, **36**, 243–263.

Chapman, C. A. and Fedigan, L. M. 1990. Dietary differences between neighboring *Cebus capucinus* groups: local tradition, food availability or responses to food profitability? *Folia Primatologica*, **54**, 177–186.

Cheney, D.L. and Seyfarth, R.M. 1990. *How Monkeys See the World*. Chicago, IL: University of Chicago Press.

Clayton, D.A. 1978. Socially facilitated behavior. *Quarterly Review of Biology*, **53**, 373–392.

Coussi-Korbel, S. and Fragaszy, D. 1995. On the relation between social dynamics and social learning. *Animal Behaviour*, **50**, 1441–1453.

Custance, D., Whiten, A., and Fredman, A. 1999. Social learning of an artificial fruit task in capuchin monkeys (*Cebus apella*). *Journal of Comparative Psychology*, **113**, 13–23.

de Waal, F.B.M. 2000. Attitudinal reciprocity in food sharing among brown capuchin monkeys. *Animal Behaviour*, **60**, 253–261.

de Waal, F.B.M., Luttrell, L.M., and Canfield, M.E. 1993. Preliminary data on voluntary food sharing in brown capuchin monkeys. *American Journal of Psychology,* **29**, 73–78.

Dunbar, R.I.M. 1988. *Primate Social Systems*. London: Croom Helm.

Fedigan, L.M. 1982. *Primates Paradigm*. Montreal: Eden Press.

Fedigan, L.M. 1991. History of the Arashiyama West Japanese macaques in Texas. In *The Monkeys of Arashiyama*, ed. L. M. Fedigan and P. J. Asquith, pp. 54–73. Albany, NY: State University of New York Press.

Fletemeyer, J.R. 1978. Communication about potentially harmful foods in free-ranging chacma baboons, *Papio ursinus*. *Primates*, **19**, 223–226.

Fossey, D. 1979. Development of the mountain gorilla (*Gorilla gorilla beringei*): the first thirty-six months. In *The Great Apes*, ed. D.A. Hamburg and E.R. McCown, pp. 138–184. Menlo Park, CA: Benjamin/Cummings.

Fragaszy, D. and Visalberghi, E. 1989. Social influences on the acquisition and use of tools in tufted capuchin monkeys (*Cebus apella*). *Journal of Comparative Psychology*, **103**, 159–170.

Fragaszy, D.M., Vitale, A.F., and Ritchie, B. 1994. Variation among juvenile capuchins in social influences on exploration. *American Journal of Primatology*, **32**, 249–260.

Fragaszy, D.M., Feuerstein, J.M., and Mitra, D. 1997a. Transfers of food from adults to infants in tufted capuchins (*Cebus apella*). *Journal of Comparative Psychology*, **111**, 194–200.

Fragaszy, D.M., Visalberghi, E., and Galloway, A. 1997b. Infant tufted capuchin monkeys' behaviour with novel foods: opportunism, not selectivity. *Animal Behaviour*, **53**, 1337–1343.

Freeland, W. J. and Janzen, D.H. 1974. Strategies in herbivory by mammals: the role of plant secondary compounds. *American Naturalist*, **108**, 269–289.

Galef, B.G., Jr. 1981. Development of olfactory control of feeding-site selection in rat pups (*Rattus norvegicus*). *Journal of Comparative Psychology*, **95**, 615–622.

Galef, B.G., Jr. 1991. Tradition in animals: field observation and laboratory analyses. In *Interpretation and Explanation in the Study of Behavior,* Vol. 1: *Interpretation, Intentionality and Communication*, ed. M. Bekoff and D. Jamieson, pp. 3–28. Hillsdale, NJ: Erlbaum.

Galef, B.G., Jr. 1993. Function of social learning about food: a causal analysis of diet novelty on preference transmission. *Animal Behaviour*, **46**, 257–265.

Galef, B.G., Jr. 1996. Food selection: problems in understanding how we choose foods to eat. *Neuroscience and Biobehavioral Reviews*, **20**, 67–73.

Galef, B.G., Jr., Marczinski, C.A., Murray, K.A., and Whiskin, E.E. 2001. Food stealing by young Norway rats (*Rattus norvegicus*). *Journal of Comparative Psychology*, **115**, 16–21.

Garcia, J. and Hankins, W.G. 1975. The evolution of bitter and the acquisition of toxiphobia. In *Olfaction and Taste: Fifth International Symposium*, ed. D.A. Denton and J.P. Coghlan, pp. 39–45. New York: Academic Press.

Garcia, J. and Koelling, R.A. 1966. Relation of cue to consequence in avoidance learning. *Psychonomic Science*, **4**, 123–124.

Garcia, J., Kimeldorf, D.J., and Koelling, R.A. 1955. A conditioned aversion towards saccharin resulting from gamma radiation. *Science*, **122**, 157–158.

Giraldeau, L.A. 1997. The ecology of information use. In *Behavioral Ecology*, 4th edn, ed. J. Krebs and N. Davies, pp. 42–68. Oxford: Blackwell Science.

Goodall, J. 1973. Cultural elements in the chimpanzee community. In *Precultural Behavior*, ed. E.W. Menzel, pp. 144–184. Basel: Karger.

Goodall, J. 1986. *The Chimpanzees of Gombe*. Cambridge, MA: Harvard University Press.

Hearst, E. 1991. Psychology of nothing. *American Scientist*, **79**, 432–443.

Heyes, C. and Galef B.G., Jr. 1996. *Social Learning in Animals: The Roots of Culture*. San Diego, CA: Academic Press.

Hikami, K. 1991. Social transmission of learning in Japanese monkeys (*Macaca fuscata*). In *Primatology Today*, ed. A. Ehara, T. Kimura, O. Takemaka, and M. Iwamoto, pp. 343–344. Amsterdam: Elsevier.

Hikami, K., Hasegawa, Y., and Matsuzawa, T. 1990. Social transmission of food preferences in Japanes monkeys (*Macaca fuscata*) after mere exposure or aversion training. *Journal of Comparative Psychology*, **104**, 233–237.

Hiraiwa-Hasegawa, M. 1990. A note on the ontogeny of feeding. In *The Chimpanzees of the Mahale Mountains: Sexual and Life History Strategies*, ed. T. Nishida, p. 277–283. Tokyo: University of Tokyo Press.

Hladik, C.M. and Simmen, B. 1996. Taste perception and feeding behavior in non-human primates and human populations. *Evolutionary Anthropology*, **5**, 58–72.

Izawa, K. 1978. Frog-eating behavior of wild black-capped capuchin (*Cebus apella*). *Primates*, **19**, 633–642.

Izawa, K. 1980. Social behavior of the wild black-capped capuchin (*Cebus apella*). *Primates*, **21**, 443–467.

Izawa, K. and Mizuno, A. 1977. Palm-fruit cracking behavior of wild black-capped capuchin (*Cebus apella*). *Primates*, **18**, 773–792.

Jacobs, G.H. 1995. Variations in primate color vision: mechanisms and utility. *Evolutionary Anthropology*, **3**, 196–205.

Janson, C.H. 1996. Toward an experimental socioecology of primates: examples from Argentine brown capuchin monkeys (*Cebus apella nigrivittatus*). In *Adaptive Radiations of Neotropical Primates*, ed. M. Norconk, P. Garber, and A. Rosenberger. New York: Plenum Press.

Janson, C.H. and van Schaik, C.P. 1993. Ecological risk aversion in juvenile primates: slow and steady wins the race. In *Juvenile Primates: Life History, Development, and Behavior*, ed. M.E. Pereira and L.A. Fairbanks. Oxford: Oxford University Press.

Johnson, E. 2000. Food-neophobia in semi-free ranging rhesus macaques: effects of food limitation and food sources. *American Journal of Primatology*, **50**, 25–35.

Jones, D.A., Keymer, R.J., and Ellis, W.M. 1978. Cyanogenesis in plants and animal feeding. In *Biochemical Aspects of Plant and Animal Co-Evolution*, ed. J. B. Harborne, pp. 21–34. London: Academic Press.

Jouventin, P., Pasteur, G., and Cambefort, J. P. 1976. Observational learning of baboons and avoidance of mimics: exploratory test. *Evolution*, **31**, 214–218.

King, B. J. 1994a. *The Information Continuum: Social Information Transfer in Monkeys, Apes, and Humans*. Santa Fe, NM: SAR Press.

King, B. J. 1994b. Primate infants as skilled information gatherers. *Pre- and Perinatal Psychology Journal*, **8**, 287–307.

King, B. J. 1999. New directions in the study of primate learning. In *Mammalian Social Learning: Comparative and Ecological Perspectives*, ed. H. O. Box and K. R. Gibson, pp. 17–32. Cambridge: Cambridge University Press.

Kinzey, W. G. 1997. *New World Primates: Ecology, Evolution and Behavior*. Hawthorne, NY: Aldine.

Kummer, H. 1971. *Primate Societies*. Arlington Heights, IL: Harlan Davidson.

Laska, M. 1997. Taste preferences for five food-associated sugars in the squirrel monkey (*Saimiri sciureus*). *Journal of Chemical Ecology*, **23**, 659–672.

Laska, M. and Metzker, M. 1998. Food avoidance learning in squirrel monkeys and common marmosets. *Learning and Memory*, **5**, 193–203.

Laska, M., Carrera Sanchez, E., and Rodriguez Luna, E. 1998. Relative taste preferences for food-associated sugars in the spider monkey (*Ateles geoffroyi*). *Primates*, **39**, 91–96.

Laska, M., Hernandez Salazar, L. T., and Rodriguez Luna, E. 2000. Food preferences and nutrient composition in captive spider monkeys, *Ateles geoffroyi*. *International Journal of Primatology*, **21**, 671–683.

Lee, P. C. 1994. Social structure and evolution. In *Behaviour and Evolution*, ed. P. J. B. Slater and T. R. Halliday, pp. 266–302. Cambridge: Cambridge University Press.

Lefebvre, L. 1995. Culturally-transmitted feeding behaviour in primates: evidence for accelerating learning rates. *Primates*, **36**, 227–239.

Matsuzawa, T., Hasegawa, Y., Gotoh, S., and Wada, K. 1983. One-trial long-lasting food-aversion learning in wild Japanese monkeys (*Macaca fuscata*). *Behavioral and Neural Biology*, **39**, 155–159.

Menzel, C. R. 1997. Primates' knowledge of their natural habitat: as indicated in foraging. In *Machiavellian Intelligence II: Extensions and Evaluations*, ed. A. Whiten and R. W. Byrne, pp. 207–239. Cambridge: Cambridge University Press.

Miklósi, Á. 1999. The ethological analysis of imitation. *Biological Review*, **74**, 347–374.

Mineka, S. and Cook, M. 1993. Mechanisms involved in the observational conditioning of fear. *Journal of Experimental Psychology: General*, **122**, 23–38.

Nishida, T. 1987. Local traditions and cultural transmission. In *Primate Societies*, ed. B. B. Smuts, D. L. Cheney, R. M. Seyfarth, R. W. Wrangham, and T. T. Struhsaker, pp. 462–474. Chicago, IL: University of Chicago Press.

Nishida, T., Wrangham, R. W., Goodall, J., and Uehara, S. 1983. Local differences in plant-feeding habits of chimpanzees between the Mahale Mountains and Gombe National Park, Tanzania. *Journal of Human Evolution*, **12**, 467–480.

Panger, M., Perry, S., Rose, L., Gros-Louis, J., Vogel, E., MacKinnon, K.C., and Baker, M. 2002. Cross-site differences in the foraging behavior of white-faced capuchins (*Cebus capucinus*). *American Journal of Physical Anthropology* **119**, 52–66.

Perry, S. and Rose, L. 1994. Begging and transfer for coati meat by white-faced capuchin monkeys, *Cebus capucinus*. *Primates*, **35**, 409–415.

Preuschoft, S. 2000. Primate faces and facial expressions. *Social Research*, **67**, 245–271.

Provenza, F. D. 1995. Postingestive feedback as an elementary determinant of food preference and intake in ruminants. *Journal of Range Management*, **48**, 2–17.

Robinson, J. G. 1986. Variation in group size of wedge-capped capuchin monkeys (*Cebus olivaceus*): effects on survival, fecundity, and social structure. *Primate Report*, **14**, 67–68.

Rozin, P. 1996. Sociocultural influences on human food selection. In *Why We Eat What We Eat: The Psychology of Eating*, ed. E. D. Capaldi, pp. 233–263. Washington, DC: American Psychological Association.

Schafe, G. E. and Bernstein, E. L. 1996. Taste aversion learning. In *Why We Eat What We Eat: The Psychology of Eating*, ed. E. D. Capaldi, pp. 31–51. Washington, DC: American Psychological Association.

Siemers, B. M. 2000. Seasonal variation in food resource and forest strata use by brown capuchin monkeys (*Cebus apella*) in a disturbed forest fragment. *Folia Primatologica*, **71**, 181–184.

Simmen, B. and Hladik, C. M. 1998. Sweet and bitter discrimination in primates: scaling effects across species. *Folia Primatologica*, **69**, 129–138.

Srivastava, A. 1991. Cultural transmission of snake-mobbing in free-ranging hanuman langurs. *Folia Primatologica*, **56**, 117–120.

Sussman, R. W. 2000. *Primate Ecology and Social Structure*, Vol. 2: *New World Monkeys*. Boston, MA: Pearson Custom Publishing.

Terborgh, J. 1983. *Five New World Primates*. Princeton, NJ: Princeton University Press.

Thierry, B., Wunderlich, D., and Gueth, C. 1989. Possession and transfer of objects in a group of brown capuchins (*Cebus apella*). *Behaviour*, **110**, 294–305.

Thomas, O. 1903. *Annals and Magazine of Natural History, 7th Series.*, **XI**, p. 396.

Tomasello, M. and Call, J. 1997. *Primate Cognition*. Oxford: Oxford University Press.

Visalberghi, E. 1993. Tool use in a South American monkey species: an overview of characteristics and limits of tool use in *Cebus apella*. In *Tool Use in Human and Nonhuman Primates*, ed. A. Berthelet and J. Chavaillon, pp. 119–131. Oxford: Oxford University Press.

Visalberghi, E. 1994. Learning processes and feeding behavior in monkeys. In *Behavioural Aspects of Feeding: Basic and Applied Research on Mammals*, ed. B. G. Galef Jr., M. Mainardi, and P. Valsecchi, pp. 257–270. Chur: Harwood Academic.

Visalberghi, E. 2000. Tool use behaviour and the understanding of causality in primates. In *Comparer ou Prédire: Exemples de Recherches en Psychologie Comparative Aujourd'hui*, ed. E. Thommen and H. Kilcher, pp. 17–35. Fribourg, Switzerland: Les Editions Universitaires.

Visalberghi, E. and Addessi, E. 2000a. Response to changes in food palatability in tufted capuchin monkeys (*Cebus apella*). *Animal Behaviour*, **59**, 231–238.

Visalberghi, E. and Addessi, E. 2000b. Seeing group members eating a familiar food enhances the acceptance of novel foods in capuchin monkeys. *Animal Behaviour*, **60**, 69–76.

Visalberghi, E. and Addessi, E. 2001. Acceptance of novel foods in *Cebus apella*: do specific social facilitation and visual stimulus enhancement play a role? *Animal Behaviour*, **62**, 567–576.

Visalberghi, E., and Fragaszy, D. M. 1990a. Do monkeys ape? In *"Language" and Intelligence in Monkeys and Apes*, ed. S. T. Parker and K. R. G. Gibson, pp. 247–273. Cambridge: Cambridge University Press.

Visalberghi, E. and Fragaszy, D. 1990b. Food-washing behaviour in tufted capuchin monkeys (*Cebus apella*) and crabeating macaques (*Macaca fascicularis*). *Animal Behaviour,* **40**, 829– 836.

Visalberghi, E. and Fragaszy, D. 1995. The behaviour of capuchin monkeys, *Cebus apella*, with novel foods: the role of social context. *Animal Behaviour*, **49**, 1089–1095.

Visalberghi, E. and Fragaszy, D.M. 1996. Pedagogy and imitation in monkeys: Yes, no, or maybe? In *The Handbook of Education and Human Development*, ed. D.R. Olson and N. Torrance, pp. 277–301. Cambridge, MA: Blackwell.

Visalberghi, E., and Fragaszy, D. 2002. Do monkeys ape? Ten years after. In *Imitation in Animals and Artifacts*, ed. K. Dautenhahn, and C.L. Nehaniv, pp. 471–499. Cambridge, MA: MIT Press.

Visalberghi, E., Valente, M., and Fragaszy, D. 1998. Social context and consumption of unfamilar foods by capuchin monkeys (*Cebus apella*) over repeated encounters. *American Journal of Primatology*, **45**, 367–380.

Watts, D.P. 1985. Observation on the ontogeny of feeding behavior in mountain gorillas (*Gorilla gorilla beringei*). *American Journal of Antropology*, **8**, 1–10.

Whitehead, J.M. 1986. Development of feeding selectivity in mantled howling monkeys, *Alouatta palliata*. In *Primate Ontogeny, Cognition and Social Behaviour*, ed. J.G. Else and P.C. Lee, pp. 105–117. Cambridge: Cambridge University Press.

Whiten, A., Goodall, J., McGrew, W.C., Nishida, T., Reynolds, V., Sugiyama, Y., Tutin, C.E.G., Wrangham, R.W., and Boesch, C. 1999. Cultures in chimpanzees. *Nature*, **399**, 682–685.

Wrangham, R.W. 1980. An ecological model of female–bonded primate groups. *Behaviour*, **75**, 262–300.

Yamamoto, M.E., Lopes, F.A., Leite, T.S., and Azevêdo, S.D. 2000. Exploration and consumption of novel and familiar food items by captive common marmosets. Poster. *European Federation of Primatology 2000 Meeting*. London, 28–29 November.

Zahorik, D.M. and Houpt, K.A. 1981. Species differences in feeding strategies, food hazards, and the ability to learn food aversions. In *Foraging Behavior: Ecological, Ethological and Psychological Approaches*, ed. A.C. Kamil and T.D. Sargent. New York: Garland STPM Press.

Zahorik, D.M. and Johnston, R.E. 1976. Taste aversions to food flavors and vaginal secretions in golden hamsters. *Journal of Comparative and Physiological Psychology*, **90**, 57–66.

VINCENT M. JANIK AND PETER J. B. SLATER

8

Traditions in mammalian and avian vocal communication

8.1 Introduction

The most basic definition of traditions used by biologists is the one given by Fragaszy and Perry in Ch. 1. It states that traditions are enduring behavior patterns that are shared by at least two individuals and that are acquired in part through social learning. Laland, Richerson, and Boyd (1993) distinguished between two forms of social learning. The first involves primarily horizontal information transmission (i.e., between animals of the same generation) in which information is of only transient value, as in the acquisition of foraging information in a highly variable environment. In the second, information is transmitted vertically (between generations) and results in what Laland *et al.* (1993) call stable traditions. In this definition, socially learned information has to remain in the population for a certain period of time before it can be called a tradition. These two forms appear not to be exclusive but rather are placed at different points on a continuum. However, it is useful to consider the results of social learning in this theoretical framework to demonstrate how social learning in communication systems differs from that in other domains. We will use these concepts to review vocal traditions in mammals and birds.

By definition, every form of learning about communication has to involve another individual since communication involves at least two individuals. The only exception is learning to change the quality of a signal through practicing. However, this can be recognized by observing the performance of an isolated individual as it changes. Therefore, the study of vocal traditions avoids one of the main problems in the study of social learning, namely the question of whether the trait under investigation is actually learned socially or individually. This is one of the reasons why birdsong has been a model system for the study of traditions since the early 1960s. However, not all forms of social learning that affect vocal communication lead to traditions. One requirement is that behavior patterns are shared between individuals once learning has occurred. This excludes some forms of social learning that lead to shared representations but not shared behavior patterns. Examples are comprehension learning (Janik and Slater, 2000) or the learning of song preferences (Riebel, 2000).

As with much else in science, the best evidence for the existence of a tradition comes from experiments in which social learning is demonstrated. Simple observation of differences in behavior patterns between groups of individuals can sometimes suggest the existence of traditions in each group. However, members of one group may also come to behave similarly to each other and differently from other groups as a result of genetic isolation or the influence of different environments. A good case in point here is in dialects. While those found in the vocalizations of many passerine bird species are known to have arisen and to persist through learning, differences between populations have also been described in groups where vocalizations are not known to be learnt (e.g., petrels: James, 1985; owls: Appleby and Redpath, 1997). Even where social learning is known to have a role in the development of calls or songs, we shall see that this may happen in various ways. For example, it may still be a matter for debate whether the individual sounds produced are memorized from the individual to whom they are matched or are selected from a pre-existing repertoire, those failing to match being discarded (Marler and Nelson, 1992).

8.2 Do group differences imply learning?

As with studies on other kinds of tradition, geographic or group variation can be an indicator that vocal traditions exist. Studies on birds and mammals have often taken such variation at face value as evidence for learned differences. However, learning is clearly not the only possible explanation for such variation. Genetic differences or differences in environmental factors are equally likely to cause variation. To demonstrate this, it is useful to look at animals in which learning is thought to be so unlikely that researchers have concentrated more on other possible explanations. In cricket frogs (*Acris crepitans*), substantial geographic variation in call structure can be found over just a few kilometers (Ryan and Wilczynski, 1991). These correlate clearly with genetic differences within the species and with different habitat types. To Ryan and Wilczynski, the most likely explanation for this variation was genetic differences, not learning. Many researchers working on birds or mammals would interpret the same correlations as evidence for vocal learning. However, there is unequivocal evidence that this kind of variation can be caused by genetic influences in birds and mammals as well. In the northern bobwhite (*Colinus virginianus*), differences in the call structure between familial lines are clearly related to genetic differences (Baker and Bailey, 1987). Medvin, Stoddard, and Beecher (1992) presented similar evidence for cliff swallows (*Hirundo pyrrhonota*). In these studies, individuals were raised with tutors from another location but still developed vocal patterns typical for their own families. As for geographic variation, the difference in repertoire size and the style of delivery of song between marsh wrens (*Cistothorus palustris*) from California and New York (Kroodsma and Canady, 1985), and the difference in the structure of squirrel monkey (*Saimiri sciureus*) isolation calls from two different populations (Lieblich *et al.*, 1980), are independent of social experience. In both cases, individuals showed the population-specific pattern even if raised in auditory isolation. There are several cases in which the initial description of variation in vocalizations has led to a closer genetic investigation and the description of cryptic species (review in Jones, 1997). Therefore, even though most learning about communication is social, it is still necessary to show that learning is involved in producing the variation between individuals if this variation is taken as an indicator of traditions within groups. In the following review, we will, therefore, only concentrate on cases in which this has been done. For general reviews of geographic variation in vocalizations in mammals and birds, see Janik and Slater (1997) and Catchpole and Slater (1995), respectively.

8.3 Forms of learning that can lead to traditions

There are three different aspects of communication that can be influenced by learning. These are usage, comprehension, and production (Janik and Slater, 2000). Usage learning and comprehension learning are about the context in which a signal is used and do not include the acquisition of novel signals. Usage learning occurs if an individual learns when to use a signal, that is the context in which to call. Comprehension learning is the equivalent on the receiver's side. It occurs if an individual associates receiving a signal with a novel context. Production learning does not involve the context of calling but describes the process in which an individual learns to produce a new signal; that is it refers to instances where the signal itself is modified in form or structure as a result of experience with signals of other individuals. This form of learning can lead to greater similarity between individuals but also to greater differences if it is used to avoid overlap with signals of other individuals.

Of these forms of learning, production learning can clearly lead to shared behavior patterns and thus to the formation of traditions. Comprehension learning, however, leads to a shared representation but not to shared behavior patterns. Therefore, we exclude it from our discussion of vocal traditions. The case of usage learning is more difficult. It leads to a shared pattern of signal usage. Can differences in the usage patterns of signals between groups be considered traditions? It is considered an important factor in different human cultures and, therefore, we will also include it in our discussion here.

We should note that it is often not easy to distinguish between different forms of learning. We discussed this problem in detail elsewhere (Janik and Slater, 2000). A researcher has to identify the existing repertoire of an individual before it can be decided whether production or usage learning was involved in the development of a specific call type. This is particularly difficult if we look at vocalizations that are made up of sequences of separate elements, as in birdsong. Here, we need to know what the minimal unit of production (MUP; Barlow, 1977) is in order to identify whether a new song represents a new unit and was acquired through production learning or whether each single element is a unit and existing units are recombined into new sequences through usage learning to produce a new song. Furthermore, it seems that, even among members of the same species, MUPs can be found at different levels. Adult male song sparrows, *Melospiza melodia*, for example, perceive whole songs as fundamental units (Searcy, Nowicki, and Peters, 1999), but young birds often combine elements from different tutors or song types to form new songs (Beecher, 1996; Marler and Peters, 1987). Given these complications, it is hardly surprising that in many cases it is not known what type of learning is used. We will try to point to the most likely scenarios in our review, keeping in mind that in many cases data on the learning process are sparse.

8.4 Usage learning

Vocal usage learning is widespread among birds and mammals, as demonstrated by experiments in which animals have been trained to give vocalizations in response to a conditioned stimulus (reviews in Adret, 1993; Janik and Slater, 1997). However, many animal calls are only given in specific contexts. In these cases, a strong genetic influence can often be found. Despite the variety of species that are capable of usage learning, it has hardly been studied in the wild. One well-studied case, however, is the use of alarm calls by vervet monkeys (*Cercopithecus aethiops*). Vervet monkeys give predator-specific alarm calls that distinguish between birds of prey, leopards, and snakes (Struhsaker, 1967). Infants often give alarm calls to stimuli that resemble some aspect of a predator, revealing their genetic predisposition (Seyfarth and Cheney, 1986). However, such stimuli can be very different from the actual predator. Infants have been seen to give bird of prey alarm calls to falling leaves. Only with time do they learn to distinguish between leaves and birds and later on among different birds of prey. This seems to be a tradition with little geographic variation (Struhsaker, 1970) as it is strongly influenced by the distribution of current predators. Another example of usage learning among mammals can be found in sperm whales (*Physeter macrocephalus*). They show clear matrilineal and geographic variation in the composition of click coda repertoires (Weilgart and Whitehead, 1997; Whitehead *et al.*, 1998). Click codas consist of a series of clicks of the same kind, but they differ in the number and repetition pattern of these clicks. Since it is only the temporal patterning and the number of clicks that are different, it seems this is an example of usage learning rather than production learning. However, few data are available on differences in the clicks themselves. Given the very stable matrilineal associations of these whales, variations between matrilines can be caused by different factors (Janik, 2001). However, given that sperm whales have been found to match arbitrary click rates (Backus and Schevill, 1966) it is likely that usage learning is involved.

Studies on chimpanzees (*Pan troglodytes*) (Clark Arcadi, 1996; Mitani *et al.*, 1992) and Japanese macaques (*Macaca fuscata*) (Green, 1975) have described variation in call parameters between populations. While non-human primates are clearly capable of usage learning and production learning in the temporal domain (review in Janik and Slater, 1997), there is some debate as to whether they can learn to alter frequency parameters. Geographic variation in chimpanzee pant hoots was primarily caused by differences in the frequency range of calls (Mitani and Brandt, 1994). Such differences in frequency parameters could be caused by production learning. However, there is no experimental evidence for production learning in chimpanzees. Mitani *et al.* (1992) argued that differences may be caused by usage learning stemming from the selective reinforcement within each population if an animal produces a sound that matches the population norm. Alternatively, provisioning by humans at both sites could have had a conditioning effect that resulted in the observed differences, again through usage learning. Mitani, Hunley, and Murdoch (1999) suspected that variations in frequency range could even be explained by size differences of the animals at the different sites, given that differences between populations were not exclusive but reflected the average frequency range used at each site. However, chimpanzees are clearly capable of usage learning involving pant hoots (Marshall, Wrangham, and Arcadi, 1999); given the additional differences between populations in call usage (Clark Arcadi, 1996), it seems clear that they have traditions of usage of calls within geographically isolated groups. A similar case of geographic variation in coo calls has been found for Japanese macaques (Green, 1975). A cross-fostering study on three individuals seemed to support the idea that Japanese macaques show production learning in the frequency domain (Masataka and Fujita, 1989). However, a subsequent more detailed cross-fostering study could find no evidence of production learning having any influence on coo call development (Owren *et al.*, 1992). As in the chimpanzee study, usage learning may explain the differences in call structure between populations (for a review of unexplained group differences in vocalizations of other primate species, see Janik and Slater (1997)).

Until a few years ago, birdsong learning was thought to be entirely a matter of memorizing songs, in some cases well before adulthood, which were then reproduced when the adult was mature, and examples of usage learning were few. A nice exception was the study by Spector, McKim, and Kroodsma (1989) on yellow warblers, a species that uses one song type in long series of repeats largely in the middle of the day, while at other times it has a variety of other song types that it switches rapidly between, particularly at dawn. Training birds with song types presented in these ways leads them to be more likely to use those particular songs in the same fashion.

Usage learning has come to the fore in recent years particularly through the notion of action-based learning, put forward by Marler and Nelson (1992, 1993). Many bird species produce a large variety of sounds during subsong but settle down to a much smaller repertoire once their final song has crystallized (e.g., Marler and Peters, 1982). Field observations suggest that the songs most likely to be discarded are those that are not shared with neighbors (Nelson, 1992, 2000). The birds thus appear to learn a large repertoire of songs early in life, but only to retain those in their repertoire with which they can usefully interact with neighbors later on. Hough, Nelson, and Volman (2000) showed that one of these species, the white-crowned sparrow (*Zonotrichia leucophrys*), is clearly capable of re-expressing songs that seemingly had been lost after the overproduction of song during vocal development. Therefore, usage learning can be an important factor in the formation and maintenance of local song traditions (Nelson, 2000). While no clear case has yet been described where the initial song repertoire is not memorized, this is also a theoretical possibility (Marler, 1997; Slater, Lachlan, and Riebel, 2000). The sorts of sound that some species can produce are very heavily constrained (Marler and Pickert, 1984), and in some cases a fixed and relatively limited repertoire of sounds has been proposed (Baker and Boylan, 1995). It is not easy to distinguish between the idea that these sounds are not influenced by production learning, with ones being selected for use depending on experience, and the alternative that the young bird memorizes the sounds themselves. However, if such lack of memorization exists, it must be rare. Birds can often be trained to produce sounds from beyond the normal species-specific range, which would indicate that production learning is at work. However, usage learning may be involved in the generation of new sequences of existing elements.

Another particularly interesting example of usage learning is in the brown-headed cowbird (*Molothrus ater*). While male cowbirds sing their own species-specific song, the species is a brood parasite so that opportunities for learning from adults are very limited. King and West (1983) found that young males housed with females of a different subspecies develop songs appropriate to that subspecies rather than their own. As the females do not sing, this was a perplexing finding. However, West and King (1988) found that the females have a display, wing stroking, that they perform in response to some male songs and that these songs are, therefore, more likely to be repeated. Consequently, just as is proposed to occur between territorial neighbors, the males produce a wide variety of sounds during subsong but then discard many of them, retaining only those that are most effective. In addition to this process, recent work has shown that the rate of song development, which also differs between subspecies, is influenced by the females (Smith, King, and West, 2000). Furthermore, young male cowbirds that are placed in a population with a different song learn this song, and this new song is then in turn passed on to their offspring, demonstrating that there is little genetic influence on song differences between populations (Freeberg, King, and West, 2001). Therefore, the case of the brown-headed cowbird demonstrates clearly that usage learning can lead to stable traditions.

An impressive final piece of evidence for usage (and production) learning comes from experiments in which African grey parrots (*Psittacus erithacus*) are trained in the laboratory. In addition to the well-known capacity of these animals to learn to produce sounds (Pepperberg, 1981), they also use them in the right context, for example by naming objects (Pepperberg, 1990). Unfortunately, little is known on parrot communication in the wild so that we cannot tell whether usage learning leads to vocal traditions in parrots.

8.5 Production learning

Production learning is relatively rare in mammals. It has only been found in pinnipeds, chiropterans, cetaceans, and humans. In pinnipeds, geographic variation of calls has been described in many species (Cleator, Sitrling, and Smith, 1989; Morrice, Burton, and Green, 1994; Terhune, 1994; Thomas and Golladay, 1995; Thomas and Stirling, 1983), but only one study describing a harbor seal (*Phoca vitulina*) mimicking human speech has provided evidence for production learning (Ralls, Fiorelli, and Gish, 1985). However, even though harbor seals appear to show geographic variation in their calls (van Parijs, Hastie, and Thompson, 2000) it is still unclear how learning influences call development.

In bats, production learning was first found in the greater horseshoe bat (*Rhinolophus ferrumequinum*), in which infants copy the acoustic frequency of their mother's echolocation call (Jones and Ransome, 1993). More detailed information on vocal tradition comes from another species, the greater spear-nosed bat (*Phyllostomus hastatus*). Females of this species live in stable groups of unrelated individuals and use group-specific screech calls (Boughman, 1997). If group composition is changed experimentally by adding a new individual, all bats re-adjust their calls, which results in increased similarity in calls among all group members including the new one (Boughman, 1998). Currently there are no data on the stability of these calls. However, given that they are transmitted horizontally, they seem to belong to this more variable and transient class of social learning described by Laland *et al.* (1993).

In cetaceans, production learning occurs in every species in which it has been investigated. However, only in the bottlenose dolphin (*Tursiops truncatus*), the killer whale (*Orcinus orca*), and the humpback whale (*Megaptera novaeangliae*) do we have information on vocal traditions in connection with production learning. Bottlenose dolphins develop individually distinctive signature whistles (Caldwell, Caldwell, and Tyack, 1990) that are used in the maintenance of group cohesion (Janik and Slater, 1998). However, they also copy each other's signature whistles, most likely to address specific individuals (Janik, 2000). Signature whistle development has received little study, but it seems that individuals learn whistles they hear and then modify them to develop their own signature whistle (Tyack, 1997). Accordingly, geographic variation in acoustic parameters of whistles can be found at sites only a few hundred kilometers apart (Wang, Würsig, and Evans, 1995). However, there is no information on the stability of local traditions.

Killer whales off British Columbia have been reported to use pod-specific call repertoires that are thought to be vocal traditions (Ford and Fisher, 1983). Miller and Bain (2000) found that within-pod variation in calls correlated with matrilineal relatedness. Genetic evidence shows that mating is rare within pods but frequent between different pods that do not share calls (Barrett-Lennard, 2000). Therefore, learning is the most likely cause for within-pod variation in call structure. This makes it also more likely that interpod differences are influenced by learning. Deecke, Ford, and Spong (2000) found that the acoustic structure of one shared call type produced by two different pods changed significantly over a period of 12 years. In this process, the rate of divergence between the groups was lower than the rate of modification. Such parallel changes between groups could have been caused by maturational processes. However, a second call type did not show any change in either pod, suggesting that this kind of drift is influenced by learning.

Male humpback whales produce elaborate songs in their breeding season that are clearly influenced by production learning (Janik and Slater, 1997). All males within a population sing the same song at any one time (Payne and Payne, 1985), but songs of isolated populations show hardly any similarities (Winn *et al.*, 1981). However, songs are not very stable since the common song changes considerably over just one singing season (Payne, Tyack, and Payne, 1983). The most dramatic change has been reported from Australia. Humpback whale song off the west coast differs greatly from that of the east coast. Usually there is little migration between these two populations. However, Noad *et al.* (2001) found that virtually all humpbacks from the east coast changed their song to that of west-coast animals within one season after a few individuals migrated from the west to the east. Thus, transmission is clearly horizontal.

Among birds, there are three groups in which learning plays a role in song development: hummingbirds (Apodiformes), parrots (Psittaciformes), and the true songbirds (oscine Passeriformes). Between them, these amount to more than half the current species of birds, around 5000. Production learning, in which sounds of other individuals are memorized, has been found in all cases in these groups that have been analyzed in detail, except for the grey catbird (*Dumetella carolinensis*), where individuals develop highly varied songs but these seem to be invented rather than based on ones they have experienced (Kroodsma *et al.*, 1997). Where sounds copied from other individuals are memorized and later produced, a vocal tradition may become established, its longevity depending on the probability that that particular song is copied and on the accuracy of copying. The identification of such traditions is eased where learning takes place after dispersal, for example from territorial neighbors, so that particular traditions tend to persist in a given locality. A good example here is the village indigobird (*Vidua chalybeata*), in which birds form leks within which the numerous song types are shared by all the members. On the one hand, occasional birds that move from one lek to another in adulthood alter their songs to match those of the group that they are joining (Payne, 1985). On the other hand, where birds learn as juveniles and then disperse before breeding, they may sing songs which, though accurately copied elsewhere, bear little similarity to those round about them (e.g., Slater and Ince, 1979).

Many birds have repertoires of song types and in some cases whole repertoires are learnt as a package (e.g., corn bunting, *Miliaria calandra*: McGregor, 1980; McGregor and Thompson, 1988; but see Latruffe *et al.*, 2000). More usually, birds copy different songs from different individuals so they may end up with a mixture of song types that differs from the repertoire of any other bird in the population (Slater, Ince, and Colgan, 1980). In short-toed treecreepers (*Certhia brachydactyla*), birds have been described as learning each song, not from a single other individual but by blending the characteristics of several (Thielcke, 1987). Such an averaging process would lead to greater conservatism.

Most of the evidence we have points to traditions in birdsong deriving from random processes (e.g., Chilton and Lein, 1996; Payne, Payne, and Doehlert, 1988). Occasional transcription errors or immigration lead to new song types being introduced into the population, while other song types fail to be copied and thus become extinct. These two processes of introduction and extinction balance out so that there is a gradual turnover of the songs present but the variety remains the same. As most birds learn their songs only as juveniles or young adults, the rate of change depends very much on turnover in the population though changes within and between seasons, similar to those described above for humpback whales, have been described in thrush nightingales (*Luscinia luscinia*) by Sorjonen (1987). The most detailed studies, by Lynch *et al.* (1989) on chaffinches and Payne (1996) on indigo buntings (*Passerina cyanea*), find no evidence for any systematic change, but simply turnover of the songs present. The only case described so far of directional change, suggesting some songs are favored over others, is in the Darwin's medium ground finches (*Geospiza fortis*) studied by Gibbs (1990). Over the course of six years, the commonest song type in the population became rarer, while the three less-common types became commoner. It appeared that males singing the rarer types survived better and produced more male offspring that survived to join the breeding population. Darwin's finches are among the few species of birds in which sons normally learn their songs from their fathers. Just why possessing a rare song should lead to greater longevity and fecundity remains obscure: if this were generally true, presumably rare songs would become commoner, and common ones rarer, until all were equal in frequency, and this was certainly far from the case at the start of Gibbs' (1990) study. The population being a closed one, we can discount immigration and, while rarity on its own could easily be achieved by innovation, no new song types were recorded in the course of Gibbs' study. Again this raises the issue of what it was about the rarer song types in this population that gave them an advantage.

Despite the apparent role of random processes in the development of traditions, there are cases where traditions seem to be connected to functional aspects of communication. As in mammals, some bird species form group-specific calls, which are shared by all group members but also show change over time. The contact calls of male budgerigars (*Melopsittacus undulatus*) placed in a group converge over the course of a few weeks (Farabaugh, Linzenbold, and Dooling, 1994). When a new bird is added to an established group, its call changes to match that shared by the others (Bartlett and Slater, 1999). The calls of female budgerigars also converge in groups (Hile and Striedter, 2000), and pairs also match their calls, but here because the male call is modified to match that of the female (Hile, Plummer, and Striedter, 2000) The horizontal transmission of calls and the flexibility of contact call systems are likely to make their traditions rather unstable, but little is known about long-term changes in calls. Another case of group-specific calls is in the yellow-rumped cacique (*Cacicus cela*) (Feekes, 1982). Here, songs change rapidly both within and between seasons, as in humpback whales (Trainer, 1989). That the changes are gradual and in constant directions suggests a benefit to adopting certain songs, rather than drift or error in copying.

8.6 The rate of change in vocal traditions

As we have seen, vocal traditions can have very different rates of change. They can be an important factor in the survival of individuals, as in the appropriate use of alarm calls. Such highly relevant information only changes if the predator distribution changes, for the obvious reason that the relatives of an inappropriate user would not be around for very long. However, other information does not seem to be as vital, and this is where we can observe change over time for no apparent reason. If copying errors do not lead to a decrease in reproductive success, they can explain the origins and some of the changes in vocal traditions that are observed in bird and mammal populations. Another factor that may influence stability is the extent to which learning has an influence on call development.

Unfortunately, there is very little information on the rate of change in vocal traditions of mammals. Terhune (1994) recorded harp seals (*Phoca groenlandica*) at the same location on occasions 18–20 years apart. A comparison only showed very slight differences in calls, which may have been caused by sampling errors. Similarly, Deecke *et al.* (2000) showed very slight changes in one shared call type for killer whales over a period of 12 years, while another call type did not change at all. In neither of these species has production learning been demonstrated, even though it seems likely for the killer whale (see above). The low rate of change may indicate that production learning is of little importance in call development. However, killer whales and harp seals are very long lived and the seemingly long intervals between recordings may have not been long enough to capture changes from one generation to the next. Payne and Payne (1985) provided a detailed study of changes in humpback whale song over 19 years. One song is made up of three to nine different themes, which consist of repeated sequences of elements called phrases. If songs from different years are compared, some interesting patterns emerge. Each song consists of material that is unique to that year, but most songs also include some thematic material that is not unique to that year. Humpback whales also sing old song themes more slowly than newer ones. While songs from subsequent years tend to share a lot of song themes, drastic changes from one year to the next can occur, with the most drastic case being the one described for Australia above. On average, 63% of all themes were shared in songs of subsequent years, 20% in songs separated by 2 to 11 years, and nothing was shared in songs from 15 or more years apart.

Vocal traditions in birds vary immensely in their longevity. In some cases, substantial changes have been recorded from year to year (Avery and Oring, 1977; Sorjonen, 1987; Trainer, 1989). Equally, the songs in an area may remain substantially the same over a decade or more (Bradley, 1994; Dixon, 1969; Thielcke, 1987). At an even greater extreme, Sorjonen (2001) describes two dialects of the chaffinch rain call on the border between Finland and Russia that seemed much the same in structure and distribution as those described in the same area over a century earlier. This is a male call and he attributed this persistence to site fidelity of males. In the white-crowned sparrow, Harbison, Nelson and Hahn (1999) described four populations in two of which there had been little change in song over 26 years while in the other two it had changed a great deal. The latter involved small populations in which it is argued syllables are more likely to go extinct and newly introduced ones to spread. A final, remarkably persistent, tradition is that of rufous-collared sparrow (*Zonotrichia capensis*) song in agricultural areas of the pampas of Argentina. This species occurs in a wide variety of habitats and its song tends to be matched to the habitat: in particular, the trill that it includes is slower in densely wooded areas, where faster ones would tend to be distorted by reverberation. In agricultural areas, however, trill rate varies considerably, and Handford (1981) found that they were appropriate to the habitat that had been present in the area 100 or more years ago, before the introduction of agriculture. One reason Handford suggests for this conservatism is that rufous-collared sparrows are among the few species that breed in farmland, and they occur there at very high densities. This habitat does not present such constraints on trill rate as woodland, and as the birds are close together the habitat characteristics are less important to sound transmission. The young birds will also have little difficulty in hearing models to copy. The persistence of a particular song, therefore, probably stems from lack of pressure to change combined with very high fidelity of copying.

Copying fidelity is likely to be a very important factor in the duration of traditions. Laboratory experiments suggest that song types can be copied accurately from a tutor even without a great many repeats. Petrinovich (1985) found that white-crowned sparrows would not learn from less than 120 repeats, but two birds did learn songs heard only 256 times, which is not a large number compared with the hundreds of repetitions per day common in singing birds. More striking, however, is the finding of accurate copying by nightingales (*Luscinia megarhynchos*) of songs heard only 10–20 times (Hultsch and Todt, 1989a,b). Therefore, the potential is there for extremely persistent traditions. However, the situation in the wild may be quite different. First, provided the song conforms to species-specific constraints, the pressure to get the copy absolutely right may not be great. Second, opportunities to copy may be much more limited. Young chaffinches sometimes hatch in early July when the adult males around them are no longer singing, so they may have no opportunity to memorize song in their first summer. While they can memorize either at this stage or in the following spring (Slater and Ince, 1982), opportunities in this next season may also be limited, for example if they set up territory in a small patch of habitat where there are no neighbors. It is perhaps remarkable that it is rare to hear a bird singing an untutored song.

Detailed studies of particular populations over time give some indication of the rate of change that is occurring. Payne *et al.* (1981) described a population of indigo buntings in terms of the "half life" of particular song types. Indigo bunting songs consist of a variable number of phrases within which a particular syllable is repeated a number of times. Traditions exist where the same series of phrases occurs repeatedly, something that would be very unlikely to happen by chance. As the number of phrases varies between songs, Payne *et al.* (1981) decided to look at strings of three and found that these averaged a half-life of 3.8 years, though some persisted for the full 15 year duration of the study. While this is impressive evidence for a persistent tradition, the exact result depends on their choice of three phrases as the cultural unit for analysis: shorter modules would presumably have persisted for longer and have been more likely to arise by chance, while longer modules would have certainly not lasted so long.

In chaffinches, songs fall into clear types of fixed structure; consequently a traditional unit is easier to define. Two separate lines of evidence converge on the idea that 85% of songs are accurate copies of songs already present in the neighborhood while 15% are either introduced from elsewhere or miscopies. Because of the possibility that new song types arise by immigration, the 15% figure is a maximum cultural mutation rate. The first evidence for it came from studies of the distribution of song types in a population (Slater *et al.*, 1980). Many songs are unique to the individual singing them, while, at the other extreme, some song types are shared by around half the birds in the wood. Computer simulations with various error rates pointed to 15% being that most likely to lead to such a distribution. At 10%, song types tended to be more widely shared, and at 20% a higher proportion were unshared. The second line of evidence comes from a snapshot of the same population in two summers 18 years apart (Ince, Slater, and Weismann, 1980). Only three song types were common to the two sets of recordings, out of a total of 22 recorded in one year and 35 in the other. This is closely matched to the rate of change expected from a 15% mutation rate.

As with these examples, most studies of birdsong suggest that changes over time are attributable to the gradual accumulation of copying errors (see, for example, the cases reviewed by Lynch (1996)). In some cases, however, like humpback whales or caciques, change is so rapid that it seems unlikely to be caused by error. The reasons for such accelerated change are still unclear. A likely explanation is some sort of run-away process. Examples are if intruders learn a group-specific call rapidly or if conformity in mating signals is necessary to stimulate females but slight differences bring a reproductive advantage for individual males.

8.7 The biological significance of vocal traditions

Traditions have been a major research interest of ornithologists ever since Thorpe (1958) discovered the large extent to which learning influences song development in chaffinches. In one of the first studies on animal traditions, Nicolai studied family traditions in songs of bullfinches as early as 1959. The research on vocal traditions in birds has mainly focused on their description and the mechanisms involved in their development and maintenance. Given this long history, we know a lot more about vocal traditions in birds than we do for most other animal traditions. These data allow us to look at patterns beyond those exhibited by single species.

Traditions can be split into those that concern social behavior and those that do not. Traditions of nonsocial behavior patterns, like tool use or dietary habits, enable a species to conquer an otherwise inaccessible habitat by allowing the exploitation of new food sources (e.g., Terkel, 1996) or make feeding more time efficient, which in turn can increase population density and/or free up time for other activities. This extra time may be an important support for the evolution of more complex social interactions (Byrne, 1995). Traditions of social behavior including vocal traditions can isolate individuals or groups from outsiders or improve transmission in specific habitats. This isolating mechanism can act within populations on the group level or on a larger scale between populations. The best descriptions of group-specific calls come from bats (Boughman, 1998) and budgerigars (Bartlett and Slater, 1999; Farabaugh *et al.*, 1994). These traditions are carefully maintained and adjusted to include new group members. Unfortunately, we know little about reactions to outsiders in these groups. On a population level, vocal traditions used in mate choice may help to maintain co-adapted gene complexes that represent local adaptations (Baker and Cunningham, 1985; Nottebohm, 1969). This idea has received a lot of attention in birdsong research (review in Catchpole and Slater, 1995). In some species, there are correlations between genetic and cultural variation (Balaban, 1988), while in others there are not (Lougheed and Handford, 1992; Lougheed, Handford, and Baker, 1993). However, even Balaban (1988) pointed out that such a correlation need not indicate a causal relationship. Consequently, even though the idea is intriguing, there is no good evidence supporting it. Finally, group differences may arise because of errors in the copying process. Once two groups of animals are sufficiently isolated over time, such errors can lead to progressive and divergent change in their vocal repertoires. It has been argued that this was the main reason for vocal traditions in some bird species (Andrew, 1962; Bitterbaum and Baptista, 1979; Wiens, 1982). It may also explain some of the mammalian cases, such as the group-specific repertoires of killer whales (Ford and Fisher, 1983). However, even such by-product traditions can eventually lead to reproductive isolation if they diverge far enough. In killer whales, this has been proposed for the so-called transient and resident groups, which are sympatric but do not interbreed (Baird, Abrams, and Dill, 1992). A similar argument has been put forward for the evolution of Darwin's finches (Grant and Grant, 1996). Interestingly, computer simulations have shown that once vocal learning has evolved it is very unlikely to disappear again, even if it loses its original function and lowers the average fitness of the population (Lachlan and Slater, 1999). The fact that functional explanations for vocal traditions are often hard to find makes the idea that some species are currently in this cultural trap even more appealing. Studies on vocal traditions should consider this possibility even if it will be difficult to establish.

8.8 Acknowledgements

This chapter was written with the support of a Marie Curie Fellowship of the European Community program *Improving Human Research Potential and the Socio-economic Knowledge Base* under contract number HPMF-CT-2000-00510 and a Royal Society University Research Fellowship to VMJ.

References

Adret, P. 1993. Vocal learning induced with operant techniques: an overview. *Netherlands Journal of Zoology*, **43**, 125–142.

Andrew, R. J. 1962. Evolution of intelligence and vocal mimicking. *Science*, **137**, 585–589.

Appleby, B. M. and Redpath, S. M. 1997. Variation in the male territorial hoot of the tawny owl *Strix aluco* in three English populations. *Ibis*, **139**, 152–158.

Avery, M. and Oring, L. W. 1977. Song dialects in the bobolink (*Dolichonyx oryzivorus*). *Condor*, **79**, 113–118.

Backus, R. H. and Schevill, W. E. 1966. Physeter clicks. In *Whales, Dolphins, and Porpoises*, ed. K. S. Norris, pp. 510–528. Berkeley, CA: University of California Press.

Baird, R. W., Abrams, P. A., and Dill, L. M. 1992. Possible indirect interactions between transient and resident killer whales: implications for the evolution of foraging specializations in the genus *Orcinus*. *Oecologia*, **89**, 125–132.

Balaban, E. 1988. Cultural and genetic variation in swamp sparrows (*Melospiza georgiana*) I. Song variation, genetic variation, and their relationship. *Behaviour*, **105**, 250–291.

Baker, J. A. and Bailey, E. D. 1987. Sources of phenotypic variation in the separation call of northern bobwhite (*Colinus virginianus*). *Canadian Journal of Zoology*, **65**, 1010–1015.

Baker, J. A. and Cunningham, M. A. 1985. The biology of bird-song dialects. *Behavioral and Brain Sciences*, **8**, 85–133.

Baker, M. C. and Boylan, J. T. 1995. A catalog of song syllables of indigo and lazuli buntings. *Condor*, **97**, 1028–1040.

Barlow, G. W. 1977. Modal action patterns. In *How Animals Communicate*, ed. T. A. Sebeok, pp. 98–134. Bloomington, IA: Indiana University Press.

Bartlett, P. and Slater, P. J. B. 1999. The effect of new recruits on the flock specific call of budgerigars (*Melopsittacus undulatus*). *Ethology, Ecology and Evolution*, **11**, 139–147.

Barrett-Lennard, L. G. 2000. Population structure and mating patterns of killer whales (*Orcinus orca*) as revealed by DNA analysis. PhD Thesis, University of British Columbia.

Beecher, M. D. 1996. Birdsong learning in the laboratory and field. In *Ecology and Evolution of Acoustic Communication in Birds*, ed. D. E. Kroodsma and E. H. Miller, pp. 61–78. Ithaca: Comstock Publishing.

Bitterbaum, E. and Baptista, L. F. 1979. Geographical variation in songs of California house finches (*Carpodacus mexicanus*). *Auk*, **96**, 462–474.

Boughman, J. W. 1997. Greater spear-nosed bats give group-distinctive calls. *Behavioral Ecology and Sociobiology*, **40**, 61–70.

Boughman, J. W. 1998. Vocal learning by greater spear-nosed bats. *Proceedings of the Royal Society of London, Series B*, **265**, 227–233.

Bradley, R. A. 1994. Cultural change and geographic variation in the songs of the Belding's savannah sparrow (*Passerculus sandwichensis belgingi*). *Bulletin of the South Californian Academy of Science*, **93**, 91–109.

Byrne, R. 1995. *The Thinking Ape: Evolutionary Origins of Intelligence*. Oxford: Oxford University Press.

Caldwell, M. C., Caldwell, D. K., and Tyack, P. L. 1990. Review of the signature-whistle-hypothesis for the Atlantic bottlenose dolphin. In *The Bottlenose Dolphin*, ed. S. Leatherwood and R. R. Reeves, pp. 199–234. San Diego, CA: Academic Press.

Catchpole, C. K. and Slater, P. J. B. 1995. *Bird Song: Biological Themes and Variations*. Cambridge: Cambridge University Press.

Chilton, G. and Lein, M. R. 1996. Long-term changes in songs and song dialect boundaries of Puget sound white-crowned sparrows. *Condor*, **98**, 567–580.

Clark Arcadi, A. 1996. Phrase structure of wild chimpanzee pant hoots: patterns of production and interpopulation variability. *American Journal of Primatology*, **39**, 159–178.

Cleator, H. J., Stirling, I., and Smith, T. G. 1989. Underwater vocalizations of the bearded seal (*Erignathus barbatus*). *Canadian Journal of Zoology*, **67**, 1900–1910.

Deecke, V. B., Ford, J. K. B., and Spong, P. 2000. Dialect change in resident killer whales: implications for vocal learning and cultural transmission. *Animal Behaviour*, **60**, 629–638.

Dixon, K. L. 1969. Patterns of singing in a population of the plain titmouse. *Condor*, **71**, 94–101.

Farabaugh, S. M., Linzenbold, A., and Dooling, R. J. 1994. Vocal plasticity in budgerigars (*Melopsittacus undulatus*) – evidence for social factors in the learning of contact calls. *Journal of Comparative Psychology*, **108**, 81–92.

Feekes, F. 1982. Song mimesis within colonies of *Cacicus c. cela* (Icteridae, Aves). A colonial password? *Zeitschrift für Tierpsychologie*, **58**, 119–152.

Ford, J. K. B. and Fisher, H. D. 1983. Group-specific dialects of killer whales (*Orcinus orca*) in British Columbia. In *Communication and Behavior of Whales*, ed. R. Payne, pp. 129–161. Boulder, CO: Westview Press.

Freeberg, T. M., King, A. P., and West, M. J. 2001. Cultural transmission of vocal traditions in cowbirds (*Molothrus ater*) influences courtship patterns and mate preference. *Journal of Comparative Psychology*, **115**, 201–211.

Gibbs, H. L. 1990. Cultural evolution of male song types in Darwin's medium ground finches, *Geospiza fortis*. *Animal Behaviour*, **39**, 253–263.

Grant, B. R., and Grant, P. R. 1996. Cultural inheritance of song and its role in the evolution of Darwin's finches. *Evolution*, **50**, 2471–2487.

Green, S. 1975. Dialects in japanese monkeys: vocal learning and cultural transmission of locale-specific vocal behavior? *Zeitschrift für Tierpsychologie*, **38**, 304–314.

Handford, P. 1981. Vegetational correlates of variation in the song of *Zonotrichia capensis*. *Behavioral Ecology and Sociobiology*, **8**, 203–206.

Harbison, H., Nelson, D. A., and Hahn, T. P. 1999. Long-term persistence of song dialects in the mountain white-crowned sparrow. *Condor*, **101**, 133–148.

Hile, A. G. and Striedter, G. F. 2000. Call convergence within groups of female budgerigars (*Melopsittacus undulatus*). *Ethology*, **106**, 1105–1114.

Hile, A. G., Plummer, T. K., and Striedter, G. F. 2000. Male vocal imitation produces call convergence during pair bonding in budgerigars, *Melopsittacus undulatus*. *Animal Behaviour*, **59**, 1209–1218.

Hough, G. E., Nelson, D. A., and Volman, S. F. 2000. Re-expression of songs deleted during vocal development in white-crowned sparrows, *Zonotrichia leucophrys*. *Animal Behaviour*, **60**, 279–287.

Hultsch, H. and Todt, D. 1989a. Memorization and reproduction of songs in nightingales: evidence for package formation. *Journal of Comparative Physiology A*, **165**, 197–203.

Hultsch, H. and Todt, D. 1989b. Song acquisition and acquisition constraints in nightingales, *Luscinia megarhynchos*. *Naturwissenschaften*, **76**, 83–85.

Ince, S. A., Slater, P. J. B., and Weismann, C. 1980. Changes with time in the songs of a population of chaffinches. *Condor*, **82**, 285–290.

James, P. C. 1985. Geographical and temporal variation in the calls of the manx shearwater *Puffinus puffinus* and the British storm petrel *Hydrobates pelagicus*. *Journal of Zoology*, **207**, 331–344.

Janik, V. M. 2000. Whistle matching in wild bottlenose dolphins (*Tursiops truncatus*). *Science*, **289**, 1355–1357.

Janik, V. M. 2001. Is cetacean social learning unique? *Behavioral and Brain Sciences*, **24**, 337–338.

Janik, V. M. and Slater, P. J. B. 1997. Vocal learning in mammals. *Advances in the Study of Behavior*, **26**, 59–99.

Janik, V. M. and Slater, P. J. B. 1998. Context-specific use suggests that bottlenose dolphin signature whistles are cohesion calls. *Animal Behaviour*, **56**, 829–838.

Janik, V. M. and Slater, P. J. B. 2000. The different roles of social learning in vocal communication. *Animal Behaviour*, **60**, 1–11.

Jones, G. 1997. Acoustic signals and speciation: the roles of natural and sexual selection in the evolution of cryptic species. *Advances in the Study of Behavior*, **26**, 317–354.

Jones, G. and Ransome, R. D. 1993. Echolocation calls of bats are influenced by maternal effects and change over a lifetime. *Proceedings of the Royal Society of London, Series B*, **252**, 125–128.

King, A. P. and West, M. J. 1983. Epigenesis of cowbird song – a joint endeavour of males and females. *Nature*, **305**, 704–706.

Kroodsma, D. E. and Canady, R. A. 1985. Differences in repertoire size, singing behavior, and associated neuroanatomy among marsh wren populations have a genetic basis. *Auk*, **102**, 439–446.

Kroodsma, D. E. Houlihan, P. W., Fallon, P. A., and Wells, J. A. 1997. Song development in grey catbirds. *Animal Behaviour*, **54**, 457–464.

Lachlan, R. F. and Slater, P. J. B. 1999. The maintenance of vocal learning by gene–culture interaction: the cultural trap hypothesis. *Proceedings of the Royal Society of London, Series B*, **266**, 701–706.

Laland, K. N., Richerson, P. J., and Boyd, R. 1993. Animal social learning: toward a new theoretical approach. *Perspectives in Ethology*, **10**, 249–277.

Latruffe, C., McGregor, P. K., Tavares, J. P., and Mota, P. G. 2000. Microgeographic variation in corn bunting (*Miliaria calandra*) song: quantitative and discrimination aspects. *Behaviour*, **137**, 1241–1255.

Lieblich, A. K., Symmes, D., Newman, J. D., and Shapiro, M. 1980. Development of the isolation peep in laboratory-bred squirrel monkeys. *Animal Behaviour*, **28**, 1–9.

Lougheed, S. C. and Handford, P. 1992. Vocal dialects and the structure of geographic variation in morphological and allozymic characters in the rufous-collared sparrow, *Zonotrichia capensis*. *Evolution*, **45**, 1443–1456.

Lougheed, S. C., Handford, P., and Baker, A. J. 1993. Mitochondrial DNA hyperdiversity, vocal dialects and subspecies in the rufous-collared sparrow (*Zonotrichia capensis*). *Condor,* **95**, 889–895.

Lynch, A. 1996. The population memetics of birdsong. In *Ecology and Evolution of Acoustic Communication in Birds*, ed. D. E. Kroodsma and E. H. Miller, pp. 181–197. Ithaca: Comstock Publishing.

Lynch, A., Plunkett, G. M., Baker, A. J., and Jenkins, P. F. 1989. A model of cultural evolution of chaffinch song derived with the meme concept. *American Naturalist,* **133**, 634–653.

Marler, P. 1997. Three models of song learning: evidence from behavior. *Journal of Neurobiology,* **33**, 501–516.

Marler, P. and Nelson, D. 1992. Neuroselection and song learning in birds: species universals in culturally transmitted behavior. *Seminars in the Neurosciences,* **4**, 415–423.

Marler, P. and Nelson, D. A. 1993. Action-based learning: a new form of developmental plasticity in bird song. *Netherlands Journal of Zoology,* **43**, 91–103.

Marler, P. and Peters, S. 1982. Developmental overproduction and selective attrition: new processes in the epigenesis of bird song. *Developmental Psychobiology,* **15**, 369–378.

Marler, P. and Peters, S. 1987. A sensitive period for song acquisition in the song sparrow, *Melospiza melodia*: a case of age-limited learning. *Ethology,* **76**, 89–100.

Marler, P. and Pickert, R. 1984. Species-universal microstructure in a learned birdsong: the swamp sparrow (*Melospiza georgiana*). *Animal Behaviour,* **32**, 673–689.

Marshall, A. J., Wrangham, R. W., and Arcadi, A. C. 1999. Does learning affect the structure of vocalizations in chimpanzees? *Animal Behaviour,* **58**, 825–830.

Masataka, N. and Fujita, K. 1989. Vocal learning of Japanese and rhesus monkeys. *Behaviour,* **109**, 191–199.

McGregor, P. K. 1980. Song dialects in the corn bunting (*Emberiza calandra*). *Zeitschrift für Tierpsychologie,* **54**, 285–297.

McGregor, P. K. and Thompson, D. B. A. 1988. Constancy and change in local dialects of the corn bunting. *Ornis Scandinavia,* **19**, 153–159.

Medvin, M. B., Stoddard, P. K., and Beecher, M. D. 1992. Signals for parent–offspring recognition: strong sib–sib call similarity in cliff swallows but not barn swallows. *Ethology,* **90**, 17–28.

Miller, P. J. O. and Bain, D. E. 2000. Within-pod variation in the sound production of a pod of killer whales, *Orcinus orca*. *Animal Behaviour,* **60**, 617–628.

Mitani, J. and Brandt, K. 1994. Social factors influence the acoustic variability in the long-distance calls of male chimpanzees. *Ethology,* **96**, 233–252.

Mitani, J. C., Hunley, K. L., and Murdoch, M. E. 1999. Geographic variation in the calls of wild chimpanzees: a reassessment. *American Journal of Primatology,* **47**, 133–151.

Mitani, J. C., Hasegawa, T., Gros-Louis, J., Marler, P., and Byrne, R. 1992. Dialects in wild chimpanzees? *American Journal of Primatology,* **27**, 233–243.

Morrice, M. G., Burton, H. R., and Green, K. 1994. Microgeographic variation and songs in the underwater vocalization repertoire of the Weddell seal (*Leptonychotes weddelli*) from the Vestfold Hills, Antarctica. *Polar Biology,* **14**, 441–446.

Nelson, D. A. 1992. Song overproduction and selective attrition lead to song sharing in the field sparrow (*Spizella pusilla*). *Behavioral Ecology and Sociobiology,* **30**, 415–424.

Nelson, D. A. 2000. Song overproduction, selective attrition and song dialects in the white-crowned sparrow. *Animal Behaviour,* **60**, 887–898.

Nicolai, J. 1959. Familientradition in der Gesangsentwicklung des Gimpels (*Pyrrhula pyrrhula* L.). *Journal für Ornithologie,* **100**, 39–46.

Noad, M. J., Cato, D. H., Bryden, M. M., Jenner, M. N., and Jenner, K. C. S. 2001. Cultural revolution in whale song. *Nature,* **408**, 537.

Nottebohm, F. 1969. The song of the chingolo (*Zonotrichia capensis*) in Argentina: description and evaluation of a system of dialects. *Condor,* **71**, 299–315.

Owren, M. J., Dieter, J. A., Seyfarth, R. M., and Cheney, D. L. 1992. "Food" calls produced by adult female rhesus (*Macaca mulatta*) and Japanese (*M. fuscata*) macaques, their normally-raised offspring, and offspring cross-fostered between species. *Behaviour,* **120**, 218–231.

Payne, K. and Payne, R. 1985. Large scale changes over 19 years in songs of humpback whales in Bermuda. *Zeitschrift für Tierpsychologie,* **68**, 89–114.

Payne, K., Tyack, P., and Payne, R. 1983. Progressive changes in the songs of humpback whales (*Megaptera novaeangliae*): a detailed analysis of two seasons in Hawaii. In *Communication and Behavior of Whales*, ed. R. Payne, pp. 9–57. Boulder, CO: Westview Press.

Payne, R. B. 1985. Behavioral continuity and change in local song populations of village indigobirds *Vidua chalybeata*. *Zeitschrift für Tierpsychologie,* **70**, 1–44.

Payne, R. B. 1996. Song traditions in indigo buntings: origin, improvisation, dispersal, and extinction in cultural evolution. In *Ecology and Evolution of Acoustic Communication in Birds*, ed. D. E. Kroodsma and E. H. Miller, pp. 198–220. Ithaca: Comstock Publishing.

Payne, R. B., Thompson, W. L., Fiala, K. L., and Sweany, L. L. 1981. Local song traditions in indigo buntings: cultural transmission of behavior patterns across generations. *Behaviour,* **77**, 199–221.

Payne, R. B., Payne, L. L., and Doehlert, S. M. 1988. Biological and cultural success of song memes in indigo buntings. *Ecology,* **69**, 104–117.

Pepperberg, I. M. 1981. Functional vocalizations by an African grey parrot. *Zeitschrift für Tierpsychologie,* **55**, 139–160.

Pepperberg, I. M. 1990. Some cognitive capacities of an African grey parrot (*Psittacus erithacus*). *Advances in the Study of Behavior,* **19**, 357–409.

Petrinovich, L. 1985. Factors influencing song development in the white-crowned sparrow (*Zonotrichia leucophrys*). *Journal of Comparative Psychology,* **99**, 15–29.

Ralls, K., Fiorelli, P., and Gish, S. 1985. Vocalizations and vocal mimicry in captive harbor seals, *Phoca vitulina*. *Canadian Journal of Zoology,* **63**, 1050–1056.

Riebel, K. 2000. Early exposure leads to repeatable preferences for male song in female zebra finches. *Proceedings of the Royal Society of London, Series B,* **267**, 2553–2558.

Ryan, M. and Wilczynski, F. 1991. Evolution of intraspecific variation in advertisement calls. *Biological Journal of the Linnean Society,* **44**, 249–271.

Searcy, W. A., Nowicki, S., and Peters, S. 1999. Song types as fundamental units in vocal repertoires. *Animal Behaviour,* **58**, 37–44.

Seyfarth, R. and Cheney, D. 1986. Vocal development in vervet monkeys. *Animal Behaviour,* **34**, 1640–1665.

Slater, P. J. B. and Ince, S. A. 1979. Cultural evolution in chaffinch song. *Behaviour,* **71**, 146–166.

Slater, P. J. B. and Ince, S. A. 1982. Song development in chaffinches: what is learnt and when? *Ibis*, **124**, 21–26.

Slater, P. J. B. Ince, S. A., and Colgan, P. W. 1980. Chaffinch song types: their frequencies in the population and distribution between the repertoires of different individuals. *Behaviour*, **75**, 207–218.

Slater, P. J. B., Lachlan, R. F., and Riebel, K. 2000. The significance of learning in signal development: The curious case of the chaffinch. In *Animal Signals: Signalling and Signal Design in Animal Communication*, ed. Y. Espmark, T. Amundsen, and G. Rosenqvist, pp. 341–352. Trondheim, Norway: Tapir.

Smith, V. A., King, A. P., and West, M. J. 2000. A role of her own: female cowbirds, *Molothrus ater*, influence the development and outcome of song learning. *Animal Behaviour*, **60**, 599–609.

Sorjonen, J. 1987. Temporal and spatial differences in traditions and repertoires in the song of the thrush nightingale (*Luscinia luscinia*). *Behaviour*, **102**, 196–212.

Sorjonen, J. 2001. Long-term constancy of two rain-call dialects of the chaffinch *Fringilla coelebs* in Finnish and Russian Karelia: a consequence of site fidelity? *Ornis Fennica*, **78**, 73–82.

Spector, D. A., McKim, L. K., and Kroodsma, D. E. 1989. Yellow warblers are able to learn songs and situations in which to use them. *Animal Behaviour*, **38**, 723–725.

Struhsaker, T. T. 1967. Auditory communication among vervet monkeys (*Cercopithecus aethiops*). In *Social Communication among Primates*, ed. S. A. Altmann, pp. 281–324. Chicago, IL: Chicago University Press.

Struhsaker, T. T. 1970. Phylogenetic implications of some vocalizations of *Cercopithecus* monkeys. In *Old World Monkeys: Evolution, Systematics, and Behavior*, ed. J. R. Napier and P. H. Napier, pp. 365–444. New York: Academic Press.

Terhune, J. M. 1994. Geographical variation of harp seal underwater vocalizations. *Canadian Journal of Zoology*, **72**, 892–897.

Terkel, J. 1996. Cultural transmission of feeding behavior in the black rat (*Rattus rattus*). In *Social Learning in Animals: The Roots of Culture*, ed. C. M. Heyes and B. G. Galef Jr., pp. 17–47. San Diego, CA: Academic Press.

Thielcke, G. 1987. Langjährige Dialektkonstanz beim Gartenbaumläufer (*Certhia brachydactyla*). *Journal für Ornithologie*, **128**, 171–180.

Thomas, J. A. and Golladay, C. L. 1995. Geographic variation in leopard seal (*Hydrurga leptonyx*) underwater vocalizations. In *Sensory Systems of Aquatic Mammals*, ed. R. A. Kastelein, J. A. Thomas and P. E. Nachtigall, pp. 201–221. Woerden: De Spil.

Thomas, J. A. and Stirling, I. 1983. Geographic variation in the underwater vocalizations of Weddell seals (*Leptonychotes weddelli*) from Palmer Peninsula and McMurdo Sound, Antarctica. *Canadian Journal of Zoology*, **61**, 2203–2212.

Thorpe, W. H. 1958. The learning of song patterns by birds, with especial reference to the song of the chaffinch *Fringilla coelebs*. *Ibis*, **100**, 535–570.

Trainer, J. M. 1989. Cultural evolution in song dialects of yellow-rumped caciques in Panama. *Ethology*, **80**, 190–204.

Tyack, P. L. 1997. Development and social functions of signature whistles in bottlenose dolphins *Tursiops truncatus*. *Bioacoustics*, **8**, 21–46.

van Parijs, S. M., Hastie, G. D., and Thompson, P. M. 2000. Individual and geographic variation in display behaviour of male harbour seals in Scotland. *Animal Behaviour*, **59**, 559–568.

Wang, D., Würsig, B., and Evans, W. E. 1995. Whistles of bottlenose dolphins: comparisons among populations. *Aquatic Mammals*, **21**, 65–77.

Weilgart, L. and Whitehead, H. 1997. Group-specific dialects and geographical variation in coda repertoire in South Pacific sperm whales. *Behavioral Ecology and Sociobiology*, **40**, 277–285.

West, M. J. and King, A. P. 1988. Female visual displays affect the development of male song in the cowbird. *Nature*, **334**, 244–246.

Whitehead, H., Dillon, M., Dufault, S., Weilgart, L., and Wright, J. 1998. Non-geographically based population structure of South Pacific sperm whales: dialects, fluke-markings and genetics. *Journal of Animal Ecology*, **67**, 253–262.

Wiens, J. A. 1982. Song pattern variation in the sage sparrow (*Amphispiza belli*): dialects or epiphenomena. *Auk*, **99**, 208–229.

Winn, H. E., Thompson, T. J., Cummings, W. C., Hain, J., Hudnall, J., Hays, H., and Steiner, W. W. 1981. Song of the humpback whale – population comparisons. *Behavioral Ecology and Sociobiology*, **8**, 41–46.

JANET MANN AND BROOKE SARGEANT

9

Like mother, like calf: the ontogeny of foraging traditions in wild Indian ocean bottlenose dolphins (*Tursiops* sp.)

9.1 Introduction

In this chapter, we identify aspects of delphinid socioecology and life history that relate to the probability and utility of socially aided learning. We also present new findings from our on-going research with dolphins at Shark Bay, Australia that address the possibility that the acquisition of specialized foraging techniques by young dolphins is aided by their affiliation with their mothers and, thus, may be viewed as likely traditions. Studies of bottlenose dolphins (*Tursiops* spp.) in captive and field settings over the last four decades indicate that this genus shows remarkable plasticity and convergent features with primates. Similar to primates, bottlenose dolphins have a long period of dependency and juvenile development (Mann *et al.*, 2000), large brains for body size (Marino, 1998; Ridgway, 1986), complex alliance formation (Connor *et al.*, 2000a), and social learning (reviewed in Janik, 1999; Janik and Slater, 1997; Rendell and Whitehead, 2001). Unlike nonhuman primates, bottlenose dolphins also show vocal learning in call production (Janik and Slater, 1997, 2000; see also Ch. 8); they produce individually distinctive "signature whistles" (Sayigh *et al.*, 1995, 1999; Tyack, 2000) and can also match each other's whistles in natural contexts (Janik, 2000).

Recently, several cetacean biologists have claimed that cetaceans have culture (Deecke, Ford, and Spong, 2000; Noad *et al.*, 2000; Rendell and Whitehead, 2001; Whitehead, 1998). The strongest evidence for social learning comes from bottlenose dolphins studied in captive settings (reviewed by Rendell and Whitehead, 2001). Field data are weaker, but the best field evidence for social learning is in the acoustic domain (e.g., see Deecke *et al.*, 2000; Janik and Slater, 1997; Noad *et al.*, 2000); evidence for social learning of gestures and motor movements has been limited to captive studies of bottlenose dolphins (Janik, 1999).

In this chapter, we intentionally do not use the term "culture," nor do we address whether this term accurately describes cetacean intraspecific behavioral variation (but see Mann, 2001). Where appropriate, we use the terms *social learning* and *tradition* (as defined by Fragaszy and Perry in Ch. 1). Our goals are to assess the role of maternal social influence in producing variation in Shark Bay bottlenose dolphin foraging techniques, and to show that such questions can be addressed in wild cetaceans (see Rendell and Whitehead, 2001).

Foraging presents an appropriate avenue for investigating social learning and traditions in cetaceans. First, bottlenose dolphins exhibit a diversity of foraging techniques both within and between populations (Connor *et al.*, 2000a; Shane, 1990). Second, foraging specializations within the Shark Bay dolphin population have been identified (e.g., Connor *et al.*, 2000b; Smolker *et al.*, 1997). Intrapopulation variation may provide the means for evaluating the role of experiential factors in behavioral development. Third, detailed long-term study of the Shark Bay population of bottlenose dolphin behavior and ecology allows us to identify matrilineal patterns of foraging, the ontogeny of foraging among calves, and foraging patterns of the larger population.

Much of the literature regarding primate foraging techniques is based on different methods of manipulating or processing food items (e.g., Chs. 10–13). Bottlenose dolphins cannot easily manipulate prey (except to beat it with their tails or on the water surface, or to break the fish on the seafloor). Rather, they vary in hunting technique rather than processing. For example, many of the foraging strategies identified in the bottlenose dolphins of Shark Bay are characterized by distinct dive or surfacing patterns (see Tables 9.1 and 9.2). Because most foraging occurs several meters below the surface of the water, we describe the most overt distinctions between foraging types. More subtle characteristics are difficult to observe and to identify reliably.

9.1.1 Flexibility in foraging

Although bottlenose dolphins have been characterized as catholic, opportunistic hunters that feed predominantly on fish, cephalopods, and crustaceans (e.g., Corkeron, Bryden, and Hedstrom, 1990; Cockcroft and Ross, 1990; Connor *et al.*, 2000b), and occasionally stingrays, sharks, eels, and mollusks (J. Mann, personal observation; Mead and Potter, 1990),

Table 9.1. *Twelve types of foraging strategy used by bottlenose dolphins*

Foraging strategy	Characteristics
Bird milling	Dolphins are surfacing within or around a tight feeding group of cormorants (and usually pelicans); this typically occurs in shallow water (< 4 m)
Leap and porpoise feeding	Dolphins are multidirectional (milling) and leaping continuously within an area, which may be relatively small or spread out over as much as a kilometer. This activity usually occurs in closely spaced bouts with abrupt starts, stops, and changes in direction. The group as a whole is often travelling rapidly
Bottom grubbing	Dolphin sticks its beak to the seafloor or sea grass to ferret something out while in a vertical position. This can only be viewed in shallow water. Regular dive types characterize surfacing
Milling	Dolphin forages and changes direction with virtually every surface and breath; breathing intervals tend to be irregular
Tail-out/peduncle dive foraging	The predominant dive types during foraging include tail-out or peduncle dives (Table 9.2). Dolphins typically stay submerged for 1–3 minutes after a tail-out or peduncle dive; once surfacing, they typically take 1–12 breaths before diving again
Rooster tailing	The predominant dive type during foraging is a rooster tail, which is a kind of fish chase with a fast swim along the surface of the water in which a sheet of water trails off the dorsal fin; After the rooster tail, the dolphin dives to the bottom, often back-tracking the direction of the fast swim
Sponge carrying	Dolphin forages wearing a sponge on its rostra while doing tail-out dives and staying down in the water for 2–3 minutes (Smolker *et al.*, 1997). The dolphin also tends to change directions often. This occurs almost exclusively in channels 8–12 m
Snacking	Characterized by a belly-up chase and capture of fish trapped at the surface. Calves typically have prolonged circular belly-up swims during the fish chase; adult snacking tends to be brief
Trevally hunting	Begins with tail-out diving, but once the trevally (always golden trevally (*Gnathanodon speciosus*)) is located, there are directed leaps after the fish. The fish is then processed in a particular way (see text)
Beaching	Shallow-water feeding involving chasing fish close to the shore-line such that the ventrum is on the seafloor or beach; fish are often trapped onto the shore, with the dolphin launching partially or fully out of the water onto the beach. The dolphin turns sharply to return to the water head first
Boat begging	Dolphin approaches stationary or slow-moving boats within 1–2 m and opens jaw or brings the head out of the water
Provisioning[a]	Dolphin receives fish handouts (thawed, dead fish) from humans standing in shallow water

Calves are not provisioned, so the category "provisioning" was excluded from data analysis. Further, all observations in the present study were conducted away from shore, where provisioning does not occur (although a few dolphins beg from boats).

Table 9.2. *Definition of surface and dive types associated with foraging*

Dive type	Description
Tail-out dive	Deep dive, flukes out of the water
Peduncle dive	Peduncle or tail-stock arched at dive, flukes partially submerged
Rooster tail	Not really a dive type; more a kind of fish chase. A fast swim along the surface in which a sheet of water trails off of the dorsal fin. Following this type of swim, the dolphin descends rapidly, often opposite to the direction of the swim
Rapid surface	A rapid surface in which the dolphin maintains a horizontal posture and the dolphin's ventrum does not clear the water surface
Porpoise	A rapid surface in which the dolphin maintains a normal horizontal posture but the entire ventrum does not clear the water surface at once; the dolphin's entire body does leave the water surface in the course of the dive
Leap	A rapid surface in which the dolphin maintains a normal horizontal posture and the dolphin completely clears the water surface
Regular dive	The dolphin sinks down at the end of a breath series without arching the peduncle or raising the flukes out of the water. Regular dives are typical of infants
Humping surface	A normal speed surface in which the dolphin "humps up" its posterior half to break its forward motion as it descends. Often seen when dolphins are driving or pursuing a fish school in shallow water but also seen in aggressive contexts
Fast swim	A dolphin rapidly accelerates and/or swims fast along or below the water surface

a number of distinct population-specific foraging techniques have been described. These include sponge carrying to ferret prey from the sea floor (Smolker *et al.*, 1997; Fig. 9.1); corkscrewing into the sand after fish (Rossbach and Herzing, 1997); belly-up chasing of fish at the surface (Bel'kovich *et al.*, 1991; Mann and Smuts, 1999); strand feeding on mud-banks in Portugal (dos Santos and Lacerda, 1987), Georgia and South Carolina, USA (Hoese, 1971; Petricig, 1993), and on beaches in Shark Bay, Australia (Berggren, 1995; Fig. 9.2); stunning or killing fish with a tail-hit (Shane 1990; Wells, Scott, and Irvine, 1987); or tail-whacking the water surface to scare up fish (Connor *et al.*, 2000b). As a coastal cosmopolitan species, bottlenose dolphins have also learned to take advantage of human activity. For example, bottlenose dolphins have learned to feed on fish drawn to a garbage barge and were predictably found based on the schedule of the garbage barge (Norris and Dohl, 1980). They follow shrimp trawlers (e.g., Caldwell and Caldwell, 1972; Corkeron *et al.*, 1990; Leatherwood, 1975; Norris and Prescott, 1961) and steal bait from lines or crab pots (Noke and Odell, 1999). In Laguna, Brazil, fishermen and

Fig. 9.1. Photograph of "Original Spongemom", who was observed sponging in the late 1980s and continues to carry sponges in 2001. Her surviving offspring, Grunge, still carries a sponge, two years after weaning, at age five.

dolphins appear to net mullet cooperatively, with the dolphins herding the fish into the nets and feeding easily off the remains (Pryor *et al.*, 1990). Historical accounts of Australian aboriginal cooperative fishing with dolphins have also been reported (Corkeron *et al.*, 1990). Provisioned females in Monkey Mia, Shark Bay, Australia beg for fish from boats and tourists (Connor and Smolker, 1985; Mann and Smuts, 1999). The Monkey Mia and Laguna "traditions" have continued across at least three generations.

Although these studies describe foraging strategies that differ between populations, in Shark Bay, the intrapopulation variation is remarkable and distinguishes Shark Bay bottlenose dolphins from other populations and species studied to date. In particular, individual females and their offspring have distinctive foraging strategies ranging from one to seven foraging types out of the 11 that we have studied and 13 that have been documented at our field site (Table 9.1). For example, one technique, sponge carrying, is clearly a form of tool use and is restricted to a limited number of animals (Fig. 9.1).

9.1.2 Is the duration of lactation related to calf foraging skill?

It is clear that dolphins are precocious and well developed at birth but maintain a long period of dependency. Bottlenose dolphins typically

Fig. 9.2. Photograph of beaching behavior (triple sequence): an adult female, Rhythm, lunges out of the surf (a), catches a mullet (b), and turns back into the water (c).

nurse for three to six years in Shark Bay (Mann *et al.*, 2000). A calf must be able to forage successfully before being completely weaned. Learning to forage appears to be a slow process, warranting the overlap between nursing and foraging for the first years of life. This contrasts with most mammals, where independent foraging does not begin until late lactation. Compared with toothed whales, baleen whales have shorter periods of maternal investment, less overlap between nursing and calf foraging, and fewer, less-complex feeding strategies (e.g., Clapham, 2000; Whitehead and Mann, 2000).

Johnston (1982) proposed that parental investment is likely to be intensive and prolonged for species with complex foraging skills (high dependence on learning). Although this seems likely, we suggest that social learning would reduce the mother's lactation costs by decreasing the period of nursing or by increasing the overlap of nursing and foraging. A longer period of dependence allows for the infant to learn specialized foraging skills from its mother or on its own while still nutritionally dependent on and protected by her. Dolphin calves maintain roughly the same home ranges as their mother after weaning (Mann *et al.*, 2000); a similar habitat would favor similar hunting strategies. Thus, selection should favor social learning from the mother. Consistent with this hypothesis, Laland and Kendal (Ch. 2) propose that moderately low environmental variability will favor vertical transmission.

Unlike carnivores (e.g., felids, canids, mustelids) and primates, cetacean mothers generally do not share prey with young. One exception is the killer whale, where prey sharing between mother and offspring has been well documented (Baird, 2000). Despite several thousand hours of observation of bottlenose dolphin calves in Shark Bay by J. Mann, prey sharing has not been observed. Nevertheless, calves seem quite interested in fish caught by other individuals. They frequently approach and inspect prey caught by others and will sometimes travel tens of meters to observe (unpublished data). Even with the fish or pieces of the fish floating in the water, calves have never been observed taking fish caught by another.

9.1.3 Dolphin social structure and foraging strategies

Because bottlenose dolphins live in a fission–fusion society with flexible group membership (Smolker *et al.*, 1992) and travel costs are low (Williams *et al.*, 1992; Williams, Friedl and Haun, 1993), dolphins can likely enjoy the benefits of group living without the costs of direct feeding competition (Connor *et al.*, 2000a). Individuals have the opportunity to associate in a number of small groups or to travel alone, allowing individuals to benefit from the group structure as well as from individual foraging success. Although most hunting is a solitary affair, schooling fish may attract groups and individual dolphins may benefit by collective balling of fish. We predict that some of the group-foraging techniques (feeding on large schools) are more widely shared (less specialized) across individuals, as large prey patches may attract all dolphins more readily.

Males and females differ in their social affiliations. Females, especially those with calves, are usually in larger groups than are males (Scott, Wells, and Irvine, 1990), although group size is variable. Some females remain fairly solitary while others are quite social (Mann *et al.*, 2000). Females tend to associate with their mothers after weaning; males do not disperse, but they do form coalitions with other males within the community (Connor *et al.*, 2000a; Wells *et al.*, 1987). Sons are weaned at an earlier age than daughters (Mann, 1998); consequently, daughters have a longer time to learn specific foraging skills from their mothers and could be expected to have a higher degree of similarity with their mothers for the specialized types of foraging. There is also a difference in the movement patterns between the sexes, with the females covering smaller areas than males (Bearzi, Notarbartolo-Di-Sciara, and Politi, 1997). The differences in social affiliation and use of space for male and female dolphins suggest that there may also be differences in foraging techniques and degree of specialization, with females more likely than males to acquire specialized techniques.

Female dolphins spend, on average, 19–36% of the daytime foraging. Their hunting strategy and choice of habitat are likely to affect their fitness. Shallow-water habitats in Shark Bay are associated with higher female reproductive success than deep-water habitats (Mann *et al.*, 2000), possibly because of differences in food density (Heithaus and Dill, 2002) or in fish species or distribution.

9.1.4 Research questions

In the following sections, the study of Shark Bay dolphins is used to examine foraging techniques and their dissemination. Three specific areas are discussed.

1. The diversity and distribution of foraging techniques used by mothers and their calves are examined, specifically to identify foraging techniques shared widely by members of the population and techniques that are more specialized (restricted to a few members).

2. Associations between the foraging strategies used by mothers and their calves are determined to see if similarities in foraging technique between mothers and calves increases with the calf's age.
3. The evidence that some specialized foraging techniques qualify as "traditions" (Ch. 1), passing from one generation to the next via vertical social transmission, is outlined. Widely shared (generalized) foraging techniques may also be socially influenced, but it will be more difficult to document their status as traditions *per se*.

9.2 Methods

9.2.1 Background and field site

The study incorporated a 130 km^2 area east of the Peron Peninsula, which bisects Shark Bay (25° 47′ S, 113° 43′ E), Western Australia. A longitudinal field study was established in 1984 (Connor and Smolker, 1985). By 2000, over 600 dolphins had been identified and 200 animals were sighted regularly. Dolphins are identified by their fin shape, nicks, and other natural markings. Calves have been sexed using views of the genital region. A mother–calf study was initiated in 1988 by Janet Mann and Barbara Smuts. Observations of mothers and calves by the former has continued for two to six months every year since, except 1995.

Since the early 1960s, 6 to 11 dolphins (at a time) have been provisioned by tourists and fishers at a small fishing camp, turned resort, called Monkey Mia. Since the mid-1980s, the feeding has been controlled and monitored by rangers currently employed by the Department of Conservation and Land Management (CALM) of Western Australia. Since 1995 feeding from boats has been firmly restricted by CALM although it still occasionally occurs. At present, three adult females (Nicky, Puck, and Surprise) and their offspring visit the Monkey Mia beach up to three times per day and receive up to 2 kg of fish per day. To discourage dolphins from spending too much time near the provisioning area, no dolphins are fed after 1 p.m. Nicky, Puck, and Surprise visit daily, with only a few absences per annum. During their visit, the mothers remain in shallow water near people and make frequent contact with the rangers until the feeding, which occurs approximately 30–60 minutes after their arrival at the beach. Calves typically remain in deeper water until the feed is over. The dolphins leave almost immediately after each feeding.

Offshore focal observations involved following individual animals in small boats (4–5 m dinghies equipped with 6–45 hp motors) for up to

Table 9.3. *Observation record for dolphin calves at each age*

Age class	Calf age (months)	Total focal calf observation minutes (hours)	No. Days observed	No. calves with focal data
Newborn	0–3	4721(78.68)	41	6
1st year	4–11	20931(348.85)	119	34
2nd year	12–23	21247(354.12)	135	35
3rd year	24–35	14534(242.23)	93	25
4th year	36–47	11192(186.53)	67	11
5th year	48–59	3617(60.28)	25	4
6th year	60+	566(9.43)	4	1

10 hours at a time (Smolker, Mann, and Smuts, 1993; Mann and Smuts, 1998). Observers typically remained < 50 m from the mother or calf. Between 1996 and 1998, and during 2000 field seasons, two boats, a 5 m fiberglass dinghy and a 10 m catamaran (*Nortrek*) were used for observations and acoustic recordings and localization. When mother and calf were together (< 10 m), only one boat stayed with the focal pair. When separated (> 10 m), one boat would stay within 100 m of each member of the dyad. *Nortrek* typically remained > 50 m from the mother or calf.

9.2.2 Subjects

The study, conducted between 1989 and 2001, incorporated 1280.1 hours of focal observations on 58 calves (18 males, 22 females and 18 of unknown sex) born to 37 mothers. We used 1781 calf and 3020 maternal foraging bouts for these analyses. Of the 37 mothers in the current sample, five visit (or visited) the provisioning beach. The remainder, to our knowledge, have had no contact with humans. The data include information about the infant's and mother's time spent foraging overall, the types of foraging, the dive type, the depth, and group membership. "Group" is defined using a 10 m chain rule: any animal that is within 10 m of any animal within the group is in the group. The total number of hours observed for the calves at each age is detailed in Table 9.3. The approximate date of birth of the infant is known for most subjects.

9.2.3 Focal sampling

Data were collected with a focal-animal procedure using several different observational methods including continuous, scan, and point sampling (Altmann, 1974; Mann, 1999). In addition to boat-based observations, similar focal methods were applied during five days of shore-based observations from cliffs and beaches of Point Peron to study the "beaching behavior". For follows from Point Peron, we also sampled the mother's and the calf's distance from the beach every minute and during beaching events.

Activity data were gathered using continuous or point sampling, with duration and/or frequencies of behaviors maintained in the sampling record. This method was used to record the duration of foraging (bout length and frequency) and dive types (frequency). Every 5 minutes we measured water depth (using a depth sounder); water depths were further classified as shallow (≤ 4 m), moderate (4–7 m), and deep (> 7 m). Group composition for mother and calf were determined every minute (post-1996) or every 5 minutes (pre-1996). Latitude and longitude were determined every 15 or 30 minutes using the Magellan Pro-Mark X or (pre-1996) using compass bearings on landmarks.

Foraging was recorded when there was reasonable evidence that the animals were actually searching for, catching, processing, and eating prey. Foraging is a regular and more or less exclusive search for prey. It is difficult to diagnose because foraging occurs below the surface and is not always successful. Further, successful prey capture often eludes observers since prey are typically swallowed whole immediately. Specific types of foraging were identified and given names, as listed in Table 9.1, although in some cases, the foraging type was defined *post hoc* based on absence or presence of defining features. *Post hoc* coding was done "blind" to dolphin identification. Foraging that could not be classified was placed in a generic category of "foraging". Foraging types were determined, in part, by dive types, which are indicated in Table 9.2.

9.2.4 Data reduction and analysis

A foraging bout was defined as each onset and offset of foraging. When point sampling was used, or if it was impossible to determine the exact time of onset or offset, the midpoint between point samples was used as the onset or offset. Each bout was classified as a type in Table 9.1. Percentage time foraging for each calf for each age class observed was determined by dividing the total minutes foraging that year by the total time observed that year. Similarly, the rate of foraging (bouts per hour) was determined by dividing the number of foraging bouts by the total time observed. For all calves that foraged ($n = 51$), percentage of foraging bouts by type was determined overall for each calf and for each age class by dividing the number of bouts of foraging type by the total number of foraging bouts.

Chi-square (Yates corrected) analysis was used to determine the association between mother and calf in foraging types in two ways. First, each mother and calf was coded according to whether or not they engaged in a specific foraging type. Some pseudoreplication was inevitable given that 13 of 37 mothers had more than one calf in the sample. Seven calves were not used in this analysis because they did not forage; in one case, the mother was not observed foraging. If the foraging type was not indicated or could not be coded using descriptions of dive types and other information, these cases were excluded from the "foraging type" analyses but included in time budgets and bout rates. Second, each mother and calf pair was coded according to the calf's age. For this analysis, the earliest year was used to characterize each mother's foraging type(s) (first year of her calf's focal data). The calf's foraging type(s) was coded for subsequent years (see Table 9.1). This way, independent datasets were obtained for mothers and calves. Hence, this analysis is more conservative because foraging similarity could be demonstrated across years. The Fisher exact test (two-tailed) was used for this analysis because the expected values for cells were less than five. The second analysis reduced our sample size to 31 mother–calf pairs because some calves were only observed in one year.

9.3 Results and discussion

9.3.1 Diversity of foraging types

Twelve foraging types were identified in this study and the distribution and mother–calf similarity of 11 were analyzed (provisioning was excluded). The mothers' predominant (most common) foraging type was tail-out/peduncle dive foraging (Fig. 9.3). The calves' predominant foraging type was snacking, followed by tail-out/peduncle dive foraging. Most mothers and calves used only a few of the foraging types available (Fig. 9.4).

The *number* of techniques employed ranged from one to seven for mothers and calves. This variable was significantly correlated for mother and calf (Pearson $r = 0.66$; $p < 0.001$; $n = 51$). That is, mothers who engaged in multiple foraging types had calves who tended to do the same. However, the number of foraging tactics used by mothers strongly correlated with the number of hours she was observed (Pearson $r = 0.71$;

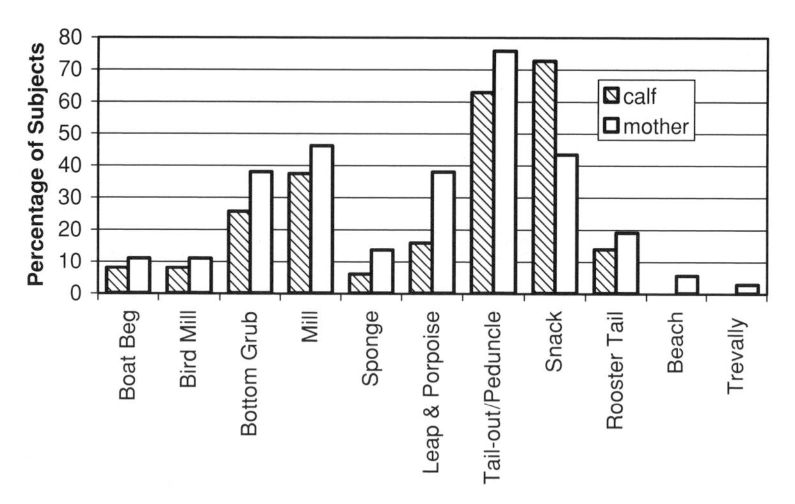

Fig. 9.3. Proportion of mothers and calves engaging in different foraging types. The predominant foraging tactics used were tail-out and peduncle dive foraging, milling, and snacking (especially for calves). Only a few subjects engaged in behaviors deemed "traditions".

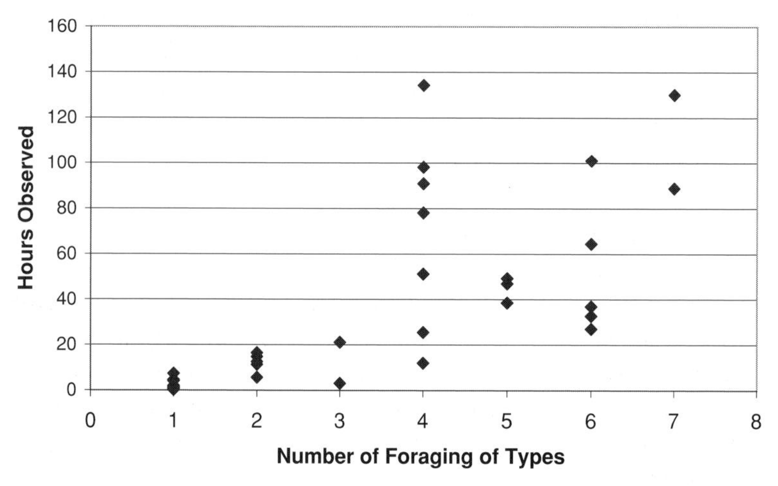

Fig. 9.4. Correlation between the number of maternal foraging types per female and observation hours (Pearson $r = 0.711$; $p < 0.01$; $n = 33$). Females who were observed for more hours exhibited more foraging types, but some females who were observed 50 hours or more still exhibited only four types. The two females who exhibited the most foraging types were provisioned. Provisioning has been associated with innovation and behavioral flexibility at other mammalian research sites (Ch. 10).

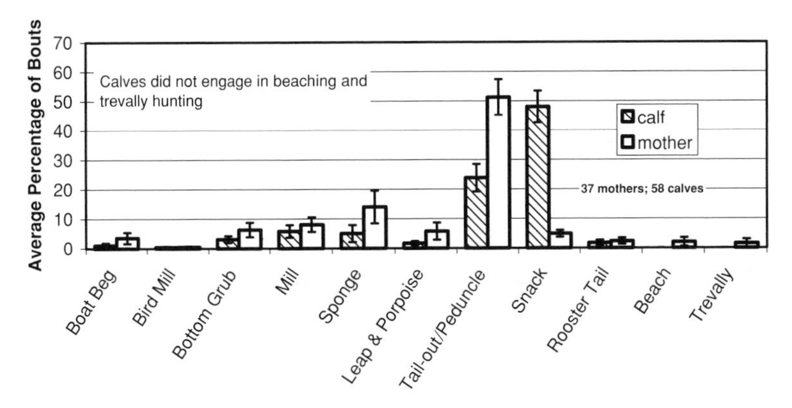

Fig. 9.5. Average proportion of bouts by foraging type across all subjects. This shows the mean proportion ($\pm$ SE) of foraging types mothers and calves engaged in. Tail-out and peduncle dive foraging clearly made up the greatest proportion of foraging bouts, averaging approximately 51% for mothers. For calves, snacking clearly was the predominant foraging type, averaging 49% of calf foraging bouts.

$p < 0.01; n = 33$) and this variable accounted for 50% of the variance in diversity of foraging types in mothers (Fig. 9.4). Across mother–calf pairs, the average proportion of foraging bouts by type illustrates the preponderance of snacking for calves and tail-out and peduncle dive foraging for mothers and calves. Other types of foraging occurred at low rates across our sample, although they may represent a high proportion of an individual mother's foraging bouts (see Fig. 9.5).

9.3.2 Development of foraging in calves

Calves increased both bout rate and proportion of time foraging with age (Fig. 9.6). *Post hoc* comparisons revealed significant differences between the newborn period (birth to three months) and the third year, and between the first and third year. The dip in foraging rate and percentage time foraging during the calf's fourth year is not significant. Calves did not forage (chase and catch fish) during the first three months. Maternal foraging did not significantly change as a function of calf age.

Among the 34 calves observed in the first year, all foraging types were observed except beaching, trevally hunting, and sponge carrying. Milling and snack foraging were first observed at 3.4 months of age. Leap feeding was first observed at 6.4 months. At seven to eight months, four foraging types were observed: boat begging, bird milling, rooster tail and tail-out/peduncle dive foraging. Calves have not been observed beaching or

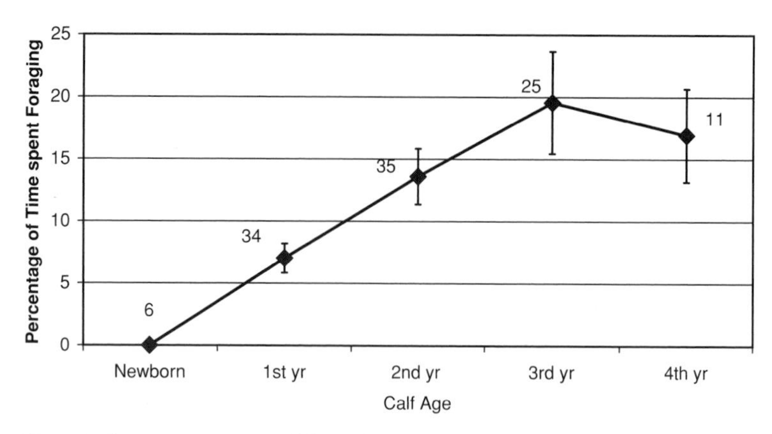

Fig. 9.6. The mean percentage of time calves spent foraging from the newborn period (0–3 months) until their fourth year. Sample sizes are indicated above the average. Calves increased the proportion of time and rate of foraging with age and no calves foraged during the newborn period. (Forage rate (bouts per hour): Kruskal–Wallis = 16.68; $p = 0.002$; Bonferroni post-hoc pairwise comparisons for newborn versus third year, $p < 0.016$; first versus third year, $p = 0.006$. Percentage time calf foraging: Kruskal–Wallis = 23.87, $p < 0.001$; Bonferroni post-hoc pairwise comparisons for newborn versus third year, $p = 0.016$; first year versus third year, $p = 0.006$.)

trevally hunting up until their third year (no fourth year observations have been conducted for calves born to mothers who engage in these behaviors). Sponge carrying was first observed at 20 months for one calf (Grunge), and at 31 months for another (Demi). The third calf was only observed during the fourth year of life and was already sponge carrying. Both Demi and Grunge have continued to sponge carry after weaning (Fig. 9.1).

Snacking, the most common calf foraging behavior, declined with calf age, suggesting that calves increase nonsnack foraging tactics with age. Snacking is also the first type of foraging to appear developmentally and is practiced in the newborn period (Mann and Smuts, 1999).

9.3.3 Correspondence between maternal and calf foraging style

Mother–calf similarity in foraging was evident for nearly all foraging types, with calves almost exclusively engaged in techniques used by their mothers. With the exception of snacking, there were only five cases where calves engaged in a foraging type not seen in their mothers. Three of those cases were calves born to sponge-carrying females, and the calves were not (yet) sponging themselves. One case was Whoops, offspring of the "trevally hunter" Wedges. The remaining case referred to a single bout of leap and porpoise feeding by a calf. The biggest difference between mothers and calves related to snack foraging. Eight calves snacked, although their mothers did not. In contrast, mothers engaged in two foraging techniques not observed in calves, beaching and trevally hunting.

9.3.3.1 Boat begging and its relationship to provisioning

Boat begging typically occurred near shore when boats were stationary (fishing) or on return from fishing trips. Average water depth of boat begging was 3.07 ± 0.15 m; four focal calves and four of five provisioned mothers engaged in boat begging.

All calves in the database that engaged in boat-begging behavior had mothers who were provisioned. One calf born to a nonprovisioned female (and not included in this dataset) was observed begging from a boat once. This calf spent the majority of his time associating with provisioned mothers and their calves (unpublished data). No nonprovisioned mothers begged from boats. In addition, one provisioned mother did not beg at boats and neither did her two calves. Boat begging is significantly associated between mothers and calves (chi-square Yates corrected = 16.92; $p < 0.001$; $n = 51$ calves). If the presence or absence of boat begging in the mother's first year of observation is compared with subsequent years of calf observation, the association between mothers and calves is weaker but approaches significance (Fisher exact test, $p = 0.060$, $n = 31$ calves). Most begging by calves occurred close to the mother, when she too was begging at the boat (average distance was < 10 m). The only calf who was provisioned (by CALM) did the most begging (84% of all calf begging bouts, at a rate of 0.5 per hour), whereas the three nonprovisioned calves of provisioned mothers begged at low rates of 0.009–0.117 per hour. The sole provisioned calf became dependent on provisioning and this was probably the cause of his death at the age of four years.

The association patterns and foraging techniques of the offspring of provisioned females indicate why foraging traditions may be more likely to be transmitted to daughters than to sons. The two nonprovisioned sons of provisioned females (now aged 6 and 13 years) rarely visit the provisioning area. In contrast, the four daughters of provisioned females (now aged 7, 9, 25, and 26 years) frequently visit the provisioning area with their mothers, and all but the youngest have been offered and have accepted fish handouts after being weaned. Although the sample is small, these observations are consistent with the pattern away from the beach, which is that mother–son association declines after weaning and likely inhibits sharing of foraging tactics that require small and specific habitats such as the provisioning beach. This pattern would reinforce foraging traditions for daughters that were limited to specific areas but would work against sons adopting their mothers' specializations (Peron beach, Monkey Mia beach, channels).

9.3.3.2 Bird milling

Large groups of pelicans (*Pelecanus conspicullatus*) and pied cormorants (*Phalacrocorax varius*) gather in shallow water and co-feed with bottlenose dolphins, typically in shallow water < 4 m (average depth of bird milling 3.51 ± 0.56 m). The surfacing patterns and numbers of dolphins attracted to bird-milling groups suggest that they are foraging on large schools of fish. However, not all individuals are attracted to bird-milling groups. In contrast to "leap and porpoise foraging", (see below), dolphins do not travel a kilometer or more to join bird-milling groups. Bird milling was seen in four (8%) of focal calves and four (11%) of focal mothers. All calves who engaged in bird milling had mothers who did so. The behavior is strongly associated for mother–calf pairs (chi-square Yates corrected = 16.93; $p < 0.001$; $n = 51$ calves). The presence or absence of the mother's bird milling in the calf's first year of observation compared with calf's bird milling in subsequent years indicates no relationship (Fisher exact test, $p = 1.0$, $n = 31$ calves). However, this type of foraging is infrequent, making up only 0.6% of calf foraging bouts and 0.6% of maternal foraging bouts.

9.3.3.3 Bottom grubbing

Grubbing in the sea grass or seafloor to ferret out fish probably occurs in all habitat types, but observers can only be certain of bottom grubbing in shallow water, when the behavior can be clearly seen. Fourteen females (38%) and 13 calves (25%) use this foraging technique, typically in 2.9 ± 0.11 m of water. One calf bottom grubbed, although his mother was not observed doing this behavior. Bottom grubbing was strongly associated for mother–calf pairs (chi-square Yates corrected $= 7.96$; $p = 0.005$; $n = 51$ calves). This association remained significant when the mother's first year of bottom-grub foraging was compared with her calf's bottom grubbing in subsequent years (Fisher exact test, $p = 0.008$; $n = 31$).

9.3.3.4 Milling

Milling, surfacing repeatedly in different directions, involves feeding on schooling fish (mid-water). This behavior occurs in both shallow and deep water (average depth 4.73 $\pm$ 0.21 m). A large proportion of mothers (46%) and calves (37%) engaged in mill foraging and there was significant mother–calf similarity (chi-square Yates corrected $= 6.50$; $p = 0.011$; $n =$ 51 calves). When the mother's mill foraging was compared in her first year of observation with that of the calf's subsequent years, the relationship approached significance (Fisher exact test, $p = 0.056$; n $= 31$). This foraging technique ranked third in proportion of subjects using the technique and may be considered a generalized or widely shared foraging tactic.

9.3.3.5 Sponge carrying

Sponge carrying, the only known example of tool use in any wild dolphin or whale, was observed in five adult females (14%) and three of their calves in this dataset. Of 141 identified mothers in the Shark Bay population, 15(11%) carry sponges. All five sponge carriers use this foraging technique almost exclusively (100, 100, 96, 90, and 75% of their foraging bouts, respectively) and tend to forage in specific deep water channels (> 8 m). Sponge carrying shows a clear female bias. Of 25 sexed animals known to carry a sponge at least once, 20 are female. Only one of the males that carried a sponge is an adult. Out of our total population of sexed animals, (192 females, 166 males), females were more likely to carry sponges than males (chi square Yates corrected $= 7.52$; $p = 0.006$; $n = 358$).

The occurrence of sponge carrying is clearly associated for mothers and calves (chi-square Yates corrected $= 15.73$; $p < 0.001$; $n = 51$ calves), and relationship between the presence or absence of the mother's sponge carrying in the first year of observation and her calf's sponge carrying in subsequent years approaches significance (Fisher exact test, $p = 0.065$, $n = 31$). Of the three calves observed sponge carrying in our focal sample, two were sexed as female. The sex of the third calf was not known. Of the three calves who did not carry sponges (although their mothers did), one died in the second year (too early to begin sponge carrying), another (not sexed) began sponging in the fourth year of life (after the analyses described here were completed). The third, a male, was observed sponge-carrying once postweaning.

The strong female bias in sponge carrying could be related to several factors. First, since sponge carrying occurs mainly in deep channels, male offspring may be unable to maintain such a specialized technique and still range widely enough to herd adult females (see Smolker *et al.*, 1997). Further, males might be unable to find other sponge-carrying males and thus maintain the behavior with their alliance partners. As yet, we know of no dolphins who became "spongers" who were not born to sponge carriers. Some males born to spongers have been observed carrying sponges, but only on a few occasions. It is unclear what developmental mechanisms might inhibit males from sponge carrying before weaning.

Sponge carrying appears latest developmentally, in the second rather than first year of life. This may be because of its difficulty, involving long dives (typically 2–3 minutes) to tear off and hunt with sponges along the seafloor. Calves under one year of age are capable of remaining submerged for 3 minutes, but this diving pattern may be difficult to maintain, or conduct appropriately, while wearing a sponge. Alternatively, the prey may be quite difficult to catch, with or without a sponge. We doubt that the prey are difficult to process, since females appear to swallow these quickly, rarely bringing prey to the surface.

As mentioned above sponge carrying tends to occur in deep channels (8–12 m), but not exclusively, and sponge carriers occasionally sponge in other areas. Further, many dolphins regularly forage in the same channels without sponging. Four of our focal females regularly used the "sponge channels" but did not sponge. Thus, the behavior does seem largely habitat specific, but use of channel habitats is not sufficient to explain the development of this foraging tactic. Recent genetic data (Krützen *et al.*, unpublished) suggest that nearly all spongers share the same mitochondrial DNA haplotype, which is rare in the rest of the (nonsponging) population. This lends further support to the suggestion that sponge carrying is transmitted through matrilines. It is unclear why other dolphins do not, at least occasionally, try to sponge. Perhaps there is some 'sensitive period' during which exposure to foraging is most likely to lead to a young dolphin acquiring similar practices.

9.3.3.6 Leap and porpoise foraging

Leap and porpoise foraging typically attracts dolphins from large distances (several kilometers) to feed on large schools of fish. Even sponge-carrying females drop their sponges and travel some distance to join leap-foraging groups. Dolphins do not appear to "specialize" in this technique; rather they take opportunistic advantage of large schools that periodically occur in the bay. More than a third of mothers (38%) and 16% of calves (Fig. 9.3) engaged in leap and porpoise foraging. Leap and porpoise foraging occurs at variable depths (4.62 ± 0.28 m), typically in moderate or deep water (> 4 m). This behavior was not associated for mothers and calves (chi-square Yates corrected $= 1.97$; $p = 0.16$; $n = 51$ calves);

although calves nearly always accompanied their mothers long distances to leap-foraging groups, they did not forage in these but appeared to concentrate on tracking their mothers when so many animals (often more than 20) were present. In comparing leap feeding across years, no association between mothers and calves was found (Fisher exact test, $p = 1.0; n = 31$).

9.3.3.7 Tail-out and peduncle dive foraging

Tail-out and peduncle dive foraging was the most common foraging technique. It was exhibited by 76% of the mothers and 63% of the calves. This type of foraging occurs in moderate and deep water, averaging 6.75 ± 0.12 m. It is not significantly associated for mother–calf pairs (chi-square Yates corrected$= 1.29; p = 0.256; n = 51$). Two calves engaged in tail-out and peduncle dive foraging although their mothers did not. Both calves were born to spongers and did not sponge themselves. In comparing tail-out and peduncle foraging for the mother's first year of observation with the calf's subsequent years, no relationship was found (Fisher exact test, $p = 1.0; n = 31$). Tail-out and peduncle dive foraging may be considered a generalized or shared foraging tactic.

9.3.3.8 Snacking

Snacking was clearly the predominant foraging type for calves, accounting for nearly half (48 ± 5%) of all calf foraging bouts. We observed 73% of calf subjects snacking. Snack foraging was the only foraging type that calves did more often than mothers. Only 43% of mothers snack foraged and only 5 ± 1% of their foraging bouts were snacking. Eight calves snacked although their mothers did not. Most maternal snacking involved single belly-up chases of fish, rather than the repeated circular swims belly-up to chase fish that are characteristic of calves. There was a significant association between maternal and calf snacking (chi-square Yates corrected $= 6.51; p = 0.011; n = 51$ calves). If the mother's snacking in the first year is compared with the calf's snacking in subsequent years, the relationship approaches significance (Fisher exact test, $p = 0.059$; $n = 31$). Snacking occurs in all water depths, averaging 4.92 ± 0.22 m.

Based on the observations of newborns, who appear to practice snacking for several months before actually catching a fish (Mann and Smuts, 1999), it may be that snack foraging allows the calf to coordinate visual images (backlit when belly-up towards the water surface) and motor activity with developing echolocation skills. Dolphins see most acutely in the ventral direction. Therefore, by swimming belly-up, calves may optimize visual and acoustic (amodal) perception. Although object play is rare in dolphins, newborn calves repeatedly belly-up "chase" and "capture" sea grass in the first months of life (Mann and Smuts, 1999). Between four and six months, calves begin capturing small minnow-sized fish, and snacking may be the easiest way for them to catch such small fish. When mothers snack, it is on much larger fish, sometimes 60 cm long (e.g., longtoms, either *Strongylura leiura* or *Tylosurus gavialoides*). Because of the early appearance of snack foraging, its apparent "practice" with seagrass, the relative lack of snack models (especially the mother), and its predominance as a calf foraging technique, we propose that snack foraging, unlike other techniques, is predominantly individually learned. Further, snacking disproportionately declines with age. Since snacking occurs in all habitat types, the decline in snacking is unlikely to be strictly caused by habitat changes during development.

9.3.3.9 Rooster-tail foraging

Seven females (19%) and seven calves (14%) rooster-tail foraged. For those seven mothers, rooster tailing made up a small to moderate proportion of their foraging bouts, ranging from 3 to 27% (chi-square Yates corrected 10.2 ± 3.9). Two of the adult females who rooster tailed were also mother and daughter. This behavior has spanned at least three generations and is significantly associated for mother and calf, with all seven rooster-tailing calves having a mother that rooster tailed (chi-square Yates corrected = 5.53; $p = 0.019$; $n = 51$ calves.) The association between rooster tailing for mothers and calves remained significant when the mother's first year was compared with the calf's subsequent years of observation (Fisher exact test, $p = 0.001$, $n = 31$). Rooster tailing usually occurs in water of shallow to moderate depth (4.23 ± 0.09 m). Because similar habitats and presumably similar prey occur throughout the bay, we would expect more dolphins to rooster tail. The complex aspect of the foraging technique is that the dolphin appears intentionally to overshoot the prey at the surface, often, but not always, back-tracking for the capture. Since a rooster-tail swim is always followed by a dive to the seafloor, it is interesting that the dolphins do not just dive immediately and pursue the fish at depth.

9.3.3.10 Beaching

Two mother–calf pairs were observed for 8.6 hours at Point Peron, just north of our main study area, specifically for their beaching behavior. These observations were conducted from cliffs and so only the foraging types used by the Peron females close to shore could be observed.

Beaching was first described by Berggren (1995) although locals have known about it since the 1980s. One of the beaching females, Reggae, has been observed beaching regularly in the period from 1991 to 2001 (J. Mann, personal observation; Berggren, 1995). The behavior pattern appears to be restricted to one to three matrilines. One of Reggae's offspring beached as a juvenile (Berggren, 1995), although no one has observed dependent calves beaching. Three adult females in the current study were observed beaching fish (typically mullet, *Mugil cephalus*) on a 1 km stretch of beach. Two had calves. The calves, one in its first year and another in its third, did not stay near the mother during beaching (they were typically > 50 m from the mother) and they did not participate in any type of beaching behavior. The technique may be risky and calves are likely to be in the way.

9.3.3.11 Golden trevally hunting

One female, Wedges, engaged in trevally hunting. Wedges begins with tail-out dive foraging in deep water (over 6–7 m) and then begins a high-speed chase, always leaping (typically 3–17 leaps), to catch golden trevally (*Gnathanodon speciosus*). Once she catches the fish, she will first take several deep dives with the fish, perhaps to strike the head against the bottom (thus killing or stunning the fish). The fish, still whole, is then carried to shallow water (< 4 m). The head is broken off and the fish is eaten in shallow water. Her calf, Whoops, nurses or stays in infant position (in contact under the mother) as she carries the fish to shallow water. Once she begins breaking up the fish in shallow water, the calf moves away and forages independently, sometimes traveling several hundred meters away, but staying in shallow water to bottom grub or snack. The calf does not regain infant position for another 30–60 minutes while Wedges breaks up the trevally. (Calves are usually out of infant position for 10 minutes or less.) The calf remains 50–300 m away from the mother during the catching and eating phases but takes the opportunity to nurse or be in infant position during the carrying phase of trevally hunting. We have seen Wedges catch seven golden trevally, six during focal observations (16.4 hours of focal observation, or one trevally every 2.7 hours.; 50% of Wedges' foraging bouts are trevally hunting). It takes nearly 1 hour to break up and eat fish this size. The remarkable aspect of this phenomenon is the size of the fish, which can reach up to 111 cm in length and 15 kg (Allen and Swainston, 1988). In 17 years of long-term study, no one has observed other Shark Bay dolphins catch fish this size.

9.4 Conclusions

A set of common foraging techniques and a set of individually distinctive foraging techniques can be identified for both Shark Bay bottlenose dolphin females and their calves. Tail-out and peduncle dive foraging and milling are common in both mothers and calves, and snacking is common amongst calves in particular. Nearly all females have been observed to use fewer than half the foraging techniques observed in the population. Such a high degree of intrapopulation variation in foraging style has not been documented elsewhere.

Some types of foraging strategy are restricted to a few animals (e.g., rooster tailing, boat begging, sponge carrying, trevally hunting, and beaching). No calves developed these foraging techniques unless their mothers engaged in them. Further, when the mother's first year of observation is compared with subsequent years of calf observation, similarity between mother–calf foraging types generally remained despite the small sample size (most values for $p < 0.10$). The pattern of increasing mother–calf similarity and clear examples of lifetime stability in some foraging techniques within matrilines are strong evidence that these are traditions that are vertically transmitted (e.g., Demi, born to a sponger, has sponged all her life; Crooked-fin, her daughter, and grand-offspring have all been rooster tailers). Future analyses will focus more directly on the degree of similarity by examining the proportion of foraging time devoted to different techniques with age. The period of dependency, which ranges from 2.7 to more than eight years (Mann *et al.*, 2000), could be related to the complexity of acquiring specialized foraging skills, but we cannot test this directly with the current data.

9.4.1 Foraging traditions

Two types of foraging (bottom grubbing and rooster tailing) meet our stringent criteria for traditions (as defined in Ch. 1) by showing a statistically significant relationship between the mother's foraging technique during the first year of the calf's life and the calf's techniques during subsequent years. However, variation in use of at least one of these techniques may reflect variation in habitat use rather than social influence (i.e., it is only possible to observe bottom grubbing in shallow water, and mother–calf pairs share habitat types). There were also significant associations between mother and calf foraging patterns for sponge carrying and boat begging, though this could be the result of sampling biases from simultaneous data collection on mother and offspring. It is possible that beaching and provisioning/interactions with humans are also vertically transmitted, but data on possible vertical transmission of beaching are only anecdotal (based on one mother–offspring pair), and the provisioning/human interaction data are not presented here. Six remaining foraging types (bottom grubbing, bird milling, leap and porpoise feeding, milling, tail-out and peduncle dive foraging) do not apparently require extended exposure to an adult model for their development. Further, all of these foraging techniques have been reported at other *Tursiops* spp. study sites (e.g., Connor *et al.*, 2000a; Shane, 1990), suggesting that these tactics are widely shared and social influence is relatively less important. Finally trevally hunting may be an "innovation", specific to one female.

Of the foraging techniques that could potentially be labeled "traditions", some calves and mothers are clearly exposed to these foraging types but do not engage in them. For example, Demi (a sponger) regularly associates with the majority of our focal females, but few of her associates sponge. About eight females regularly visit the provisioning area and have access to fishing boats, but they do not attempt to take fish. One female, Joy, was born to Holeyfin, a provisioned female, but avoided the provisioning beach as soon as she was weaned and has never accepted fish handouts. At the Peron beach, other animals clearly observe the beaching behavior but do not attempt it. A large number of dolphins forage in the "sponge channels", but they do not sponge. Several of our focal females regularly tail-out and peduncle dive in the sponge channel, but never pick up sponges. There might be some inhibition to development of foraging tactics that are not exhibited by one's mother or such tactics may require some threshold of exposure, perhaps during a sensitive period.

Other foraging techniques observed in our study population but not in our focal animals (e.g., kerplunking, see Connor *et al.*, 2000b) indicate that there may be other foraging and prey specializations of which we are yet unaware. For example, Square, one of the focal mothers in this study, forages for an average of only 10% of her time during daylight hours, yet during one night-time follow, she foraged for approximately 55% of the time. Therefore, we suspect that some females may be "nocturnal" specialists. Further, we know little about the diversity of prey consumed, and there may be specializations in this domain as well. Although Wedges' consumption of golden trevally is obvious, smaller prey are difficult to identify. We have observed only one female in our sample catching and eating stingrays (blue spotted fantail, *Taeniura lymma*). Another female frequently catches flathead (possibly *Sorsogono tuberculata*); only her daughter has been observed catching the same type of fish.

9.4.2 Vertical transmission and developmental mechanisms

We suggest that social learning, especially between mothers and calves (vertical transmission), plays an important role in the calf's foraging development. Laland and Kendal (Ch. 2) suggest that predominantly vertical transmission is adaptive when environmental change occurs relatively slowly. They make two predictions for conditions favorable to social learning that are relevant for Shark Bay dolphins (see Ch. 2). First, social learning is favored when the observer and demonstrator experience the same environment. It would be expected that specialized foraging types in dolphins would be passed on from generation to generation only when the environmental conditions were similar. This is also relevant to the discussion of sex differences in the adoption of the mother's foraging strategy; females are more likely to associate with their mothers and, therefore, experience the same environmental conditions. We predict that daughters are more likely to adopt foraging strategies similar to their mothers than are sons, especially when the strategies are highly habitat specific. This appears to be the case for sponge carrying, but we have insufficient data to test this hypothesis more broadly.

The second prediction from Laland and Kendal relevant to dolphins in Shark Bay is that information regarding resources that are relatively static is more likely to be socially learned than information regarding resources that are rapidly changing. At Shark Bay, specialized foragers do not shift foraging techniques seasonally, suggesting that prey are static or specialists can exploit multiple prey types with one technique.

Laland and Kendal's third prediction is that costly skills are more likely to be socially learned. Our findings relevant to this prediction concern the restriction of sponge carrying to mothers and female calves, coupled with its delayed appearance in the calves. Given the length of time apparently necessary for the development of sponge foraging, it appears to be a difficult strategy to learn.

Why do female calves, but not male calves, readily adopt and appear to maintain foraging traditions within matrilines? It seems clear that daughters, who maintain strong ties with their mothers after weaning, would clearly benefit by developing similar foraging tactics so long as the mother's foraging tactics are adequate. Peglet, Square's fully grown daughter, appears, like her mother, to forage little during the daytime

(< 10%, J. Watson, unpublished data), although night-time follows have not been conducted. After weaning, sons may be less prone to maintain the same tactics as their mothers. Although their ranging overlaps extensively with that of their mothers after weaning, the development of strong bonds with males is likely to take precedence over feeding sites. Consequently, we would expect males to become more opportunistic and eclectic in foraging with age. By the time they reach adulthood, a male's foraging tactics may depend more upon who his alliance partner is than who his mother is.

We still know little about precisely how the calves learn to forage. Although calves are often close to their mothers while she is foraging and they can obviously see and perhaps hear what she is doing, they also are more likely separate from their mothers (> 10 m) during foraging than during other activities (J. Mann and J. J. Watson, unpublished data) and at these times would not be able to see what prey items she chases or catches. However, calves may not only hear the patterning of the mother's sonar but might also hear some of the feedback from those pulses. During the first year, when most foraging techniques appear, the calf could have significant opportunities to link acoustic and visual phenomena with foraging activity when it is close to the mother. Most foraging techniques were initiated in the first year. Data from captivity suggest that adult and immature dolphins are excellent mimics, both in gestural/motor and vocal domains (e.g., Bauer and Harley, 2001). Field data offer additional support for acoustic matching (Janik, 2000) and mother–calf swimming and breathing synchrony (Mann and Smuts, 1999). Such abilities could clearly predispose calves to learn foraging tactics from their mothers, even outside of close visual proximity.

In addition to observing or hearing maternal chase and capture methods, calves may need to learn which prey are desirable. Some fish are toxic or have spines that are difficult to process. As described in Section 9.1.2, when sizeable prey are caught (> 20 cm inches), calves frequently approach and closely inspect the fish. The "owner" can even allow the fish to float at the surface and no one will attempt to steal it. In the absence of food sharing, this behavior suggests that calves learn about prey types by inspecting what others catch.

9.4.3 Future directions

Individual differences in foraging among Shark Bay dolphins are robust, consistent, and acquired by offspring. We suggest that social learning is likely to play a part in the development of most foraging tactics in young dolphins, but only two of the eleven tactics we observed meet the more stringent definitions of tradition. Most of the literature on cetacean social learning to-date has demonstrated acoustic traditions, such as killer whale dialects (Deecke *et al.*, 2000). Our data suggest that elaborate motor skills can also be socially learned and maintained across generations. This is not surprising given the importance of social living for most cetaceans. The parallels between primates and cetaceans are striking, and such comparisons will continue to provoke us. In an environment so alien to our own, dolphins have evolved flexible learning strategies that challenge our primate-centric perspectives.

9.5 Acknowledgments

We would like to thank our colleagues that contributed to the long-term Shark Bay Dolphin Project over the last 13 years: Lynne Barre, Lars Bejder, Per Berggren, David Charles, Richard Connor, Hugh Finn, Cynthia Flaherty, Mike Heithaus, Vincent Janik, Michael Krützen, Harvey Raven, Andrew Richards, Amy Samuels, Bill Sherwin, Rachel Smolker, and Barbara Smuts, the rangers of the Dolphin Information Centre and the crew of *Shotover*. We are grateful to the many research assistants who made data collection possible: Lauren Acinapura, Ellen Aspland, Ferdinand Arcinue, Heidi Barnett, Katherine Bracke, Thalia Brine, Caryann Cadman, Cynthia Davis, Julia Dorfmann, Lindsey Durbin, Karen Ertel, Julie Gros-Louis, Kristina Habermann, Leila Hatch, Krista Irwin, Christian Johannsen, Courtney Kemps, Ann Klocke, Ryan LeVasseur, Justin Marquis, Brenda McGowan, Caitlin Minch, Kristen Nienhaus, Lynn Posse, Kylie Reid, Catherine Samson, Jana Watson, Kristel Wenziker, and Gabrielle Xena-Hailu. Research support was provided to J. Mann and her assistants by the Eppley Foundation for Research, the National Science Foundation grant 9753044, the Helen V. Brach Foundation, Georgetown University, American Cetacean Society, and the New York Explorers' Club. Support for long-term study was provided by the Dolphins of Shark Bay Research Foundation (USA) and the Dolphins of Monkey Mia Research Foundation (Australia). The Department of Conservation and Land Management of Western Australia, Ron Swann, and the University of Western Australia Department of Anatomy and Human Biology also provided logistical support. The Monkey Mia Dolphin Resort, Dean and Leisha Massie, and Graeme Robertson have been extraordinarily supportive of our efforts and donated accommodation and office space to our research team. We also thank Richard Holst and Jo Heymans for their kindness and assistance over the years.

References

Allen, G. R. and Swainston, R. 1988. *The Marine Fishes of North-Western Australia*. Perth: Western Australian Museum.

Altmann, J. 1974. Observational study of behavior: sampling methods. *Behavior*, **49**, 227–267.

Baird, R. W. 2000. The killer whale: foraging specializations and group hunting. In *Cetacean Societies: Field Studies of Dolphins and Whales*, ed. J. Mann, R. C. Connor, P. L. Tyack, and H. Whitehead, pp. 127–153. Chicago, IL: University of Chicago Press.

Bauer, G. B. and Harley, H. E. 2001. The mimetic dolphin. *Behavioral and Brain Sciences*, **24**, 326.

Bearzi, B., Notarbartolo-Di-Sciara, G., and Politi, E. 1997. Social ecology of bottlenose dolphins in the Kuarneric (Northern Adriatic Sea). *Marine Mammal Science*, **13**, 650–668.

Bel'kovich, V. M., Ivanova, E. E., Yefremenkova, O. V., Kozarovitsky, I. B., and Kharitonov, S. P. 1991. Searching and hunting behavior in the bottlenose dolphin (*Tursiops truncatus*) in the Black Sea. In *Dolphin Societies: Discoveries and Puzzles*, ed. K. Pryor and K. S. Norris, pp. 38–67. Berkeley, CA: University of California Press.

Berggren, P. 1995. Foraging behavior by bottlenose dolphins (*Tursiops* sp.) in Shark Bay, Western Australia. In *Abstracts of the Eleventh Biennial Conference of the Society for Marine Mammalogy*, Orlando, FL, p. 11.

Caldwell, D. K. and Caldwell, M. C. 1972. *The World of the Bottlenose Dolphin*. Philadelphia, PA: Lippincott.

Clapham, P. 2000. The humpback whale: seasonal feeding and breeding in a baleen whale. In *Cetacean Societies: Field Studies of Dolphins and Whales*, ed. J. Mann, R. C. Connor, P. L. Tyack, and H. Whitehead, pp. 173–196. Chicago, IL: University of Chicago Press.

Cockcroft, V. G. and Ross, G. J. B. 1990. Food and feeding of the Indian Ocean bottlenose dolphin off Southern Natal, South Africa. In *The Bottlenose Dolphin*, ed. S. Leatherwood and R. R. Reeves, pp. 295–308. San Diego, CA: Academic Press.

Connor, R. C. and Smolker, R. A. 1985. Habituated dolphins (*Tursiops* sp.) in Western Australia. *Journal of Mammalogy*, **66**, 398–400.

Connor, R. C., Wells, R., Mann, J., and Read, A. 2000a. The bottlenose dolphin, *Tursiops* sp.: social relationships in a fission–fusion society. In *Cetacean Societies: Field Studies of Dolphins and Whales*, ed. J. Mann, R. C. Connor, P. L. Tyack, and H. Whitehead, pp. 91–126. Chicago, IL: University of Chicago Press.

Connor, R. C., Heithaus, M. R., Berggren, P., and Miksis, J. L. 2000b. "Kerplunking": surface fluke-splashes during shallow-water bottom foraging by bottlenose dolphins. *Marine Mammal Science*, **16**, 646–653.

Corkeron, P. J., Bryden, M. M., and Hedstrom, K. E. 1990. Feeding by bottlenose dolphins in association with trawling operations in Moreton Bay, Australia. In *The Bottlenose Dolphin*, ed. S. Leatherwood and R. R. Reeves, pp. 329–336. San Diego, CA: Academic Press.

Deecke, V. B., Ford, J. K. B., and Spong, P. 2000. Dialect change in resident killer whales: implications for vocal learning and cultural transmission. *Animal Behaviour*, **60**, 629–638.

dos Santos, M. E. and Lacerda, M. 1987. Preliminary observations of the bottlenose dolphin (*Tursiops truncatus*) in the Sado estuary (Portugal). *Aquatic Mammals*, **13**, 65–80.

Heithaus, M. R. and Dill, L. M. 2002. Food availability and tiger shark predation risk influence dolphin habitat use. *Ecology*, 83, 480–491.

Hoese, H. D. 1971. Dolphin feeding out of water on a salt marsh. *Journal of Mammalogy*, **52**, 222–223.

Janik, V. 1999. Origins and implications of vocal learning in bottlenose dolphins. In *Mammalian Social Learning: Comparative and Ecological Perspectives*, ed. H. O. Box and K. R. Gibson, pp. 308–326. Cambridge: Cambridge University Press.

Janik, V. 2000. Whistle matching in wild bottlenose dolphins (*Tursiops truncatus*). *Science*, **289**, 1355–1357.

Janik, V. and Slater, P. J. B. 1997. Vocal learning in mammals. *Advances in the Study of Behavior*, **26**, 59–99.

Janik, V. and Slater, P. J. B. 2000. The different roles of social learning in vocal communication. *Animal Behaviour*, **60**, 1–11.

Johnston, T. D. 1982. Selective costs and benefits in the evolution of learning. *Advances in the Study of Behavior*, **12**, 65–106.

Leatherwood, S. 1975. Some observations of feeding behavior of bottle-nosed dolphins (*Tursiops truncatus*) in the northern Gulf of Mexico and (*Tursiops* cf. *T. Gilli*) off southern California, Baja California, and Nayarit, Mexico. *Marine Fisheries Review*, **37**, 10–16.

Mann, J. 1998. The relationship between bottlenose dolphin mother–infant behavior, habitat, weaning and infant mortality. In *Abstracts of the Twelfth Biennial Conference of the Society for Marine Mammalogy*, Monaco, p. 85.

Mann, J. 1999. Behavioral sampling methods for cetaceans: a review and critique. *Marine Mammal Science*, **15**, 102–122.

Mann, J. 2001. Cetacean culture: definitions and evidence. *Behavior and Brain Sciences*, **24**, 343.

Mann, J. and Smuts, B. B. 1998. Natal attraction: allomaternal care and mother–infant separations in wild bottlenose dolphins. *Animal Behaviour*, **55**, 1097–1113.

Mann, J. and Smuts, B. B. 1999. Behavioral development in wild bottlenose dolphin newborns (*Tursiops* sp.). *Behaviour*, **136**, 529–566.

Mann, J., Connor, R. C., Barre, L. M., and Heithaus, M. R. 2000. Female reproductive success in wild bottlenose dolphins (*Tursiops* sp.): life history, habitat, provisioning, and group-size effects. *Behavioural Ecology*, **11**, 210–219.

Marino, L. 1998. A comparison of encephalization between odontocete cetaceans and anthropoid primates. *Brain Behavior and Evolution*, **51**, 230–238.

Mead, J. G. and Potter, C. W. 1990. Natural history of bottlenose dolphins along the central Atlantic coast of the United States. In *The Bottlenose Dolphin*, ed. S. Leatherwood and R. R. Reeves, pp. 165–195. San Diego, CA: Academic Press.

Noad, M. J., Cato, D. H., Bryden, M. M., Jenner, M. N., and Jenner, K. C. S. 2000. Cultural revolution in whale songs. *Nature*, **408**, 537.

Noke, W. D. and Odell, D. K. 1999. Interaction between the Indian River Lagoon blue crab fishery and the bottlenose dolphin (*Tursiops truncatus*). In *Abstracts of the Seventh Annual Atlantic Coastal Dolphin Conference*, Virginia Beach, p. 12.

Norris, K. S. and Dohl, T. P. 1980. The structure and functions of cetacean schools. In *Cetacean Behavior: Mechanisms and Functions*, ed L. M. Herman, pp. 211–261. New York: Wiley.

Norris, K. S. and Prescott, J. H. 1961. Observations of cetaceans of Californian and Mexican waters. *University of California Publications in Zoology*, **63**, 291–402.

Petricig, R. O. 1993. Diel patterns of strand-feeding behavior by bottlenose dolphins in South Caroline salt marshes. In *Abstracts of the Tenth Biennial Conference of the Society for Marine Mammalogy*, Galveston, TX, p. 86.

Pryor, K., Lindbergh, J., Lindbergh, S., and Milano, R. 1990. A human–dolphin fishing cooperative in Brazil. *Marine Mammal Science*, **6**, 77–82.

Rendell, L. and Whitehead, H. 2001. Culture in whales and dolphins. *Behavior and Brain Sciences*, **24**, 309–382.

Ridgway, S. H. 1986. Physiological observations on dolphin brains. In *Dolphin Cognition and Behavior: A Comparative Approach*, ed. R. J. Schusterman, J. A. Thomas, and F.G. Wood, Hillsdale, NJ: Erlbaum.

Rossbach, K. A. and Herzing, D. L. 1997. Underwater observations of benthic feeding bottlenose dolphins (*Tursiops truncatus*) near Grand Bahamas Island, Bahamas. *Marine Mammal Science*, **13**, 498–503.

Sayigh, L. S., Tyack, P. L., Wells, R. S., Scott, M. D., and Irvine, A. B. 1995. Sex differences in signature whistle production of free-ranging bottlenose dolphins, *Tursiops truncatus*. *Behavioral Ecology and Sociobiology*, **36**, 171–177.

Sayigh, L. S., Tyack, P. L., Wells, R. S., Solow, A., Scott, M. D., and Irvine, A. B. 1999. Individual recognition in wild bottlenose dolphins: a field test using playback experiments. *Animal Behaviour*, **57**, 41–50.

Scott, M. D., Wells, R. S., and Irvine, A. B. 1990. A long-term study of bottlenose dolphins on the West Coast of Florida. In *The Bottlenose Dolphin*, ed. S. Leatherwood and R. R. Reeves, pp. 235–244. San Diego, CA: Academic Press.

Shane, S. H. 1990. Comparison of bottlenose dolphin behavior in Texas and Florida, with a critique of methods for studying dolphin behavior. In *The Bottlenose Dolphin*, ed. S. Leatherwood and R. R. Reeves, pp. 541–558. San Diego, CA: Academic Press.

Smolker, R. A., Richards, A. F., Connor, R. C., and Pepper, J. W. 1992. Sex differences in patterns of association among Indian Ocean bottlenose dolphins. *Behaviour*, **123**, 38–69.

Smolker, R. A., Mann, J., and Smuts, B. B. 1993. The use of signature whistles during separations and reunions among wild bottlenose dolphin mothers and calves. *Behavioral Ecology and Sociobiology*, **33**, 393–402.

Smolker, R. A., Richards, A. F., Connor, R. C., Mann, J., and Berggren, P. 1997. Sponge-carrying by Indian Ocean bottlenose dolphins: possible tool-use by a delphinid. *Ethology*, **103**, 454–465.

Tyack, P. L. 2000. Functional aspects of cetacean communication. In *Cetacean Societies: Field Studies of Dolphins and Whales*, ed. J. Mann, R. C. Connor, P. L. Tyack, and H. Whitehead, pp. 270–307. Chicago, IL: University of Chicago Press.

Wells, R. S., Scott, M. D., and Irvine, A. B. 1987. The social structure of free-ranging bottlenose dolphins. In *Current Mammalogy,* Vol. 1, ed. H. Genoways, pp. 247–305. New York: Plenum Press.

Whitehead, H. 1998. Cultural selection and genetic diversity in matrilineal whales. *Science,* **282**, 1708–1711.

Whitehead, H. and Mann, J. 2000. Female reproductive strategies of cetaceans: life histories and calf care. In *Cetacean Societies: Field Studies of Dolphins and Whales,* ed. J. Mann, R. C. Connor, P. L. Tyack, and H. Whitehead, pp. 219–246. Chicago, IL: University of Chicago Press.

Williams, T. M., Friedl, W. A, Fong, M. L., Yamada, R. M., Dedivy, P., and Haun, J. E. 1992. Travel at low energetic cost by swimming and wave-riding bottlenose dolphins. *Nature,* **355**, 821–823.

Williams, T. M., Friedl, W. A., and Haun, J. E. 1993. The physiology of bottlenose dolphins (*Tursiops truncatus*): heart rate, metabolic rate and plasma lactate concentration during exercise. *Journal of Experimental Biology,* **179**, 31–46.

MICHAEL A. HUFFMAN AND SATOSHI HIRATA

10

Biological and ecological foundations of primate behavioral tradition

10.1 Introduction

An interest in nonhuman primate behavioral traditions has existed since the beginning of primatology, with some of the earliest details coming from the Japanese macaque (*Macaca fuscata*). When Kyoto University researchers began their investigations in 1948, under the leadership of Denzaburo Miyadi and Kinji Imanishi (Asquith, 1991), animals were considered to act on instinct and such concepts as tradition or culture were considered to be a uniquely human trait (de Waal, 2001; Kroeber and Kluckhohn, 1952). Imanishi (1952) predicted the presence of "culture" in animals even before the results of these observations had begun to be published. He emphasized that, unlike instinct, culture in animals should be viewed as the expression of developmentally labile behaviors. He reasoned that, if one defines culture as behavior transmitted to offspring from parents, differences in the way of life of members of the same species, whether they are human, monkey, or wasp, belonging to different social groups could be attributed to culture. Imanishi's general argument still holds today, albeit with greater refinements in our overall view of the phenomenon (e.g., Avital and Jablonka, 2000; de Waal, 2001; McGrew, 2001). Currently, healthy debate over whether culture or tradition in humans and animals is really the same is ongoing (e.g., Boesch and Tomasello, 1998; Galef, 1992; Tuttle, 2001; see also Ch. 6).

We use the term behavioral tradition in this chapter to denote those behaviors for which social context contributes to their acquisition by new practitioners and which are maintained within a population through social means (as defined by Fragaszy and Perry in Ch. 1; McGrew, 2001). Operationally, we define a behavioral innovation as any single (or set of) species-typical voluntary action performed in a novel context that has not previously been observed to be performed by members of that group. Thus, new behavioral traditions arise from behavioral innovations (typically made up of existing behaviors) and can diffuse within a group through any of several possible processes, including local enhancement, social facilitation, observational learning, and imitation (Galef, 1976; Whiten, 2000). Visalberghi and Fragaszy (1990) have pointed out the importance of knowing the history of a behavioral innovation to assess the contribution of any one of these possible processes in its diffusion within a group. In nature, only under the best of long-term observational conditions is it possible to know with any degree of certainty whether or not one is documenting a behavioral innovation. Likewise, the diffusion of a new behavioral innovation into a group can be a long process. Rarely have such events been observed and documented in detail. Controlled experimentation sometimes allows us to avoid such difficulties and gain a better appreciation of the situation in nature (e.g., Hirata and Morimura, 2000; Inoue-Nakamura and Matsuzawa, 1997; Matsuzawa, 1994). Laboratory studies have also provided useful insights into the details of social diffusion, providing various parsimonious interpretations of the learning processes involved (e.g., Lefebvre, 1995; Visalberghi and Fragaszy, 1990; Chs. 6 and 7). Even under the best of experimental conditions, however, it is difficult to reconstruct faithfully the complex ecological and social conditions under which innovation and its subsequent diffusion may occur in natural populations.

Innovation and diffusion of new behaviors within a troop, and the establishment of group-specific behavioral traditions, have been topics of great interest from the beginning of research on Japanese macaques (e.g., Huffman, 1984, 1996; Huffman and Quiatt, 1986; Itani, 1958; Itani and Nishimura, 1973; Kawai, 1965; Kawamura, 1959; Watanabe, 1994). By the early 1950s, at well-known sites such as Koshima, Takasakiyama, Arashiyama, and Minoo, provisioning and individual recognition of all troop members was accomplished, starting off the practice of long-term comparative research of troops across the country (see Huffman, 1991; Takahata *et al.*, 1999; Yamagiwa and Hill, 1998). Provisioning provided the first outdoor laboratory situation for recording the process of behavioral innovation and diffusion of behaviors in a novel environment. Research at these sites has contributed much to our understanding of the patterns of diffusion of innovative behavior in monkeys (see Itani and Nishimura, 1973; Nishida, 1987; Thierry, 1994). A growing number of long-term studies of chimpanzee populations have also revealed an array of behavioral variation between populations that has been attributed to social learning. Currently, a set of behavioral traditions within a group and differences in such sets of tradition between groups is given by some as evidence for chimpanzee culture (see Whiten *et al.*, 1999).

In chimpanzees, the uniqueness of a particular behavior to a given group is often highlighted (e.g., McGrew and Tutin, 1978; Nakamura *et al.*, 2000; Nishida *et al.*, 1983). As more sites have reached long-term study status, however, a number of these behaviors, shown to be practiced regularly within a group and socially transmitted to each new generation, are frequently found to occur in more than one population. In both Japanese macaques and chimpanzees, there are examples of behavioral traditions that occur in geographically isolated groups of the same species and or among different subspecies. They include foraging skills (ant dipping, leaf sponge, honey dipping), self-medication (leaf swallowing), social conventions, communicative signals (hand clasp grooming, leaf clipping) and a form of object play (stone handling) (see Boesch, 1996; Huffman, 1996, 1997).

While it is accepted that behavioral innovations can be passed to future generations via social learning, the foundations of behavioral innovations themselves, which form the basis of behavioral traditions in any species, including our own, have been little discussed. How can it be that behavioral innovations socially transmitted amongst members of one group can also occur in other groups for which social diffusion of behavior cannot possibly occur? A historical explanation would assume that such behavioral traditions are extremely old, implying that ecological, geographical, or even subspecies barriers now close previously open pathways of intergroup behavioral transmission. While plausible, in some cases this may often be difficult to demonstrate, and it does not explain cases where a particular behavioral innovation is observed to emerge simultaneously in more than one group under geographically isolated conditions. Alternatively, a biological explanation assumes that members of the same or closely related species possess common behavioral propensities, leading to a greater than random probability of a behavioral innovation based on them to arise independently in more than one group. This can happen simultaneously or at greatly different points in time. The biological explanation helps to explain why the same or similar behavioral traditions, which apparently arise from innovation, can occur in more than one group.

This strongly suggests that aspects of species-typical behavior can and do indeed shape the pattern of behavioral innovation among members of a taxonomic group. Here, both historical and biological explanations can be complimentary, and indeed important for a clear understanding of the phenomena. However, we have excluded the overly parsimonious assumption that behavioral innovations depend only upon species-wide, latent tendencies and require only the appropriate stimuli to bring them out independently in all individuals of a group or species.

This chapter addresses the various factors supporting behavioral innovation and discusses the interaction between the biological and environmental variables (both social and ecological) that influence the diffusion of such innovations in free-living populations. We synthesize research on Japanese macaque behavioral traditions to discuss the possible effects of group size and behavior type on the rate of diffusion and the pathways of transmission. New information is also presented from long-term multisite comparative studies of two behavioral traditions: stone handling in Japanese macaques (Huffman, 1984, 1996; Huffman and Quiatt, 1986) and leaf swallowing in the African great apes (Huffman, 1997; Huffman *et al.*, 1996; Huffman and Caton, 2001). This exploration of the biological and ecological foundations of animal traditions is intended to improve our understanding of fundamental aspects of social learning, and the role behavioral traditions may play in the survival of the organism.

10.2 Biological basis of behavioral innovation: behavioral predispositions

Given enough time to familiarize oneself with a particular social group of animals, it is apparent that each individual has its own unique personality. Further time spent comparing two or more groups will invariably lead one to the conclusion that different groups in different regions of the distribution of a species can differ strikingly from one another in some details of their social or feeding habits. These population differences in the overall behavioral repertoire of a species are what we most readily identify as behavioral traditions. At the same time, there are inescapable similarities between groups and the individuals within groups, which make them recognizable as members of the same species. That is, the better you get to know the behavior of individuals in one group, the easier it becomes to predict with a relatively high level of accuracy just how any individual of that species will respond in a particular social situation elsewhere. An individual's unique qualities, based on personal experience and other attributes, coexists with a relatively high species-level predictability (behavioral predisposition) (see also Mendoza and Mason, 1989). We argue that this constitutes the biological foundations of behavioral traditions in any animal species, including humans.

Keeping these factors in mind, we make six basic assumptions about the role of species-level behavioral predispositions as important biological features of behavioral innovation and discuss their possible role in the emergence of behavioral traditions arising in geographically distinct populations of the same species.

1. The basic motor units of behavior evident in a species are shared by all members of that species. These basic behavioral units are the product of adaptation to social and ecological challenges in its evolutionary past and are shared by all members of a species in the present.
2. In order to survive and reproduce, animals have to be good at reading and appropriately responding to the behavior of conspecifics and to changing environmental conditions. These behavioral units are the basic building blocks of behavioral traditions.
3. The reliability with which a behavior occurs in a species is based on the predictability of the response to stimuli in the social and ecological environment and the reliability with which a particular set of environmental conditions occurs. (See Ch. 2, for a model of the rate of environmental change.)
4. Predictable behavior across individuals of a species reflects reliable production of a finite set of behavioral units. The capacity for innovation is limited to the possible number of permutations of such behaviors an individual can produce.
5. While the possible number of behavioral permutations is influenced by biological constraints (physiological and morphological), an individual may never fully exploit the full potential of its species in any given environment. This is considered to be the source of a behavioral innovation and intergroup variability upon which behavioral traditions are based.
6. A shared repertoire of behavioral units and a shared degree of predictability in the production of particular behaviors in particular situations makes it possible for common behavioral innovations to appear and common behavioral traditions to arise among socially and geographically isolated groups.

The juxtaposed predictability and unpredictability of behavior in a species becomes more apparent the more familiar we become with that species. Behavioral predispositions make members of a species interact with their environments in a relatively predictable way. At the same time, we fully recognize the dual importance of individual differences and the novelty of social and ecological contexts in which behavioral innovations arise. An innovation is likely to arise when an individual(s) or group is faced with new social or ecological challenges for which it currently has no workable solution in its existing behavioral repertoire.

Intuitively, the more generalist a species, the greater array of behavior it is likely to exhibit and, therefore, the more flexible to environmental change it should be. As a general principle then, those species found to exist in a wide range of social, climatic, and ecologically diverse environments should be expected to exhibit the greatest array of behavioral traditions. They may also be better social learners, but this is a different issue. Reader (Ch. 3) reports a significant positive correlation between the incidence of reports of social learning, innovation, and tool use and the absolute "executive" brain volume and the ratio of "executive" brain over brainstem in nonhuman primates. If these measures are a robust indicator of adaptability and intelligence, regardless of relative phylogenetic positions, behavioral traditions will be more frequent among generalist species than among specialists. This should hold true for any animal species and behavior in which social context contributes to behavioral acquisition.

10.3 Innovative behaviors in Japanese macaques

10.3.1 Phases of behavioral diffusion

In macaques, and presumably other social animals, the diffusion of a behavioral tradition can be divided into three distinct phases: transmission, tradition, and transformation (Huffman and Quiatt, 1986). The *transmission phase* is the period of early dissemination of a behavior and is typically similar from group to group and presumably species to species, at least among primates. The first individual(s) to display a behavior may do so repeatedly and perhaps for increasingly longer periods of time. The behavior is first acquired by a network of spatial–interactional associates of the innovator. The membership of this network is directly influenced by the nature of the behavior being performed and its context (e.g., feeding, resting, traveling, mating, etc.). Laland's (1999) work on the transmission of digging behavior in rats, however, suggests that the innovator of a new behavior is not always easy to detect, and multiple individuals may exhibit the behavior almost simultaneously.

A behavioral tradition need not diffuse to all members of a social group. The more specialized the functions and context of the behavior is, the more limited will be the subgroup of individuals (age, sex, rank, etc.) that will acquire it. Diffusion rate and the distribution of the behavior across age–sex classes should, therefore, vary according to the behavior in question. For example, if the behavior were a form of sexual display, like leaf clipping in Mahale chimpanzee, then we would not expect it to be acquired by sexually immature or postreproductive individuals at any phase of the diffusion process. In such cases, the behavior would never spread to 100% of a group. If, however, the context of leaf clipping were to be altered at some point to a general solicitation of intent or to attract attention of others, for example, we could expect to see an increase in the proportion of the population exhibiting the behavior starting from that point in time.

The *tradition phase* is the period in which a behavior is passed down from mother to offspring or along other multigeneration lines. At this time, the rate of diffusion will depend upon the direction of diffusion and once again upon who the target of the behavior is in the transmission phase.

The *transformation phase* is a period in which prolonged practice and acquired familiarity with a behavioral pattern is gained. Increased behavioral variety brought about by more active manipulation occurs largely among younger age groups, which naturally tend to be more physically active and explorative. This can be a period of behavioral drift or easily changing fads. An example of this is the divergence in behavioral patterns that developed for wheat washing and potato washing in the 20 or more years following its initial spread at Koshima (Watanabe, 1994). Details on the direction of diffusion at this stage, however, are not clear. In this case too, diffusion is expected to be influenced by the innovator's network of spatial–interactional associates.

Among the reported cases of the diffusion of behavioral innovations in Japanese macaques, most are in one way or another related to food or foraging activity, including the acquisition of new foods and food-processing techniques. In general, information regarding food should be of importance to all members of a group and, therefore, foraging innovations are

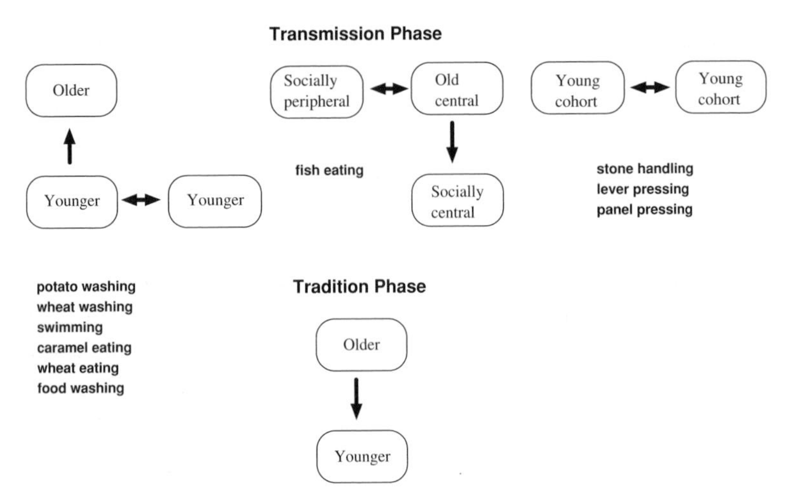

Fig. 10.1. Pathways of diffusion by behavioral type exhibited in Japanese macaques.

expected to diffuse widely. However, socio-ecological factors affecting access to novel foods or more preferred food can enhance or impede the rate and routes of diffusion (e.g., Giraldeau, 1997; Giraldeau, Caraco, and Valone, 1992; Tokida *et al.*, 1994; Watanabe, 1989).

10.3.2 Pathways of diffusion

Three basic pathways of behavioral diffusion in the transmission phase have been recorded: younger to older, peripheral to central group members, and between young of the same cohort. These pathways are shown in Fig. 10.1 along with the characteristic older to younger pathway inherent in the tradition phase. Some examples of the different behaviors associated with each pathway are also given. Other pathways are likely to exist, but to our knowledge they have yet to be reported. Pathways are dependent upon the nature of the behavior and the relationships of those individuals most likely to be in proximity with each other when the behavior is being practiced. In many, but not all, cases, this is characteristic of the transmission phase. As a general rule, behavioral type and the context of the behavioral innovation in question will strongly determine this pathway and thereby to some extent also the rate of diffusion.

Of the behaviors noted in Fig. 10.1, food-related behavioral traditions were most consistently associated with the initial lateral transmission among young, followed closely by the upwardly vertical transmission to older kin members and then across kin boundaries to other adults.

The pathway of transmission of fish eating at Koshima is different in that it first appeared in an adult male living in the troop's social periphery. An old female of the troop eventually acquired the habit next. She was reported to be the link by which the behavior then diffused to some members within the troop's social central area (Watanabe, 1989). This food habit was acquired in response to a drastic reduction in provisioned foods. Those individuals most directly affected by the food shortage, who had less access to provisioned foods, and who had social contact with the peripheral adult male were reported to be the first to acquire the behavior. Fish eating was adopted by about a third of the population and is an interesting case of a putative behavioral tradition where diffusion within the group is limited by need (Watanabe, 1989).

Stone handling was first observed being performed by a young juvenile female, Glance 6774, in December of 1979. By June 1985, stone handling was found to have diffused throughout 60% ($n = 142$) of the 236 member B troop (Huffman, 1984). Of these 142 individuals, 80% were born between 1980 and 1983; that is after stone handling was first observed in the troop. Only limited diffusion to individuals older than the first female seen stone handling occurred, suggesting the recent emergence of this behavior within the troop: at least since provisioning was first begun at this site in 1954. It also shows that the behavior did not spread to adults within the troop. Those three individuals older than Glance-6774 that acquired stone handling were two of her female cousins, Glance-6775 and Glance-6774 (sisters), and one lower-ranking non-kin-related female, Blanche-596475.

By 1985, these four females and two others (Oppress-7078, Momo-5978), all then 10 years or older, had one or more offspring of their own. All 13 of these females' offspring also acquired stone-handling behavior. In 1986, B troop divided, becoming E and F troops (Huffman, 1991). By August 1991, 12 years after the first appearance of stone handling, every individual under the age of 10 years in E troop was verified to have acquired the behavior (F troop gradually stayed away from the provisioning site and observations on them were stopped). Stone handling had spread to the young of every kin group in the troop. Unlike potato washing or wheat washing, however, no individual 5 years of age or older in 1979 (when the behavior first appeared at Arashiyama) ever acquired stone handling later on.

Long-term observations made on stone handling have revealed that the social network of diffusion has varied over time as a function of the age of individuals exhibiting the behavior and the social context of the behavior itself. In the initial transmission phase of stone handling, this network included a very small group of cousins, sisters and non-kin playmates. Very shortly thereafter, however, the behavior began to spread more widely between play groups composed of kin and non-kin as it diffused downward to younger individuals from mother to offspring, older to younger siblings, etc. in the tradition phase (Fig. 10.1).

Consequently, in the first few years, infants of stone-handling mothers were exposed to the behavior earlier than other infants. However, according to the 1985 census, even those infants whose mothers had not acquired stone-handling behavior began to pick up or scatter stones on the ground as early as 10 weeks. In all these infants, the behavior was exhibited by older siblings indicating that stone handling can also be acquired via older siblings. However, since then and up to the time of writing this paper, all infants acquire the behavior within the first 6 months of life. Multiple modes are suspected to have played a role at different stages of the behavior's history, with some form of social facilitation no doubt playing a central role.

From this, we would predict that no one particular age, sex, or rank class had a monopoly on innovation skill. Rather, the type of innovation is likely to be influenced by the unique position of each individual within its social and ecological environment.

10.3.3 Behavior type, group size, and rates of diffusion

Earlier evolutionary and population-level models of cultural transmission assumed rapid and, as discussed by Laland and Kendal (Ch. 2), temporally accelerating rates of behavioral diffusion within a group, producing a sigmoid curve (Boyd and Richardson, 1985; Cavalli-Sforza and Feldman, 1981; Pulliam, 1983). Most of these models assume that, as each new individual acquires the behavior, the rate of diffusion will increase as a function of an increase in the number of demonstrators who can influence the remaining naïve individuals. Laland and Kendal (Ch. 2) disagree and suggest that the shape of the curve of diffusion is not always consistent with the pattern of learning (social versus asocial). They conclude that the shape of the diffusion curve may not allow us to identify the learning process. Lefebvre

(1995) found supporting evidence for an increase in the rate of diffusion with the increase in number of demonstrators in an analysis of the rates of acquisition of innovative behaviors reported in the primate literature, including potato washing, wheat washing and fish eating in the Japanese macaque. He found accelerating rates of diffusion as the number of practitioners of a behavior increased in some but not all behaviors. Conversely, Lefebvre and Giraldeau (1994) also found that large group size could have a negative effect; many naïve bystanders could slow down diffusion. However, we must not assume that a behavioral innovation will be of relevance to every individual in the group. In social learning models, we cannot assume *a priori* that all behaviors will reach 100% diffusion within a group. As seen from fish eating, not even food-related innovations are totally free from such considerations.

Behavior type and group size are not typically included in models of cultural transmission. The question we ask here is, "What effect do these variables have on the rate of diffusion?" From our discussions above, we know that the pathway of diffusion is affected by behavior type and that the function of the behavior determines the type of individual and, therefore, the total number of individuals within a group that will acquire it. Based on this evidence, we predict that group size alone does not have an over-riding effect on the rate of diffusion. To test this, we calculated the theoretical rate of increase in the number of individuals performing 12 novel behaviors reported in Japanese macaques to estimate the time it would have taken each behavior to spread to 50% of the group. Here we assume a constant rate of increase. The number of days necessary to diffuse to 50% of the population in these 12 behaviors was not found to be significantly related to group size alone (Spearman rank correlation coefficient (r_s) $= 0.38$; $n = 12$; not significant). This pattern remained constant even when we excluded experimentally induced behaviors (caramel eating and lever and panel pressing; $r_s = 0.49$; $n = 7$; not significant). No consistent pattern was found with regards to troop size (Fig. 10.2). However, we did find a significant difference in the number of days to diffuse to 50% of the population when these 12 new behaviors were grouped into four behavioral types (Kruskal–Wallis ANOVA: $H_{(3,12)} = 8.1$; $p < 0.05$) (Fig. 10.3). Food processing and play were much slower to diffuse (over 1400 days) than accepting a new food and experimental tasks (less than 200 days). As predicted, behavioral type does have an important effect on the rate of diffusion and the effect of group size is inconsistent, even when using a linear model (see also Ch. 2).

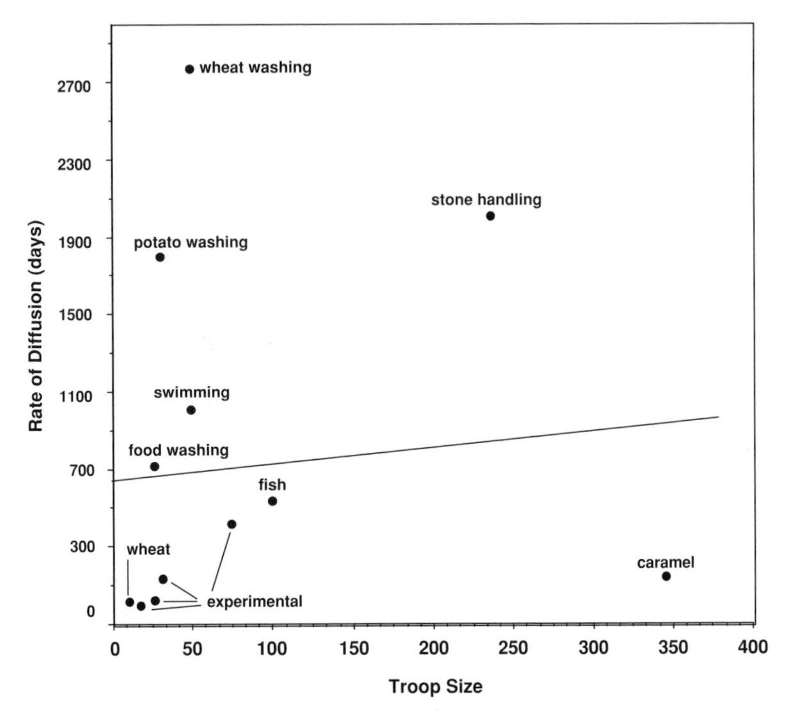

Fig. 10.2. Correlation between troop size and estimated rate of diffusion among 50% of the population for 12 behavioral traditions observed in Japanese macaques.

This difference in rates of diffusion can be explained by the fact that accepting a new food item or manipulating an experimental device for immediate food reward does not compete with an existing way of handling a problem (e.g., already available food). In the case of play, the new behavior is likely to be acquired only by a specific subset of the population, constraining the rate and defining the level of diffusion into the group.

Previously, a slow rate of diffusion has been considered an argument in favor of the more parsimonious mechanism of individual learning supporting the acquisition of some behaviors in Japanese macaques (Galef, 1991, 1992). However, in light of our empirical analysis discussed above and supported by the theoretical discussion of Laland and Kendal (Ch. 2), we conclude that variations in the rate of diffusion do not necessarily reflect more or less reliance on social context in learning.

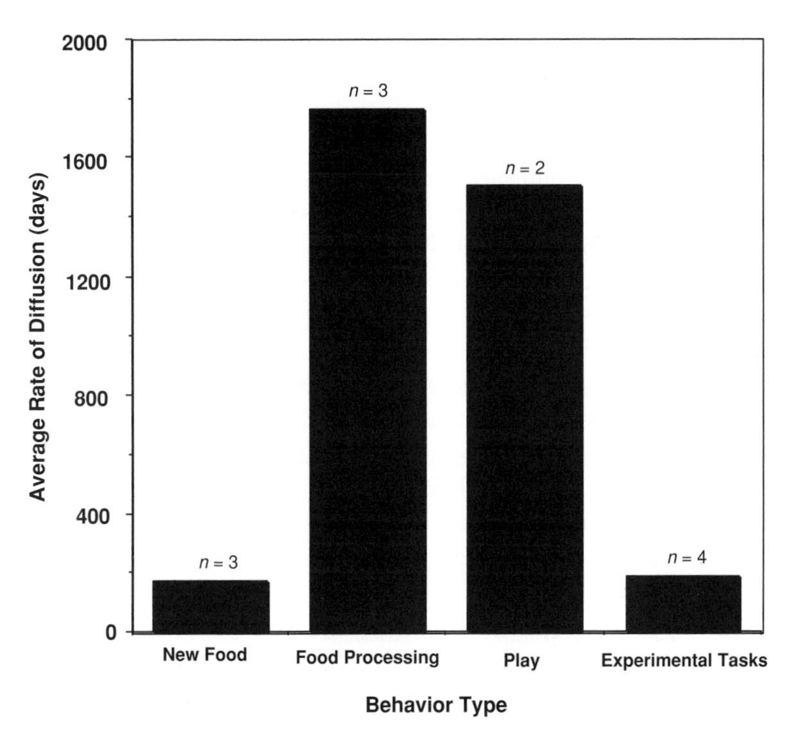

Fig. 10.3. Average rate of diffusion to 50% of the population by behavior type. New Food, acquisition and consumption of food; food processing, new way to process food; play, play-related behavior, such as stone handling and swimming; experimental tasks, new behavior induced by the introduction of a novel experimental setting.

10.4 Factors influencing the innovation, diffusion, and maintenance of primate behavioral traditions

10.4.1 Appearance and disappearance of behavioral traditions

Behavioral traditions can appear, disappear, and even reappear sometimes in slight variation within the same group over time (see Ch. 14). Very few studies have been able to document such change given the long time investment required. Some behaviors, while practiced by a few individuals in any population, may not endure very long. For example, stone handling was observed at Arashiyama and Takasakiyama for the first time in late 1979, after over 30 years of close observation at both sites by a number of researchers and park employees (Huffman, 1984, 1996). Interestingly, before provisioning was started at Arashiyama in 1954, young macaques were sometimes seen to play with inedible hard-covered citrus fruits in a fashion resembling some of the current stone-handling patterns (see Huffman, 1984). The initial circumstances bringing about the practice of stone handling and other new behaviors at these sites are unknown. Site differences in the frequency of provisioning and the size of the feeding area may have some influence on the relative daily frequency of occurrence as well as the overall opportunities for others to observe and take up the practice of the behavior habitually. Although not completely understood, some of the possible factors are discussed below.

10.4.2 Provisioning

Early research on Japanese macaques used provisioning as a way of speeding up the process of habituation and to lure them out into the open for better observation. This brought about a change in the life habits of the monkeys, providing them with access to new foods and environments previously not encountered. These types of change preceded such innovations as potato washing, wheat washing, and altered swimming behavior at Koshima (Kawai, 1965).

With provisioning comes a tendency for a more sedentary lifestyle. More time is spent around the feeding area and that too can have profound effects on behavioral and dietary innovation (see Fa and Lindburg, 1996). At Arashiyama, Huffman (1984) found, on the one hand, that the natural diet of the troop decreased from its early provisioning period (1954–1958) level of approximately 192 species of plants (Murata and Hazama, 1968) to as few as 67 species between 1979 and 1980. On the other hand, while the diversity of their natural diet decreased and dependence on provisioned foods increased, at least 17 new natural and introduced plant foods were acquired in the process of adjusting to life in the newly exploited 1 km radius of the feeding site.

Provisioning can also bring about other changes in behavior. With less time spent actively searching for food, more time is left for other activities such as play and socializing. Provisioning improves reproductive potential in females and causes a shortening of the interbirth interval. This, in turn, can have an effect on a number of behaviors, ranging from infant care practices to modification of matrilineal dominance-rank systems (Hill, 1999; Itani, 1959; Kutsukake, 2000). Little attention has yet been paid to the possible relationship of behavioral innovations and changes in population structure. This should be a fruitful area of future investigation.

In a restricted sense, provisioning can be considered to be synonymous with dramatic changes in a more natural habitat. That is to say, changes in the distribution, defensibility, and abundance of food or other sought-after resources in nature are also expected to trigger changes in social organization, group behaviors, and diet.

10.4.3 Competition

As shown in this chapter, the rate of diffusion of innovative behaviors and their longevity is a complex issue. The relative abundance of resources associated with the innovation also affects which individuals will acquire a new behavior that another practices. The type of behavior under consideration is very important. This will directly influence who is most likely to acquire the behavior, and in the end how widely a behavior will spread among members of a group. If, for example, a behavior allows an individual to obtain a resource previously denied because of sex, age, or rank, the behavior is not likely to spread widely, passing only very slowly to others in the same social situation who are tolerated by the innovator. An example of this is tool manufacture by a chimpanzee to rouse a squirrel out of its hiding place in the hole of a tree (Huffman and Kalunde, 1993). Meat is a highly prized food resource by chimpanzees, with access controlled by a few adult males of the group based upon social and sexual status of the potential recipient (Nishida *et al.*, 1992; Stanford, 1999). The manufacture and use of tools to drive a squirrel out of hiding is an extremely rare behavior at Mahale. The orphan adolescent female observed performing this behavior would normally have no chance to obtain meat from others or to hunt larger prey on her own. Hunting in the presence of others increases the likelihood of the catch being taken away from a subordinate, and, therfore, such activities tend to be done in secret (Huffman and Kalunde, 1993). Here, the lack of social tolerance (see Ch. 11) indirectly encourages efforts to obtain a meat source not highly open to competition. At the same time, this suite of characteristics of the individual and the behavior inhibits the diffusion of the behavior to more powerful individuals in the group. Consequently, although the behavior is potentially important to all, it is not likely to diffuse widely or be observed frequently, because of both limited opportunities for observation by others and the intolerance of subordinate individuals to competitors for a limited resource. This is in contrast to ant fishing, where resources are more widely distributed and abundant, resulting in less competition. By comparison, behaviors with a clear benefit to all members are more likely to spread throughout an entire group and be maintained indefinitely if the resources required for its performance are widely available (e.g., potato washing).

10.4.4 Maintenance of "neutral" behaviors

In some cases, traditions emerge that seem to have no (or even mildly negative) immediate adaptive consequences (e.g., capuchin hand sniffing, Ch. 14). Presumably these behaviors are maintained because of some internal consequences that we cannot as yet measure. Stone handling is a case in point (Huffman, 1996). Unlike the leaf-swallowing behavior exhibited by chimpanzees in the wild, the significance of stone-handling behavior to a Japanese macaque is difficult to interpret. Individually, motivation to perform the behavior simply may be the social value placed on these items by others in the group (Huffman, 1984). The immediate motivation to act as others do and the long-term motivation to continue performing a behavior may be different at both the individual and group levels. This is an especially important area for theorists to consider, because assumptions about motivation and performance should be approached from both short- and long-term perspectives. For example, during the tradition phase (Fig. 10.1) of behavioral transmission, when behaviors are acquired by the very young from their mothers or older siblings, the motivation to perform a behavior is likely to be quite different from that which induced the innovator(s), and perhaps subsequent early initiates in the propagation phase, to acquire the behavior. Furthermore, once a behavioral habit is acquired, individuals may continue to perform it even after the original conditions for promoting its adoption are no longer present if there is no cost to performing it. They are continued merely out of habit.

10.5 A behavioral tradition in multiple troops of the same species: stone handling among Japanese macaques

10.5.1 Behavioral description

Classified as a form of object manipulation or play (Candland, French, and Johnson, 1978), stone handling has so far never been observed in a nonprovisioned troop (Huffman, 1984, 1996). However, there are also several provisioned troops where stone handling has never been recorded (Fig. 10.4). In provisioned and nonprovisioned troops where stone handling has never been seen, prolonged physical contact with stones is absent. At sites where it does occur, stone handling is habitual and occurs most predictably just after feeding time. In this situation, individuals have gathered all the food (often grains of wheat or soy beans) they can

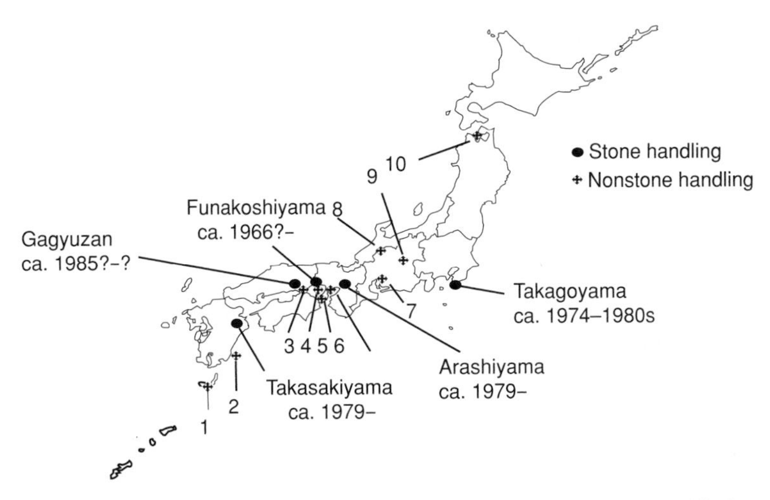

Fig. 10.4. Distribution of study sites where stone handling has been observed in free-ranging, provisioned populations of Japanese macaques. Sites denoted by numbers are provisioned sites where stone handling has not been observed; 1, Yakushima; 2, Koshima; 3, Katsuyama; 4, Shodoshima; 5, Awajishima; 6, Minoo; 7, Hanyama; 8, Hakusan; 9, Jigokudani; 10, Shimokita.

at one time and are slowly pushing the items from the cheek pouch back into the mouth to be chewed before swallowing. Stone handling can be interspersed with bouts of feeding but begins to taper off in most individuals when the food has been completely consumed. The general mood of individuals is relaxed with an intense concentration directed toward the activity, sometimes ignoring or refusing the solicitations for play or mating. There is nothing to stop them from feeding on provisioned food or from moving into the forest to feed on natural vegetation, yet they choose to manipulate stones first.

Some individuals will continue to carry stones around for several minutes after feeding time has finished, depositing the stones in piles at the feeding site, at the base of trees on slopes in the forest, or even sometimes in the fork of a tree. Based on the analyses of video surveys conducted in the winter of 1989 at Takasakiyama and the summer of 1991 at Arashiyama, a highly significant negative correlation was found between age and total handling time per stone-handling session at both sites (Arashiyama: $n = 167$; $r_s = -0.435$; $p = 0.0001$; Takasakiyama: $n = 53$, $r_s = -0.488$; $p = 0.0004$; Huffman, 1996). For both males and females, the decline in stone-handling activity with age appears to be closely correlated with social and biological life-history variables (Huffman, 1996).

Stone handling replaces other activities that normally follow or are interspersed with foraging. In a nonprovisioned troop, and indeed among individuals that do not handle stones in provisioned troops, this is a time for social grooming and play. Only the most dominant individuals are likely to continue feeding until the last bit of provisioned food has disappeared. Mothers can be seen grooming the backs of their offspring who are handling stones, while they themselves are processing the food they have stored in their own cheek pouches.

Macaques in general may have an intrinsic propensity towards manipulatory activity concurrent with foraging. If so, stone-handling activity fits nicely into that time slot. In nonprovisioned groups, forging activity is likely to take longer because food resources are less abundant in any one location and are more spread out, making it necessary to feed for a longer time to obtain the same amount of food. Interspersed with traveling between food patches, this would leave little time for such leisure activities as stone handling.

10.5.2 Intergroup and interspecies behavioral comparison of stone handling

Stone handling has been observed to occur independently in at least five free-living provisioned populations across Japan (Fig. 10.4) and in two captive groups kept at the Kyoto University Primate Research Institute (Huffman, 1984, 1996; Huffman and Quiatt, 1986).

The spatial and temporal distribution of known stone-handling sites demonstrates the behavioral tradition's independent origin in each group. There is no geographical or temporal pattern of emergence to suggest that the behavior spread between provisioned troops within a region (Fig. 10.4). The behavior does not reliably occur in neighboring populations within the same regions where it does occur. An interesting example from this perspective is that stone handling appeared in both the Arashiyama and Takasakiyama populations at around the same time in 1979, while it is thought to have started much earlier at the intermediate location of Funakoshiyama, around 1966. Separated by an ocean barrier and several hundred kilometers of land, it is implausible that the behavior was transmitted between these different populations on Kyushu and Honshu islands.

Only sketchy details are known about most of the other sites where stone handling has been observed. Hiraiwa (1975) made the first brief report of stone handling in Japanese macaques from her observations of the Takagoyama troop. The frequency of occurrence was low and only subadults, younger than 3 years of age, exhibited the behavior. Later on in 1984 when provisioning was stopped, the practice of stone handling gradually ceased (T. Fujita, personal communication). On the island of Koshima in the early 1980s, a 15–16-year-old male named Ira was frequently observed carrying and clacking stones together along the rocky shoreline, but the behavior never spread to other group members (K. Watanabe, M. Kawai, and S. Mito, personal communications). Written records do not exist for the first occurrence of stone handling at the Funakoshiyama Monkey Park, but the caretaker in charge of provisioning this free-ranging troop remembers seeing the behavior as early as 1966 (I. Narahara, personal communication). Stone handling has not spread widely within this troop with approximately 300 members. Details of stone handling at Gagyuzan are even scarcer. The behavior apparently underwent a couple of periods in which its visibility rose and subsequently dropped again in frequency (F. Fukuda, personal communication).

At Arashiyama, where the most detailed studies of stone handling have been conducted, 17 basic behavioral types have been classified (see Table 10.1 for these and the abbreviations). The first five behaviors (GA, PU, SC, RIH, and RT) are commonly exhibited by macaques in general when manipulating objects in their environment such as twigs and acorns or novel human-introduced objects with which they come into contact. The last three behaviors (TW, MP, and GW), connected with moving, are considered to reflect a growing familiarity with stones and are a product of human habitats where hard-packed ground, roofing, or concrete is available.

With the exception of one behavior (PUD), all of the Arashiyama behavioral types have been observed at Takasakiyama (Huffman, 1996), and at relatively similar frequencies (Fig. 10.5). The most common behavioral patterns observed at both Arashiyama and Takasakiyama were SC, RWH, GW, and GH. At Takagoyama, the relative frequency of occurrence was low and the behavioral patterns limited to GA, CD, PU, RT, and RIH (Hiraiwa, 1975), five of the same eight basic behaviors first observed at Arashiyama (Huffman, 1984). The general visibility of stone handling at Funakoshiyama, as observed in the mid-1990s, was much lower than at either Arashiyama or Takasakiyama, despite the troop's large size. The behavioral patterns observed (GA, RWH, RT, RIH, GH, etc.) were identical to those recorded in the other groups (K. Kaneko,

Table 10.1. *The 17 basic behavioral patterns of stone handling observed at Arashiyama*

Behavioral pattern	Characteristics
Gathering (GA)	Gathering stones into a pile in front of oneself
Pick up (PU)	Picking up and placing stones into one hand
Scatter about (SC)	Scattering stones about on the ground in front of oneself
Roll in hands (RIH)	Rolling stones in the hands
Rubbing stones together (RT)	Rubbing stones together
Clacking (CL)	Clacking two stones together
Carrying (CA)	Carrying stones from one place to another
Cuddling (CD)	Holding or cradling stones
Pick up and drop (PUD)	Pick up repeated over and over
Rub on surface (ROS)	Rubbing stones on tin roofing, cement surfaces, etc.
Flinting (FL)	Striking one stone against another held stationary
Pick up small stones (PUs)	Resembling the picking up of wheat grains or soy beans
Rub with hands (RWH)	Similar to potato-washing behavior
Grasp with hands (GH)	Clutching a pile of stones gathered and placed in front of oneself
Toss walk (TW)	Repeated tossing ahead and picking up of a stone(s) while walking
Move and push (MP)	Pushing a stone with both hands while walking forward
Grasp walk (GW)	Walking with one or more stones in the palm of one or both hands

unpublished report; J. Itani, M. A. Huffman, unpublished observations). These comparisons demonstrate that the behaviors that make up stone handling are based on the wide behavioral repertoire of the species.

The behaviors exhibited in stone handling appear to be a predisposition shared by macaques in general. Two sites where stone handling is seen in semiprovisioned troops of long-tailed macaques (*Macaca fascicularis*) in Indonesia and Thailand have recently been brought to the authors' attention. Stone handling occurs in a free-ranging, potato- and fruit-provisioned troop of long-tailed macaques inhabiting the sacred monkey forest of Padangtegal, Ubud Bali. Here, CL, SC, PUD, RWH, ROS, and RIH are the behaviors observed most often (A. Fuentes, personal communication). These macaques have also been observed to exhibit food-washing behavior similar to that on Koshima (Wheatly, 1988) and other forms of object-rubbing behavior (Fuentes, 1992). Another troop living along the coast in Prachuap Province, Thailand are opportunistically

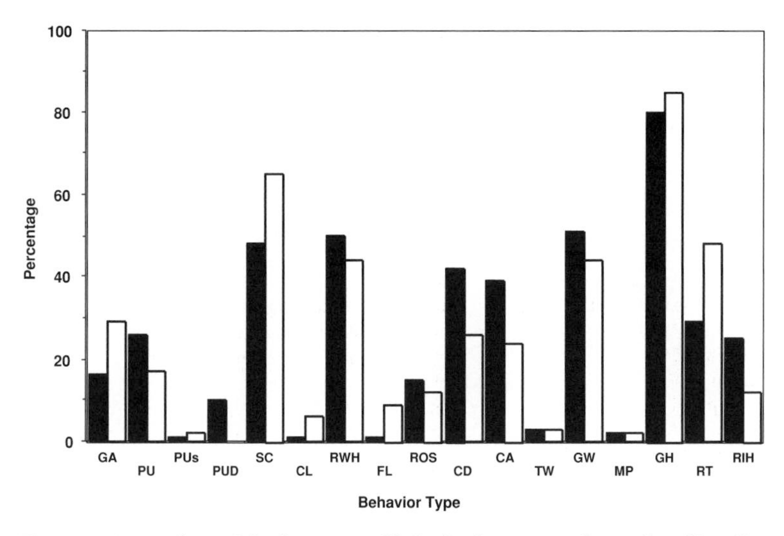

Fig. 10.5. Comparison of the frequency of behavioral patterns of stone-handling displayed by Arashiyama (■) and Takasakiyama (□) Japanese macaques. See Table 10.1 for the behavior types for these abbreviations.

fed bananas and peanuts, which are sold to tourists. At this site, the stone-handling patterns exhibited are ROS, CA, POD, TW, and MP (K. Bauers, personal communication).

In the late 1980s, at the Primate Research Institute, Inuyama Japan, the habit of clacking hard food pellets together was seen to spread from one to other individually caged rhesus macaques (A. Mikami, personal communications). Stone handling or its proximate behavior with other objects appears to be a genus level behavioral propensity associated with provisioning and a sedentary lifestyle.

10.5.3 Factors influencing the rate of diffusion of stone handling

The rate of diffusion of stone handling in Arashiyama was estimated at two time points from surveys conducted in 1983 (B troop) and 1991 (E troop) (Fig. 10.6). The natural logarithm of the yearly total number of individuals for which stone handling was observed was plotted against time to compare the slopes of the linear regression equations (Sokal and Rohlf, 1994). The rate of diffusion was significantly higher in B troop ($y = -0.9 + 0.6x$) than in E troop ($y = 2.8 + 0.2x$) ($p < 0.001$).

In 1986, B troop divided, producing E and F troops (Huffman, 1991). Regardless of the smaller size of E troop, the age–sex class structure of stone

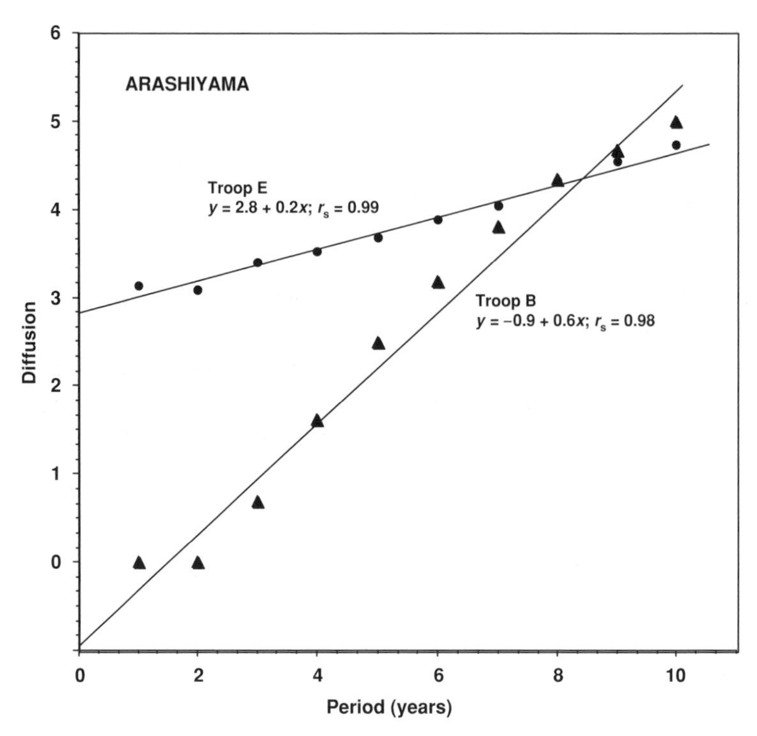

Fig. 10.6. Diffusion of stone handling in two Arashiyama troops of different sizes (troop E with 139 and troop B with 236 members). Each time period on the x axis represents one year. The number of individuals performing stone handling for each period was estimated on the basis of two surveys conducted in 1983 (B troop) and 1991 (E troop). The natural logarithm of the yearly number of individuals performing stone handling (diffusion) was calculated; the significance of the difference in slope between the two linear regression equations was $p < 0.001$.

handlers remained basically the same. Rather than the rate of diffusion being a function of group size, these estimated differences in the rate of diffusion are the result of the current phase of transmission. At this phase, acquisition of stone handling was only occurring among infants, because all individuals acquired the behavior within their first 6 months of life. Therefore, the increase in new stone handlers after this point in time is purely a function of new births.

If the period of innovation of stone-handling behavior had not been observed in the detail achieved at Arashiyama, an investigator seeing the behavior for the first time today would find it difficult to conclude that stone handling is a behavioral innovation. Although phylogenetic and group-history factors are difficult to establish in field research, it is important to keep in mind the importance of both these factors when interpreting the origin of a behavior.

10.6 Behavioral tradition in multiple groups and among subspecies: leaf swallowing, a self-medicative behavior in African great apes

10.6.1 Behavioral description and its context of performance in the wild

Attention was first brought to leaf-swallowing behavior by Wrangham and Nishida (1983). They pointed out that leaf swallowing was unlikely to provide any nutritional value as they noticed a pattern of folded, undigested leaves of *Aspilia* spp. in the dung of chimpanzees at both Gombe and Mahale.

Leaves are most commonly swallowed early in the morning or shortly after climbing out of the night nest, often by visibly ill individuals, as one of the first items ingested after waking (Huffman and Caton, 2001; Huffman *et al.*, 1996; Wrangham and Goodall, 1989; Wrangham and Nishida, 1983). Leaf swallowing is a form of animal self-medication (Huffman, 1997) and has been documented in the greatest detail in chimpanzees at four study sites in East Africa (Mahale, Gombe, Kibale and Budongo see Fig. 10.7). At these sites, the behavior is strongly associated with the expulsion of adult intestinal nematodes and or cestode proglottids (Huffman and Caton, 2001; Huffman *et al.*, 1996; Wrangham, 1995). The gastrointestinal tract responds to the swallowed leaves by rapidly expelling the undigested leaves approximately 6 hours after swallowing. Repeated periodically throughout peak periods of infection, leaf swallowing was projected to have a significant impact on the level of *Oesophagostomum* sp. infection (Huffman and Caton, 2001).

10.6.2 Species comparison and geographical distribution of leaf swallowing

The evidence that great apes practice leaf-swallowing behavior as a form of self-medication has stimulated researchers to look for this anomalous feeding habit among apes across Africa. At the writing of this paper, leaf-swallowing behavior involving the use of more than 34 different plant species has been noted in at least 22 social groups at 13 great ape study sites in Africa (Fig. 10.7). Represented by these observations are

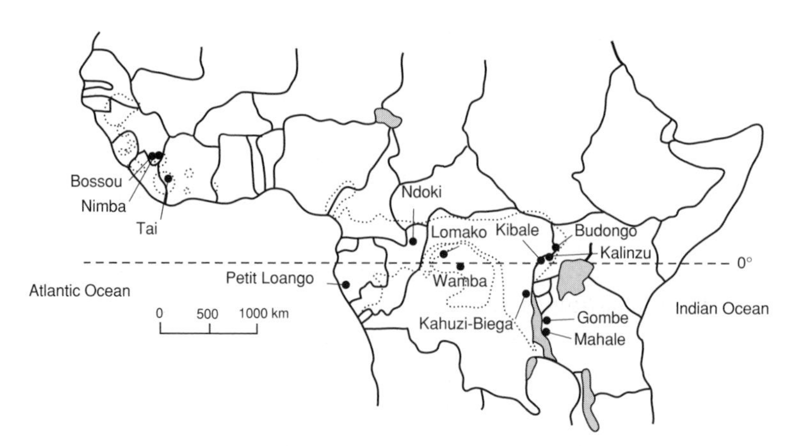

Fig. 10.7. African great ape study sites where whole leaf swallowing has been observed (Based on Huffman, 1997 and M. A. Huffman unpublished data). Species represented by each study site are as follows; bonobo (Lomako, Wamba), chimpanzee (Bossou, Nimba, Tai, Petit Loango, Ndoki, Mahale, Gombe, Kahuzi-Biega, Kalinzu Forest, Kibale, Budongo Forest), eastern and western lowland gorillas (Kahuzi-Biega, Lope).

three subspecies of chimpanzee, the bonobo, and both western and eastern lowland gorilla species. The behavioral details described above are basically the same at other sites, where leaf swallowing has been directly observed (Tai: Boesch, 1995; Bossou and Mt. Nimba: Matsuzawa and Yamakoshi, 1996; Sugiyama and Kohman, 1992; Mahale and Gombe: Huffman, 1997; Takasaki and Hunt, 1987; Wrangham and Nishida, 1983; Lomako: Dupain *et al.*, 2002).

Self-medication is most likely a very old behavior and, therefore, widespread throughout the distribution of the species that practice it. One source of behavioral variation in leaf swallowing among sites is the species of plant selected for leaf swallowing (Huffman, 1997; Huffman and Wrangham, 1994). Local variation may not be manifest in the behavioral pattern itself, but in the materials used; the species selected for use in the wild appear to be transmitted among individuals within a group. The local, regional and pan-African patterns of plant species selected for leaf swallowing suggest that transmission of information about which particular plant species are used also occurs between neighboring groups. Local and regional level similarities, not explainable by plant distribution alone, suggest that social learning and intergroup diffusion of the behavioral tradition exists for leaf swallowing (Huffman, 2001).

10.7 Future prospects and directions

To-date, the majority of behavioral traditions described in the literature have been related to food or foraging activity. While a good case for adaptive value can be made for such foraging-related behavioral traditions, it need not be an absolute criterion for the emergence or the continued existence of a behavioral tradition, as long as the behavior is not maladaptive. Pleasure seeking, stress release, even addictions are the motivation behind widespread human behavioral traditions, such as using worry beads, smoking, alcohol consumption, bungee jumping, or automobile racing. Versions of these traditions exist in practically every human culture. These behaviors are, in part, based on common propensities rooted in the evolutionary past of our species. Other species also have a great range of possibilities for traditions. New examples from emerging long-term studies on capuchin species are presented for the first time by Perry *et al.* (Ch. 14). Boinski *et al.* (Ch. 13) provide interesting new examples of social interactions, foraging techniques, and object manipulation that are suggestive of behavioral traditions.

One of our tasks for the future is to evaluate the potential impact of behavioral traditions on the survival of the individual and the group. At the same time, it will be productive to identify the ecological variability and biological foundations upon which these behaviors may be based and to look for similarities and differences among taxonomically related species. The knowledge gained from such research, when integrated into the current theoretical models used to explain the dynamics of behavioral transmission, should provide a broader understanding of the role of animal traditions in the survival of the species.

10.8 Acknowledgements

We are indebted to the many people and institutions that helped the research on which this chapter is based, in particular N. Asaba, J. Itani, T. Matsuzawa, T. Nishida, and Y. Sugiyama. We wish to give our sincere appreciation to Massimo Bardi for his help in the statistical analyses and in preparing some of the figures, for commenting on the manuscript at various stages, and for his overall technical and intellectual input. We are deeply indebted to the staff of the Arashiyama facilities, who generously provided immeasurable logistical support and friendship over the years. This study benefited in many ways from our colleagues and the facilities of the Primate Research Institute, Kyoto University. Our sincere thanks go to the people and Government of Tanzania for their long-term cooperation and commitment to chimpanzee research. In particular the Tanzanian National Scientific Research Council, Tanzanian National Parks, Serengeti Wildlife Research Institute, Mahale Mountains Wildlife Research Centre, Gombe Stream Research Centre, and the University of Dar es Salaam. To all these people and institutions, we extend our sincerest thanks.

References

Asquith, P. J. 1991. Primate research groups in Japan: orientations and East–West differences. In *The Monkeys of Arashiyama. Thirty-five Years of Research in Japan and the West*, ed. L. M. Fedigan and P. J. Asquith, pp. 81–98. Albany, NY: SUNY Press.

Avital, E. and Jablonka, E. 2000. *Animal Traditions: Behavioral Inheritance in Evolution*. Cambridge: Cambridge University Press.

Boesch, C. 1995. Innovation in wild chimpanzees (*Pan troglodytes*). *International Journal of Primatology*, **16**, 1–16.

Boesch, C. 1996. Three approaches for assessing chimpanzee culture. In *Reaching into Thought: The Minds of the Great Apes*, ed. A. E. Russon, K. Bard, and S. Taylor Parker, pp. 404–429. Cambridge: Cambridge University Press.

Boesch, C. and Tomasello, M. 1998. Chimpanzee and human cultures. *Current Anthropology*, **39**, 591– 613.

Boyd, R. and Richardson, P. J. 1985. *Culture and the Evolutionary Process*. Chicago, IL: University of Chicago Press.

Candland, D. G., French, D. K., and Johnson, C. N. 1978. Object-play: test of a categorized model by the genesis of object-play in *Macaca fuscata*. In *Social Play in Primates*, ed. E. O. Smith, pp. 259–296. New York: Academic Press.

Cavalli-Sforza, L. L. and M. W. Feldman 1981. *Cultural Transmission and Evolution: A Quantitative Approach*. Princeton, NJ: Princeton University Press.

de Waal, F. B. M. 2001. *The Ape and the Sushi Master: Cultural Reflections of a Primatologist*. New York: Basic Books.

Dupain, J., van Elsaker, L., Nell, C., Garcia, P., Ponce, F., and Huffman, M. A. 2002. *Oesophagostomum* infections and evidence for leaf swallowing in bonobos (*Pan paniscus*): indication for self-medicative behavior? *International Journal of Primatology*, **23**, 1053–1062.

Fa, J. E. and Lindburg, D. G. 1996. *Evolution and Ecology of Macaque Societies*. New York: Cambridge University Press.

Fuentis, A. 1992. Object rubbing in Balinese macaques (*Macaca fascicularis*). *Laboratory Primate Newsletter*, **31**, 14–15.

Galef, B. G., Jr. 1976. Social transmission of acquired behavior: a discussion of tradition and social learning in vertebrates. In *Advances in the Study of Behavior*, Vol. 6, ed. J. R. Rosenblatt, R. A. Hinde, E. Shaw and C. Beer, pp. 77–99. New York: Academic Press.

Galef, B. G., Jr. 1991. Tradition in animals: field observations and laboratory analyses. In *Interpretation and Explanation in the Study of Animal Behavior*, ed. M. Bekof and D. Jamieson, pp. 74–95. Boulder, CO: Westview Press.

Galef, B. G., Jr. 1992. The question of animal culture. *Human Nature*, **3**, 157–178.
Giraldeau, L.-A. 1997. The ecology of information use. In *Behavioral Ecology: An Evolutionary Approach*, 4th edn, ed. J. R. Krebs and N. B. Davies, pp. 42–68. Oxford: Blackwell Scientific.
Giraldeau, L.-A., Caraco, T., and Valone, T. J. 1992. Social foraging: individual learning and cultural transmission of innovations. *Behavioral Ecology*, **5**, 35–43.
Hill, D. A. 1999. Effects of provisioning on the social behavior of Japanese and rhesus macaques: implications for socioecology. *Primates*, **40**, 187–198.
Hiraiwa, M. 1975. Pebble-collecting behavior by juvenile Japanese monkeys. [In Japanese] *Monkey*, **19**, 24–25 .
Hirata, S. and Morimura, N. 2000. Naive chimpanzees' (*Pan troglodytes*) observation of experienced conspecifics in a tool-using task. *Journal of Comparative Psychology*, **114**, 291–296.
Huffman, M. A. 1984. Stone-play of *Macaca fuscata* in Arashiyama B troop: transmission of a non-adaptive behavior. *Journal of Human Evolution*, **13**, 725–735.
Huffman, M. A. 1991. History of Arashiyama Japanese Macaques in Kyoto, Japan. In *The Monkeys of Arashiyama. Thirty-five Years of Research in Japan and the West*, ed. L. M. Fedigan and P. J. Asquith, pp. 21–53. Albany, NY: SUNY Press.
Huffman, M. A. 1996. Acquisition of innovative cultural behaviors in non-human primates: a case study of stone handling, a socially transmitted behavior in Japanese macaques. In *Social Learning in Animals: The Roots of Culture*, ed. B. G. Galef, Jr. and C. Heyes, pp. 267–289. Orlando, FL: Academic Press.
Huffman, M. A. 1997. Current evidence for self-medication in primates: a multidisciplinary perspective. *Yearbook of Physical Anthropology*, **40**, 171–200.
Huffman, M. A. 2001. Self-medicative behavior in the African great apes: an evolutionary perspective into the origins of human traditional medicine. *BioScience*, **51**, 651–661.
Huffman, M. A. and Caton J. M. 2001. Self-induced increase of gut motility and the control of parasitic infections in wild chimpanzees. *International Journal of Primatology*, **22**, 329–346.
Huffman, M. A. and Kalunde, M. S. 1993. Tool-assisted predation by a female chimpanzee in the Mahale Mountains, Tanzania. *Primates*, **34**, 93–98.
Huffman, M. A. and Quiatt, D. 1986. Stone handling by Japanese macaques (*Macaca fuscata*): implications for tool use of stone. *Primates*, **27**, 427–437.
Huffman, M. A. and Wrangham, R. W. 1994. Diversity of medicinal plant use by chimpanzees in the wild. In *Chimpanzee Cultures*, ed. R. W. Wrangham, W. C. McGrew, F. B. deWall, P. G. Heltne, pp. 129–148. Cambridge, MA: Harvard University Press.
Huffman, M. A. Page, J. E., Sukhdeo, M. V. K., Gotoh, S., Kalunde, M. S., Chandrasiri T., and Towers, G. H. N. 1996. Leaf-swallowing by chimpanzees: a behavioral adaptation for the control of strongyle nematode infections. *International Journal of Primatology*, **72**, 475–503.
Imanishi, K. 1952. Evolution of humanity. [In Japanese] In *Man*, ed. K. Imanishi. Tokyo, Mainichi-Shinbunsha.
Inoue-Nakamura, N. and Matsuzawa, T. 1997. Development of stone tool use by wild chimpanzees (*Pan troglodytes*). *Journal of Comparative Psychology*, **111**, 159–173.
Itani, J. 1958. On the acquisition and propagation of a new food habit in the troop of Japanese monkeys at Takasakiyama. In *Japanese Monkeys: A Collection of*

Translations, ed. K. Imanishi and S. Altmann, pp. 52–65. Edmonton: University of Alberta Press.

Itani, J. 1959. Paternal care in wild Japanese monkeys, *Macaca fuscata fuscata*. *Primates*, **2**, 61–93.

Itani, J. and Nishimura, A. 1973. The study of infrahuman culture in Japan. In *Symposia of the Fourth International Congress of Primatology*, Vol. 1, ed. E. W. Menzel Jr., pp. 26–60. Basel: Karger.

Kawai, M. 1965. Newly acquired pre-cultural behavior of a natural troop of Japanese monkeys on Koshima Island. *Primates*, **6**, 1–30.

Kawamura, S. 1959. The process of sub-human culture propagation among Japanese macaques. *Primates*, **2**, 43–60.

Kroeber, A. L. and Kluckhohn, C. 1952. Culture: a critical review of concepts and definitions. *Papers of the Peabody Museum of American Archeology and Ethnology*, **47**, 41–72.

Kutsukake, N. 2000. Matrilineal rank inheritance varies with absolute rank in Japanese macaques. *Primates*, **41**, 321–336.

Laland, K. N. 1999. Exploring the dynamics of social transmission with rats. In *Mammalian Social Learning: Comparative and Ecological Perspectives*, ed. H. O. Box and K. R. Gibson, pp. 174–187. Cambridge: Cambridge University Press.

Lefebvre, L. 1995. Culturally transmitted feeding behavior in primates: evidence for accelerating learning rates. *Primates*, **36**, 227–239.

Lefebvre, L. and Giraldeau, L.-A. 1994. Cultural transmission in pigeons is affected by the number of tutors and bystanders present. *Animal Behaviour*, **47**, 331–337.

Matsuzawa, T. 1994. Field experiments on use of stone tools by chimpanzees in the wild. In. *Chimpanzee Cultures*, ed. R. W. Wrangham, W. C. McGrew, F. B. de Waal, and P. G. Hiltne, pp. 351–370. Cambridge, MA: MIT Press.

Matsuzawa, T. and Yamakoshi, G. 1996. Comparison of chimpanzee material culture between Bossou and Nimba, West Africa. In *Reaching into Thought: The Minds of the Great Apes*, ed. A. Russon, K. A. Bard and S. Taylor, pp. 211–232. Cambridge: Cambridge University Press.

McGrew, W. C. 2001. The nature of culture: prospects and pitfalls of cultural primatology. In *Tree of Origin: What Primate Behavior Can Tell Us About Human Social Evolution*, ed. F. B. M. de Waal, pp. 229–254. Cambridge, MA: Harvard University Press.

McGrew, W. C. and Tutin, C.E.G. 1978. Evidence for a social custom in wild chimpanzees? *Man*, **13**, 234–251.

Mendoza, S. and Mason, W. 1989. Primate relationships: social dispositions and physiological responses. In *Perspectives in Primate Biology*, Vol. 2, ed. P. K. Seth and S. Seth, pp. 129–143. New Delhi: Today and Tomorrow's Printers and Publishers.

Murata, G. and Hazama, N. 1968. Flora of Arashiyama, Kyoto, and plant foods of Japanese monkeys. [In Japanese] *Iwatayama Shizen Kenkyujo Chosa Kenkyu Hokoku*, **2**, 1–59.

Nakamura, M., McGrew, W. C., Marchant, L. F., and Nishida, T. 2000. Social scratching: another custom in wild chimpanzees? *Primates*, **41**, 237–248.

Nishida, T. 1987. Local traditions and cultural transmission. In *Primate Society*, ed. B. B. Smuts, D. L. Cheney, R. M. Seyfarth, R. W. Wrangham, and T. T. Struhsaker, pp. 462–474. Chicago, IL: University of Chicago Press.

Nishida, T., Wrangham, R. W., Goodall, J., and Uehara, S. 1983. Local differences in plant-feeding habits of chimpanzees between the Mahale Mountains and Gombe National Park. *Journal of Human Evolution*, **12**, 467–480.

Nishida, T., Hasegawa, T., Hayaki, H., Takahata, Y., and Uehara, S. 1992. Meat-sharing as a coalition strategy by an alpha male chimpanzee? In *Topics in Primatology*, Vol. 1, *Human Origins*, ed. T. Nishida, W. C. McGrew, P. Marler, M. Pickford, and F. D. M. de Waal, pp. 159–174. Tokyo: University of Tokyo Press.

Pulliam, H. R. 1983. On the theory of gene–culture co-evolution in a variable environment. In *Animal Cognition and Behavior*, ed. R. Melgren, pp. 427–443. Amsterdam: North Holland.

Sokal, R. R. and Rohlf, F. J. 1994. *Biometry*. New York: Freeman.

Stanford, C. B. 1999. *The Hunting Apes: Meat Eating and the Origins of Human Behavior*. Princeton, NJ: Princeton University Press.

Sugiyama, Y. and Kohman J. 1992. The flora of Bossou: its utilization by chimpanzees and humans. *African Studies Monographs*, **13**, 127–169.

Takahata, Y., Huffman, M. A., Suzuki, S., Koyama, N., and Yamagiwa, J. 1999. Male-female reproductive biology and mating strategies in Japanese macaques. *Primates*, **40**, 143–158.

Takasaki, H. and Hunt, K. 1987. Further medicinal plant consumption in wild chimpanzee? *African Studies Monographs*, **8**, 125–128.

Thierry, B. 1994. Social transmission, tradition and culture in primates: from the epiphenomenon to the phenomenon. *Techniques and Culture*, **23–24**, 91–119.

Tokida, E., Tanaka, I., Takefushi, H., and Hagiwara, T. 1994. Tool-using in Japanese macaques: use of stones to obtain fruit from a pipe. *Animal Behaviour*, **47**, 1023–1030.

Tuttle, R. H. 2001. On culture and traditional chimpanzees. *Current Anthropology*, **42**, 407–408.

Visalberghi, E. and Fragaszy, D. M. 1990. Food-washing behaviour in tufted capuchin monkeys, *Cebus apella*, and crab eating macaques, *Macaca fascicularis*. *Animal Behaviour*, **40**, 829–836.

Watanabe, K. 1989. Fish: a new addition to the diet of Japanese macaques on Koshima Island. *Folia Primatologica*, **52**, 124–131.

Watanabe, K. 1994. Precultural behavior of Japanese macaques: longitudinal studies of the Koshima troops. In *The Ethnological Roots of Culture*, ed. R. A. Gardner, A. B. Chiarelli, B. T. Gardner, and F. X. Plooji, pp. 81–94. Dordrecht, the Netherlands: Kluwer Academic.

Wheatly, B. P. 1988. Cultural behavior and extractive foraging in *Macaca fascicularis*. *Current Anthropology*, **29**, 516–519.

Whiten, A. 2000. Primate culture and social learning. *Cognitive Science*, **24**, 477–508.

Whiten, A., Goodall, J., McGrew, W. C., Nishida, T., Reynolds, V., Sugiyama, Y., Tutin, C. E. G., Wrangham, R. W., and Boesch, C. 1999. Cultures in chimpanzees. *Nature*, **399**, 682–685.

Wrangham, R. W. 1995. Relationship of chimpanzee leaf-swallowing to a tapeworm infection. *American Journal of Primatology*, **37**, 297–303.

Wrangham, R. W. and Goodall, J. 1989. Chimpanzee use of medicinal leaves. In *Understanding Chimpanzees*, ed. P. G. Heltne and L. A. Marquardt, pp. 22–37. Cambridge, MA: Harvard University Press.

Wrangham, R. W. and Nishida, T. 1983. *Aspilia* spp. leaves: a puzzle in the feeding behavior of wild chimpanzees. *Primates*, **24**, 276–282.

Yamagiwa, J. and Hill, D. 1998. Intraspecific variation in the social organization of Japanese macaques: past and present scope of field studies in natural habitats. *Primates*, **39**, 257–273.

CAREL P. VAN SCHAIK

11

Local traditions in orangutans and chimpanzees: social learning and social tolerance

11.1 Introduction

Upon sufficiently close inspection, virtually all animals will show spatial variation in their behavior. Most studies of behavioral geography have focused on local adaptation, in part based on genetic differences in learning predispositions (e.g., Foster and Endler, 1999). Only a few studies have assumed that the geographic variation in behavior was affected by social learning; that is, it was traditional rather than genetic in origin (e.g., Galef, 1976, 1992, 1998). Primate studies, however, are clearly the exception to this rule: social transmission is often thought to be important (e.g., Wrangham, de Waal, and McGrew, 1994). Unfortunately, descriptive field studies face various obstacles, making it very difficult to demonstrate unequivocally that differential invention and social transmission underlie the pattern of geographic variation.

Local variants can be defined as behaviors that show geographically patchy distribution (Table 11.1.) Following Galef (1976, 1992) and Fragaszy and Perry (Ch. 1), it appears that three criteria must be met to decide that a local variant qualifies as a tradition: (a) the local variant must be common, shown by multiple individuals (cf. McGrew, 1998); (b) it must be long lasting, probably persisting across generations; and (c) it must be maintained by some form(s) of social learning[1].

Field studies can yield information on condition (a) and, with some patience, on condition (b); however, the processes underlying the acquisition

[1]This use of the term tradition is at the level of the individual behavioral variant not at that of the total number of local behavioral variants, population-dependent clusters of variants (cf. Table 1), which one might call local tradition *repertoires*. This distinction is the equivalent of that between tool type and tool kit (e.g., McGrew 1992).

Table 11.1. *Geographic distribution of behavioral variants and their possible interpretation: distinct local clusters, consistent with limited invention and reliable local social learning*

Behavior	Individuals of population 1			Individuals of population 2				Individuals of population 3			
	a	b	c	d	e	f	g	h	i	j	k
A	1	1	1	0	0	0	0	0	0	0	0
B	1	1	1	1	1	1	1	0	0	0	0
C	1	1	1	1	1	1	1	0	0	0	0
D	0	0	0	1	1	1	1	0	0	0	0
E	0	0	0	1	1	1	1	0	0	0	0
F	0	0	0	0	0	0	0	1	1	1	1
G	0	0	0	0	0	0	0	1	1	1	1
Number of variants	3	3	3	4	4	4	4	2	2	2	2

Table 11.2. *Geographical distribution of behavioral variants and their possible interpretation: a mosaic distribution pattern consistent with limited local invention and poor social learning, limiting transmission to matrilineal inheritance*

Behavior	Individuals in population 1			Individuals in population 2				Individuals in population 3			
	a	b	c	d	e	f	g	h	i	j	k
A	1	0	1	0	0	1	1	0	1	0	0
B	0	0	1	1	0	0	0	1	0	0	1
C	1	1	0	0	0	1	1	1	1	0	0
D	0	0	0	1	0	0	1	0	0	0	1
E	0	1	0	1	0	1	1	0	0	1	0
F	0	0	1	0	0	0	0	1	1	0	0
G	1	0	1	0	0	1	0	1	1	0	1
Number of variants	3	2	4	3	0	4	4	4	4	1	3

of behavior (condition (c)) are notoriously difficult to study in the wild. Only studies that control the environment and the history of exposure to novel stimuli can produce definitive statements about the nature of the learning process. However, establishing the potential for traditions in the laboratory does not amount to demonstrating their existence in

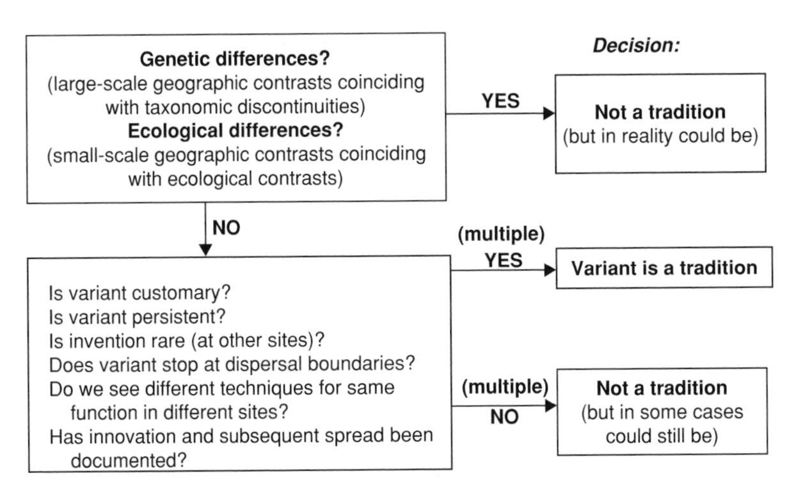

Fig. 11.1. Illustration of the "method of elimination" used to assess whether a geographically variable behavior is or is not a tradition.

the field; consequently, field studies remain essential to demonstrate the existence of the spatiotemporal patterning of behavior expected for traditions. Hence, the "ethnographic" work in nature and the detailed scrutiny of the process of behavior acquisition in the laboratory can be regarded as complementary endeavors, each necessary but insufficient on its own for a complete understanding of the behavioral geography of a species.

What can field studies do to increase the plausibility of traditions as the proper interpretation of local variants in species for which laboratory studies indicate the potential for traditions? Figure 11.1 illustrates the procedure, modeled after the one used for chimpanzees (cf. Boesch, 1996a; Boesch *et al.*, 1994; Whiten *et al.*, 1999). First, the field studies must demonstrate the ubiquity and persistence in the local community of variants that show a geographic distribution of clear local clusters (cf. Table 11.1). Second, they must eliminate simple alternative explanations. Third, they must generate other evidence consistent with the notion of traditions (cf. Whiten *et al.*, 1999) that helps to increase confidence in the conclusion that the variants are indeed traditions. As more elements are found to be consistent with the notion of tradition, the likelihood of alternative interpretations decreases, especially if the required capacity for social learning has been proven in controlled studies.

The figure also shows that this procedure is conservative. First, behavioral variants that are not customary can still be traditions: some know but do not show. Consequently, while a distribution as found in Table 11.2 is not normally considered evidence for traditions, it could, in principle, still be a product acquisition through social transmission. Unfortunately, we have no way to tell. Second, behavioral variants showing strong ecological or genetic correlates are not considered traditions, although in reality they could be. Clear correlations with ecological or genetic factors do not obviate the need for a developmental explanation for the presence of behaviors. These seemingly simpler explanations assume that these behaviors develop reliably under the given ecological conditions or the given genetic context. Third, even some behaviors that do not vary geographically could still be socially transmitted, but there is no easy way for us to identify them.

In this chapter, I examine local variants in orangutans. The focus is on tool use, which is more easily recognized as distinct variants than the potentially subtle variants in displays or techniques of feeding or nest building. Consequently, it is relatively easy to map tool-use distribution within and across groups or local populations. There are two main themes. Section 11.2 argues that local variants in tool use, such as those seen in chimpanzees, almost certainly qualify as traditions using the procedure laid out in Fig. 11.1. Section 11.3 examines the "opportunities for social learning" hypothesis (van Schaik, Deaner, and Merrill, 1999) for geographic variation in the local tradition repertoire using data from both great apes species that use tools in the wild: orangutans and chimpanzees. The implications of these findings for the understanding of human evolution are discussed in Section 11.4.

11.2 Behavioral geography of orangutans

Orangutans have been studied at about a dozen field sites, although truly long-term studies have been conducted only at about five or so (Kutai, Tanjung Puting, and Gunung Palung in Borneo, and Ketambe and Suaq Balimbing in Sumatra; see map in Delgado and van Schaik, 2000). Until a few years ago, there had been very little documentation of variation in behavior among populations of orangutans that could not be directly linked to demographic or habitat differences, even though students of chimpanzee behavior had drawn attention to such geographic variation for a long time (e.g., McGrew and Tutin, 1978).

It is now known that orangutans, like chimpanzees, have many local behavior variants that are probably traditions (van Schaik *et al.*, 2003a), although orangutans may have fewer local traditions than chimpanzees. In most locations, they are far more solitary than chimpanzees, and the only long and intensive associations are between mothers and offspring. In the orangutans, a mosaic distribution of behavioral variants would be expected, as shown in Table 11.2, linked to matrilineally transmitted inventions, and hence an absence of clear local variants. In line with this expectation, there are common incidental observations of the use of tools by individual animals at various sites, but they do not amount to systematic behavioral variants, and until recently no study had ever reported the systematic use of feeding tools in any population.

Suaq Balimbing was first surveyed in 1992, and a station was established in 1994. It is a swamp area, near Sumatra's west coast. At about seven individuals per square kilometer, orangutan densities are the highest on record (Rijksen and Meijaard, 1999), which is related to the unusual density of trees producing edible fruit and the high and relatively nonseasonal productivity (e.g., van Schaik, 1999).

It was soon discovered that the orangutans at Suaq Balimbing show manufacture and use of tools in two distinct contexts, as well as more incidental uses in other contexts (Fox, Sitompul, and van Schaik, 1999; van Schaik, Fox, and Sitompul, 1996). First, they use tools to extract social insects (ants or stingless bees, and termites) or, more commonly, their products (honey) from tree holes. Second, they use tools to dislodge the nutritious seeds from *Neesia* fruits, large, dehiscent woody fruits in which the seeds are embedded in a mass of stinging hairs. Both forms of tool use are not known from other long-term study sites in either Sumatra or Borneo, despite suitable ecological conditions: tree holes inhabited by stingless bees, ants or termites; *Neesia* seeds eaten by orangutans in at least two other study sites (van Schaik and Knott, 2001). Importantly, all known individuals that were followed long enough showed both forms of tool use, making it customary (sensu McGrew, 1998) (Fig. 11.2). Also, animals of all ages, from juvenile to old adults, show the behavior, and they have shown it for the duration of the study, making it almost certain that it persists over time. Thus, the first criterion for local traditions is fulfilled.

There are also other unique customary behaviors. During nest building in the evening, orangutans at Suaq Balimbing are known to make sputtering vocalizations very similar to the sounds made when expelling fibers that remain after a large amount of plant material, usually liana stems,

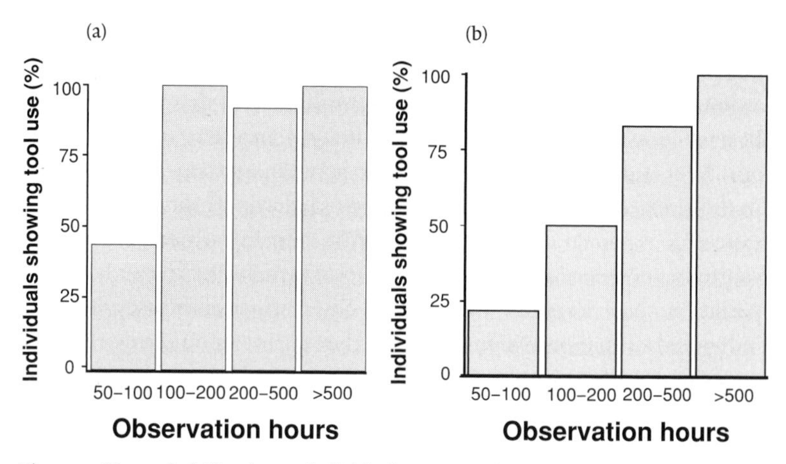

Fig. 11.2. The probability that an individual orangutan in Suaq Balimbing is known to use a particular kind of tool as a function of total follow time as a focal animal, for *Neesia* tools (a) and for tree-hole tools (b). The positive relationships show both tool uses are customary. (From data presented in van Schaik and Knott (2001) and van Schaik *et al.* (2003b).)

have been chewed thoroughly to remove all the soluble contents. Usually, just before these noises are made, the animals will move the end of a leafy twig, used to produce the soft lining of their bed, past their mouth, as if they were to bite off a piece. The vocalization seems to be symbolic and entirely nonfunctional, but so far all regularly followed orangutans at Suaq Balimbing are known to make this vocalization. At none of the other long-term study sites has there ever been any record of this phenomenon.

These observations establish that orangutans at Suaq Balimbing show behavioral variants. Because these variants are so ubiquitous at the site, yet entirely absent elsewhere, it is tempting to consider them local traditions. However, we must first exclude alternative possibilities and add more evidence for social learning (cf. Fig. 11.1). We do this by a closer examination of the tool use.

11.2.1 Are these local variants traditions?

As we noted above, other, simpler explanations to account for local variants first need to be eliminated before the more onerous concept of tradition (Galef, 1992; Williams, 1966) is invoked (Fig. 11.1). First, the geographic variation can represent local adjustment to ecological or demographic conditions that reliably arises ontogenetically through a combination of maturation and experience. For example, the variant could simply reflect the opportunities to perform the behavior, which are strictly linked to ecological conditions. Second, it is possible, although rather unlikely, that the presence or absence of the variant is a consequence of strongly genetically canalized development, for instance because the behavior depends on motor patterns tightly linked to a consistent morphological variant, which in turn is either adaptive or merely the result of drift. Obviously, this hypothesis also assumes that the behavior arises reliably during ontogeny (albeit without social inputs), largely independently of local ecology.

In the first case, one expects clear ecological correlates to the behavioral variants or local repertoire and, therefore, differences between orangutans in distinct habitat types or in relation to the presence of particular plant or animal species. In the second case, one expects that the presence or absence of the variants are associated with major genetic discontinuities such as subspecies boundaries, for example between the Bornean and Sumatran forms, which are genetically quite distinct (Xu and Arnason, 1996).

Although ecological and genetic (constitutional) contrasts often go together, they can be considered logically independent explanations. It is reasonable to expect that ecological differences can occur within a common genetic background, or that genetic differences can produce variants even where the relevant ecological variables are constant. Because these issues have also been examined in chimpanzees, I will first review the evidence for genetic or ecological discontinuities accompanying the local variants in chimpanzees, and then examine the same issue in orangutans.

11.2.1.1 Chimpanzees

Genetic explanations have been rejected for the distribution of tool-use types in chimpanzees (McGrew, 1992; Sugiyama, 1997), because subspecies do not show systematic differences in how they use tools, and even nearby sites (e.g., Gombe versus Mahale) may show appreciable differences in local variants (e.g., Whiten *et al.*, 1999). Ecological explanations could also be rejected in several cases for chimpanzees. Nut cracking is found in a variety of sites but is also absent at others, despite the occurrence of edible nuts and suitable raw materials for tools (McGrew *et al.*, 1997). Most dramatically, chimpanzees west of the Sassandra river do crack nuts, whereas those east of this river, in very similar forest, do not (Boesch *et al.*, 1994). Therefore, in chimpanzees, feeding tools show no evidence for strong genetic or ecological influences.

Indeed, there is strong circumstantial evidence for a role of social learning in the maintenance of the tool-use techniques: that is for considering these variants traditions (step 3 in Fig. 11.1). First, many of the techniques are complex enough that individuals are unlikely to invent them independently. Indeed, nut cracking is such a skilled technique that juvenile chimpanzees take many years to reach adult levels of competence (Boesch and Achermann-Boesch, 2000). Even if these changes reflect maturation only, it is likely that social factors make the immatures persist at the task in spite of the lack of reinforcement. Second, in some areas, an identical task, ant dipping, is performed in rather different ways, using tools of systematically different dimensions with dramatically different efficiencies (Boesch and Boesch, 1990; McGrew, 1992). This suggests that once a technique is established in a population, there is a strong social component in the acquisition process for the great majority of individuals, and they do not invent their own technique independently (but see Humle and Matsuzawa, 2002). Finally, many arbitrarily variable gestural behaviors show the clustered distribution of variants expected when social learning is the norm (Whiten *et al.*, 1999).

11.2.1.2 Orangutans

Orangutans are only known to use feeding tools in Sumatra, consistent with a role for genetic differences. However, within the Sumatran subspecies, two of the three populations subject to behavioral study do not use these tools. These tool-using behaviors are almost certainly not highly ontogenetically canalized (or else all populations within a region characterized by common dispersal barriers should show them). Hence, if they are based on genetic differences, these differences should be large enough to cause visible morphological differences or locomotor patterns. This is implausible, however, because variation in Sumatra is rather limited, at least compared with that found in Borneo (Courtenay, Groves, and Andrews, 1988; Uchida, 1998).

The use of *Neesia* tools has now been mapped for various swamps along Sumatra's west coast (van Schaik and Knot, 2001), as the tools can be found under trees that recently bore fruit and the mapping, therefore, does not require habituated populations. Whereas tool use was found in three large swamp areas, two of which are separated by well over 100 km, it was absent in a smaller swamp directly across an impassable river from the largest swamp area (van Schaik and Knott, 2001). Therefore, the loss of tool-use techniques is associated with a dispersal barrier that causes an immediate break in social transmission but is associated with, at most, only subtle genetic differences.

Ecological differences could, in theory, explain why orangutans ignore *Neesia* at one site but not at another, particularly if *Neesia* would not be part of the optimum diet at some sites. This possibility is remote, however, because caloric gains from feeding on *Neesia* seeds are exceptionally great. Indeed, even where the orangutans do not use tools, they still eat the *Neesia* seeds, which they can only obtain by laboriously breaking off the fruit's valves (van Schaik and Knott, 2001). At the site where *Neesia* is eaten without tools, the seeds are mainly taken by adult males, strong enough to break open the fruits, and total daily intake is far less than at Suaq Balimbing. Therefore, neither genetic nor ecological differences explain the distribution of *Neesia* tool use.

At first sight, ecological differences seem to offer a plausible explanation for the geographic pattern in tree-hole tool use. Orangutans at Suaq Balimbing inhabit swamp forest, which is a rather different habitat from the other study sites (although both Gunung Palung and Tanjung Puting study sites contain some swamp forest). The use of tree-hole tools is not reported for any of the other sites with long-term studies. Yet, ecological factors do not explain the absence of tree-hole tool use elsewhere. First, all forests have trees with abundant tree holes. A comparison of equal samples of trees of > 30 cm girth at breast height found exactly the same incidence of holes in the trunks at Suaq Balimbing's swamp forest as in the dryland riverine forest at Ketambe (C. P. van Schaik, unpublished data), although it is not known whether there are dramatic differences in the proportion of tree holes inhabited by insects that are edible or produce edible products. Second, when the swamp orangutans entered the hills during a mast fruiting event there, they occasionally used tools to exploit tree holes. General habitat conditions and forest structure of the mixed dipterocarp forest growing in the hills adjacent to the swamp are representative of a very widespread forest type.

The only plausible conclusion for orangutans, then, is that the presence or absence of tool use reflects aspects of the acquisition process, a combination of invention and social learning. However, to decide that a variant is a tradition we must show that most individuals acquire it through social transmission rather than through independent invention. The main alternative is that sites with the local variants have conditions that are very conducive to frequent, independent invention of the behavioral variant by individuals as they mature. The geographic pattern in *Neesia* has already been mentioned, but there are also other reasons that make it parsimonious to assume that social learning is mainly responsible in the orangutans, although the conditions that favor invention and successful social acquisition overlap (van Schaik *et al.* 1999, and below).

First, the three local traditions documented so far (tree-hole tool use, seed-extraction tool use, nest-building splutters) are customary at the site where they are known to occur, whereas they are absent at the other study sites. Since ecological factors cannot account for their presence, occasional tool use should have been seen at other sites as well if invention were the main determinant of a skill's presence. Yet, no instances of tool use in the tree-hole or *Neesia* context are known for any of the other long-term study sites. Second, they are persistent, having been observed throughout our study (six years) and in all age–sex classes. Third, numerous observations at Suaq Balimbing are consistent with the interpretation that social context contributes heavily to skill acquisition. Immature animals, particularly, frequently associate with others, more so than at other sites (van Schaik, 1999), often at very close range and also during foraging. They also show great inquisitiveness toward activities of others, both relatives and nonrelatives (Fox *et al.*, 1999). Both chimpanzees and orangutans can match aspects of actions on objects they see others perform (although they may attend the object more closely than the actor: Myowa-Yamakoshi and Matsuzawa, 2000).

On the whole, the orangutan pattern is remarkably consistent with the one found among chimpanzees. Both species suggest a picture of rare invention of new variants, which subsequently persist by reliable social learning, in other words exactly the process postulated by the notion of local tradition. The similarity between the two great apes suggests that local tool-using traditions are a primitive (shared) trait of great apes.

11.3 Explaining variation in the size of local tradition repertoires

11.3.1 The "opportunities for social learning" hypothesis

If the geographic distribution of local traditions mostly reflected the history of invention, with perhaps subtle ecological or demographic influences on the likelihood of invention, then all local populations would have roughly the same tradition repertoire size. However, there is clear variation in the size of local repertoires, and an explanation for this is needed as well. Orangutans at Suaq Balimbing, and probably other nearby swamps, have a comparatively rich tool kit, as well as a unique vocalization. This suggests that some local traditions or cultures may be richer than others: that is, contain a larger number of customary variants. A similar pattern is seen among chimpanzees (Whiten *et al.*, 1999). Although there is an obvious effect of duration of study (e.g., Gombe has a high number of variants) and a possible effect of the interest of the observers, the data also strongly suggest other influences. For example, the number of variants described for Kibale is much lower despite intensive study and evident interest on the part of the observers. Therefore, there seems to be variation across populations in the local tradition repertoire in both orangutans and chimpanzees.

Is this variation predictably related to ecological conditions, population history, or social structure? New skills can be introduced into a local community in two distinct ways: through invention and through importation by dispersing individuals (diffusion). What is critical, however, is what happens to these skills after the first individual in a community has acquired it. Because invention is rare, social learning is essential in order for a skill to be maintained. Both the uniformity of the techniques and their ubiquity within communities are consistent with the notion that the acquisition of these skills is strongly influenced by the social context. Hence, to explain the presence of population-wide tool use in a population, we should focus on the conditions that favor the social learning of complex skills. In particular, our focus should be on horizontal and oblique transmission (not involving the mother) rather than vertical (e.g., mother to offspring) transmission. Even though in the "normal" situation of population-wide tool use most learning may be vertical, horizontal transmission is critical in the initial spread or propagation phase as well as to ensure continued presence of a skill in a population whenever vertical transmission fails. It is also the only plausible explanation for variation across populations in the size of the tradition repertoire. Another form of horizontal transmission is diffusion, when an immigrant introduces a skill in a community. Since dispersal of one sex or the other (females in chimpanzees, males in orangutans) is universal, diffusion may be an important source of introduction of new skills in great ape communities.

How can variation across populations in the reliability of horizontal transmission arise? Van Schaik *et al.* (1999) suggested that *social tolerance in a gregarious foraging context* should not only be conducive to the invention

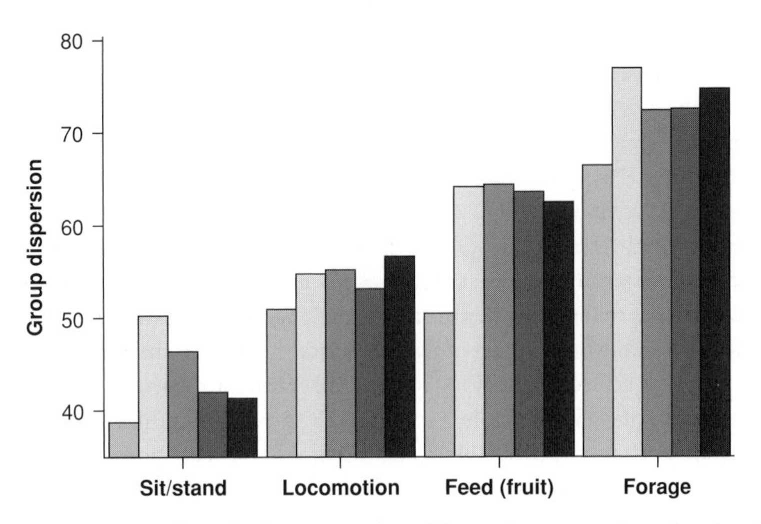

Fig. 11.3. Group dispersion (percentage time without others at < 5 m) as a function of activity for five different groups of long-tailed macaques (*Macaca fascicularis*). (Data taken from van Schaik and van Noordwijk (1986).)

of new techniques but also enhance the spread and maintenance of tool-using skills through some form of social learning. The number of technically difficult skills in the community, as indexed for instance by the size of the tool kit, should reflect the balance between the rate of introduction (through invention or diffusion) and the rate of extinction (from failed transmission). The "opportunities for social learning" hypothesis predicts that social tolerance affects both introduction and maintenance (the opposite of extinction) and should thus produce a higher equilibrium number of customary technical skills in the community. This hypothesis should hold not only for foraging skills but also more generally for other behaviors where acquisition depends on close-range social learning.

The need for social tolerance during foraging is not a trivial condition. Among group-living species, such as most cercopithecines, when groups forage on dispersed food such as insects, they spread out so there is little opportunity in the wild for the close proximity required for social learning; even during feeding on clumped resources such as fruit trees, animals tend to disperse to the maximum extent allowed by these resources (Fig. 11.3). The social situation of adult females in particular sets the stage for the learning of their juvenile or adolescent offspring. Among the tool-using great apes, adult females are semi-solitary (van Schaik, 1999; Wrangham *et al.*, 1996), and females in *Pan* spp. are also nonmatrilineal.

Therefore, close proximity of animals during foraging is not expected to be the norm, and if it occurs it would reflect social tolerance.

The hypothesis that social tolerance in gregarious foraging would be conducive to the transmission of new techniques was based on the suggestion that socially tolerant conditions should enhance the likelihood that naïve animals learn skills through some form of social learning (Coussi-Korbel and Fragaszy, 1995; Fragaszy and Visalberghi, 1990; Lee, 1994). The effect of social tolerance is based on three mechanisms. First, it allows close proximity during foraging, the only context in which observational learning of complex skills is possible. Second, the relaxed social atmosphere allows for attention on the task without the risk of agonistic interactions. Third, social tolerance allows the subordinate animal to keep the food or to share it with others rather than to lose it to kleptoparasitism ("snatching" in Hirata, Watanabe, and Kawai, 2001).

There may be several useful operational definitions of tolerance. First, a good ecological measure is close proximity during foraging. In a detailed comparison, a tolerant species showed greater foraging proximity and less kin bias than a more despotic congener (Matsumura, 1999). A second good indirect measure is the rate of food sharing, although not all species have an ecology involving highly valuable food in discrete packages, which makes food sharing likely. Third, useful social measures (especially useful in captive settings) are the rate of reconciliation and the extent of counterthreats within dyads (Aureli, Das, and Veenema, 1997; de Waal and Luttrell, 1989; Preuschoft and van Schaik, 2000). These features characterize so-called tolerant species; the terms tolerance and egalitarianism are often used interchangeably for this phenomenon, but here I use them with separate technical meanings, with egalitarian referring to overall rarity of escalated aggression and tolerance to more symmetric and less-damaging aggression, compared with despotic dominance styles (Preuschoft and van Schaik, 2000; Sterck, Watts, and van Schaik, 1997).

The hypothesis makes a straightforward prediction: more opportunities for social learning should produce richer local traditions. These traditions may involve complex learned skills or more generally a greater repertoire of communicative gestures or arbitrary acts (e.g., the spluttering sounds). Social tolerance may vary predictably across species but also intraspecifically among populations.

The rest of this section is concerned with testing this hypothesis, focusing particularly on the technical foraging skills such as the use of feeding tools but also occasionally addressing the broader hypothesis, which requires less stringent conditions. I will first examine the evidence for the underlying suggestion that invention and social learning show a strong dependence on social conditions. I will then turn to testing its main prediction: a positive correlation between social tolerance during gregarious foraging and the size of the feeding tool kit, or, more generally, the repertoire of complex (and thus also presumably socially transmitted) feeding techniques, or even all possibly socially transmitted behavioral variants.

11.3.2 Testing the assumptions of the "opportunities for social learning" hypothesis

11.3.2.1 Social tolerance and invention

Innovation is usually thought of as being affected by two main factors: opportunities for playful or relaxed exploration, and dire need (Kummer and Goodall, 1985; Woodworth, 1958). The focus here is on only one aspect, namely the influence of the number of opportunities for undisturbed object handling. The hypothesis predicts that there should be a correlation between social tolerance and the amount of object handling or the invention of novel techniques. This relationship could hold at two levels. First, individuals within the group who invent a novel technique should be more likely to be socially unconstrained, regardless of overall social structure. Second, invention should be more common in species with tolerant social systems, all other things being equal. A problem with the first prediction is that it does not control for the influence of necessity; this problem is much less acute for the second prediction because variation across species should keep "necessity" constant.

Reader and Laland (2001) combed the literature to study the effects of sex, age, and social rank on innovation (of all kinds) in primates. They found that males were more likely to innovate than females, adults more than young, and there was a trend for more innovation by low-ranking individuals in chimpanzees, but not other species. Therefore, only the sex difference clearly supports the hypothesis, perhaps because of the possible impact of necessity.

The second prediction, while more difficult to test, is less affected by the confounding effect of necessity as a factor favoring innovation. It can be tested by comparing the monkey species in which attempts were made to induce feeding tool manufacture in captivity with those that did not, and assessing whether the inventor species are more tolerant. Although a quantitative test is still far from possible, the sparse data support this assumption. First, the only monkey species to show more than incidental tool use in the wild (the use of crumpled leaves to bring up water from tree holes: Phillips, 1998), *Cebus albifrons*, is known to be socially tolerant (Janson, 1986) and to show extensive food sharing (C. P. van Schaik, personal observation). Second, another species of *Cebus*, *C. apella*, is known to learn tool use easily in captivity or in semi-free ranging conditions (Ottoni and Mannu, 2001; Westergaard and Fragaszy, 1987; Westergaard *et al.*, 1998). Individuals of this species allow others, especially immatures, to take food (i.e., share passively) and occasionally actively share food in captivity (Fragaszy, Feuerstein, and Mitra, 1997). A similar argument can be made for *Cebus capucinus* (see Ch. 14).

Macaques vary dramatically in the degree of social tolerance, as measured by the amount of counteraggression within dyads (Thierry, 1985) and the kinds of appeasement and subordination signal (Preuschoft, 1995). The most accomplished tool users among the macaques are *Macaca silenus* (Westergaard, 1988) and *Macaca tonkeana* (Anderson, 1985), both of which are remarkably tolerant (Preuschoft, 1995). Several other macaque species are more abundantly accessible to researchers yet have not generated reports of extensive tool-using skills, quite possibly because they are less likely to invent novel skills and learn them from others (Zuberbühler *et al.*, 1996). The Japanese macaque, *Macaca fuscata*, may be an exception, since some forms of combinatorial behaviors (e.g., stone handling: Huffman, 1984) are known in spite of social despotism, but one could argue that the huge provisioned groups form such large matrilines that islands of social tolerance exist. Therefore, the trend is suggestive but systematic comparisons are needed.

Another way to examine the link between social tolerance and invention is to examine the effect of social tolerance on object handling, the likely precursor to invention of actions involving objects. There is only one known systematic study, that of Thierry (1994), who compared three macaque species for the amount of novel object handling. In the order rhesus monkeys (*Macaca mulatta*), long-tailed macaques (*Macaca fascicularis*), and tonkean macaques (*Macaca tonkeana*), object handling increased, as did social tolerance measured independently using the amount of counteraggression (Thierry, 1985). This sample is too small to settle this issue convincingly, but the strong tendency of lion-tail macaques (*M. silenus*) to handle novel objects, coupled with its highly tolerant social structure (Preuschoft, 1995) suggests that the correlation would hold across a larger range of species.

In conclusion, invention, the discovery of behavioral innovations, may well be enhanced by socially tolerant conditions, but more systematic work is needed to uncover the factors influencing innovation.

11.3.2.2 Social tolerance and social learning of skills

A good test of the assumption that social learning is a function of social tolerance is to examine patterns of spread of novel techniques inside groups. The expectation is that those who learn are indeed those most likely to be tolerated near the model. Although one might argue that this procedure only tests for the effect of variation in proximity rather than the effect of tolerant proximity, in practice this amounts to the same thing: close proximity in a feeding context is actually a good measure of social tolerance. Between-species or between-population variation in social tolerance should affect the presence or absence in skills in a given group by affecting the number of individuals able to acquire any given skill through social learning, which is largely a function of proximity during the performance of the skill. If such groups are not socially tolerant, many individuals in the same group will only quite rarely find themselves near skilled tool users, and spread will be highly spatially biased (cf. Ch.2). Consequently, patterns of spread within groups of a species are a good test of the general principle, perhaps especially if they concern a relatively intolerant species.

Huffman and Hirata (Ch. 10) demonstrated that a novel feeding technique (folded leaf swallowing) in a captive group of chimpanzees spread only to those animals that had immediate social contact with the two inventors during performance. The only study on monkeys that presented quantitative measures of the incidence and timing of acquisition of the novel techniques by naïve animals as a function of their proximity to the model during moments when the model performed the task did not confirm the prediction (Fragaszy and Visalberghi, 1989). However, it is possible that short-term experiments will not reveal the patterns of spread in the absence of observational learning. Interestingly, various other, less carefully controlled but generally longer-term studies showed an unambiguous effect of tolerant proximity.

Zuberbühler *et al.* (1996) described the spread of the use of a stick to rake in pieces of fruit from outside the cage in a group of long-tailed macaques (*M. fascicularis*). The practice spread extremely slowly. By the end of the study, the only three individuals in the group of 35–37 that had mastered the technique were a peer from the same matriline, a younger brother, and a young protégé: all individuals expected to enjoy close proximity to the model (the dominant male) during the demonstration of the task.

Anderson (1985) elicited a form of tool use to obtain honey in captive tonkean macaques (*M. tonkeana*). The novel behavior spread to only one other individual, a close playmate of the discoverer, who actually learned it by co-manipulating the tool with the original inventor.

Another example concerns playful object manipulation, not a feeding technique, in Japanese macaques (*M. fuscata*): stone handling (Huffman, 1984). This novel technique initially spread in a remarkably similar pattern "to a network of spatial-interactive associates which is specific to the innovator(s)" (Huffman and Quiatt, 1986), in other words to those who could be in close proximity to the innovators in the period after feeding when the behavior generally took place (see Ch. 10).

It is also instructive to re-examine the pattern of initial spread of the famous new techniques, sweet potato washing and placer mining, that appeared in the Koshima group of Japanese macaques during the 1950s (Kawai, 1965; Nishida, 1987). Both novel skills spread quite slowly, showing the small effect of social context on learning in monkeys. Of interest here is the first stages, when social learning was still largely horizontal, that is, not involving mother–infant pairs. Sweet potato washing spread from Imo, who invented it when she was about 1.5 years old, initially to two playmates and her mother, and later to other peers and younger members of her own matriline. All of these individuals would be expected to be in frequent close proximity with Imo, and also often in somewhat relaxed proximity during foraging (in this case provisioning). This clearly echoes Kawamura's early conclusions, as reported by Hirata *et al.* (2001). The placer mining was more difficult since it involved overcoming the reluctance to "throw food away" and, consequently, it spread even more slowly (Itani and Nishimura, 1973). Initially, it spread from Imo, who was about 4 years old by then, to a small juvenile and two of her sisters, and subsequently to several relatives and younger females: again individuals with whom relatively relaxed association was likely.

By 1962, when many animals born before 1961 had learned the techniques, it is clear that members of Imo's matriline had become much more proficient in these techniques than members from other matrilines (Fig. 11.4), especially the more difficult placer-mining technique, and that those who learned in the other matrilines were mainly young animals. These latter animals may have been better at learning either because they were young (the traditional assumption: Hirata *et al.* 2001) or because they

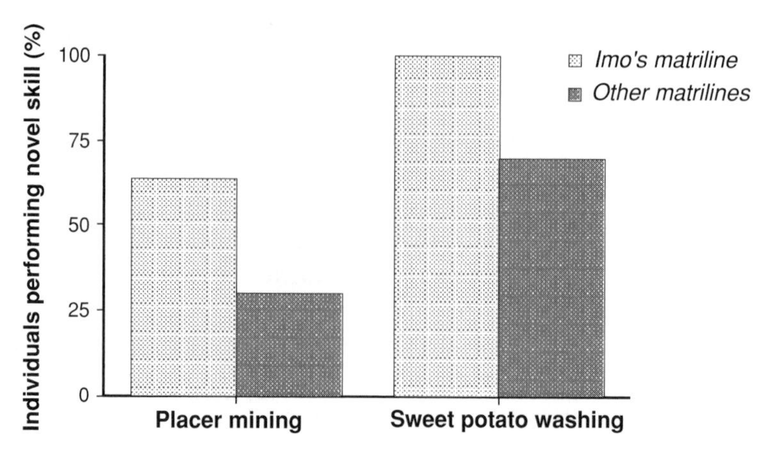

Fig. 11.4. Spread of two novel feeding techniques, following invention by one individual (Imo), to members of the same or different matrilines of the inventor, in a group of Japanese monkeys (data taken from Kawai, 1965).

had far more opportunities to learn, being members of play groups that are subject to less-intense feeding competition (Fragaszy and Visalberghi, 1990), or both.

Several other experiments with Japanese monkeys involved novel foods rather than novel techniques, and their spread was much more rapid and probably depended much less on close proximity to models. However, it is perhaps remarkable that the spread of candy eating in the artificially composed Ohirayama group, which lacked matrilines and was generally less cohesive, lagged behind that observed in other groups (Itani and Nishimura, 1973).

The studies reviewed above show that social learning is most likely or much faster in dyads within a group that show mutual tolerance at close range during situations of potential competition, for example during feeding. In practice, these are kin, play partners among immatures ("peers"), or special friends among adults. In fact, this process may be enough to explain propagation of novel skills, making it unnecessary to appeal to individual traits such as status or age (Hirata *et al.*, 2001).

Two possible caveats deserve comment. First, many examples concerned Japanese monkeys. It is possible that the high frequency of novel techniques observed in them is a by-product of provisioning: only provisioned groups have the large matrilines in which individuals are sufficiently socially relaxed to concentrate on situations long enough to achieve serendipitously some novel technique and to learn it from others.

However, this merely serves to demonstrate the role of tolerant proximity. Second, most examples concerned monkeys rather than apes, but it can be argued that the simpler mechanisms of social learning (although some observational learning is probably involved, e.g., Hirata *et al.* 2001) bring out the role of social factors even more clearly. There are few examples for apes, but Russon and Galdikas (1995) noted that orangutans learnt more readily from models with whom they had an affiliative relationship. No doubt, numerous nonprimate examples could be found as well (e.g., Laland and Reader, 1999).

In sum, this survey provides support for the hypothesis that social learning is enhanced when communities of gregarious individuals show social tolerance. Section 11.3 will test these predictions.

11.3.3 Testing the predictions of the hypothesis

11.3.3.1 Between-population variation in orangutans

The opportunities for social learning hypothesis predicts that the animals at Suaq Balimbing should show more social tolerance than orangutans elsewhere. All the data available so far suggest that this is indeed the case (see also van Schaik *et al.*, 1999). First, the orangutans at Suaq Balimbing are far more likely to move around in travel parties, regardless of their reproductive state, the only exception being nondominant flanged ("adult") males (van Schaik, 1999). They also forage in parties. Orangutans elsewhere also form parties, especially at the other Sumatran site, Ketambe, but the majority of those are feeding parties, which form in fruit trees and disband upon departure (Sugardjito, te Boekhorst, and van Hooff, 1987), rather than the travel parties seen at Suaq Balimbing. Second, food sharing among adults is commonly observed at Suaq Balimbing, whereas reports from other sites are rare (Ketambe) or it is said to be absent (reports from Borneo). Above, I noted that immatures use the social tolerance in the foraging context to observe the foraging behavior of unrelated mature animals at very close range. The same behaviors can be observed in consortships, which are both frequent and long lasting at Suaq Balimbing relative to other sites (Delgado and van Schaik, 2000). Therefore, the social system at Suaq Balimbing, with more gregarious foraging and frequent food sharing, is as expected by the hypothesis; however, as the hypothesis was developed on the basis of observations at Suaq Balimbing, this is not to be taken as critical support for the hypothesis.

The most likely reason for the higher social tolerance at Suaq Balimbing is habitat productivity, which increases association and tolerance. High productivity also should increase population density and, at the large home range size observed at Suaq Balimbing, home range overlap; as a result, each individual is likely to meet many more others at Suaq Balimbing than in areas with lower density and smaller home ranges. The high productivity of the coastal swamp forests such as Suaq Balimbing and adjacent forests, in turn, is a result of a combination of factors: the absence of mast fruiting, a rather muted seasonality in fruit production, relatively high annual fruit production, and low tree species richness, with most species producing edible fruit for orangutans (Delgado and van Schaik, 2000). The densities at the other sites with known *Neesia* tool use are also in the upper range observed for orangutans (van Schaik and Knott, 2001).

11.3.3.2 Within-population variation in orangutans

Although there is very little variation in the rate with which individual orangutans at Suaq Balimbing use tools once they feed in *Neesia* trees with dehisced fruits, there is remarkable individual variation in how often they use tree-hole tools. Recently, we examined this variation and the degree of specialization on this subsistence behavior (van Schaik, Fox, and Fechtman, 2003b). We tested four plausible hypotheses: (a) intrinsic sex differences cause especially females to specialize on rich food sources to support reproduction; (b) displacement by dominants causes subordinates to specialize more on foods that are less-abundantly available and difficult to procure; (c) different opportunities for tool use in different habitats causes variation among individuals in relation to home range location, and (d) opportunities for social learning are available during the maturation period.

We found no evidence for an effect of sex difference (unlike in chimpanzees: Boesch and Boesch, 1981; McGrew, 1992), dominance, or habitat on tool-use rates or specialization. We did, however, find a correlation between tool-use specialization and mean party size, which was used as a proxy measure for cumulative learning opportunities. This was not a direct effect of party size, because being in parties, if anything, suppressed rates of tool use (for details, see van Schaik *et al.*, 2003b). Because a female's mean party size was stable over the years, we consider it a reflection of the ecological conditions on and near the natal range. The party size effect was not found for males, as expected, because males emigrate from their natal range around sexual maturation, and their current mean party size is not thought to reflect the opportunities for social learning on their natal range. Therefore, the observations are consistent with a key role of the cumulative number of opportunities for social learning even at the level of individual proficiency or specialization of tool use. This result provides an important test of the opportunities for social learning model because it is closest to the level of mechanisms. The only limitation is that it refers to degree of specialization on particular skills, rather the presence or absence of the skill (as is examined across sites).

11.3.3.3 Between-population variation in chimpanzees

To relate tolerant gregariousness to tool-kit size in chimpanzees requires that we produce an operational definition of social tolerance in chimpanzees that can be compared across sites. Among the various measures that are available for at least some sites are traveling in (foraging) parties, food sharing, and grooming as direct measures, and interbirth intervals as an index of habitat productivity. These various measures show a remarkable concordance in their pattern across sites (van Schaik *et al.*, 1999). The one measure that is used most easily for comparisons, because it is known for many sites, is the percentage of time independent individuals spend in parties (as opposed to alone) (largely based on Boesch, 1996b; see van Schaik *et al.*, 1999). Since chimpanzee males tend to be in parties much of the time, variation across sites in this measure should mostly reflect variation in female tendency to associate during travel and foraging. Adult females tend to forage alone and will only forage together when food is abundant (as shown by intrapopulation studies, e.g., in Kibale (Chapman, Wrangham, and Chapman, 1995) and in Gombe (Wrangham, 1977)). This measure should, therefore, be a good proxy for social tolerance during foraging.

Variation across known populations in chimpanzee tool use is remarkably large (see overviews by Boesch and Tomasello, 1998; McGrew, 1992). My own compilation of these overviews (van Schaik *et al.*, 1999) shows a very strong correlation ($r = +0.962$; $n = 7$; $p < 0.001$) with the tolerance index (percentage of time in parties) (Fig. 11.5). However, this compilation can be criticized as being insufficiently rigorous, because it may include tool uses that are less than customary or habitual and because tool use at some sites may be under-reported. I, therefore, also used the recent systematic compilation for six chimpanzee sites by Whiten *et al.* (1999), which

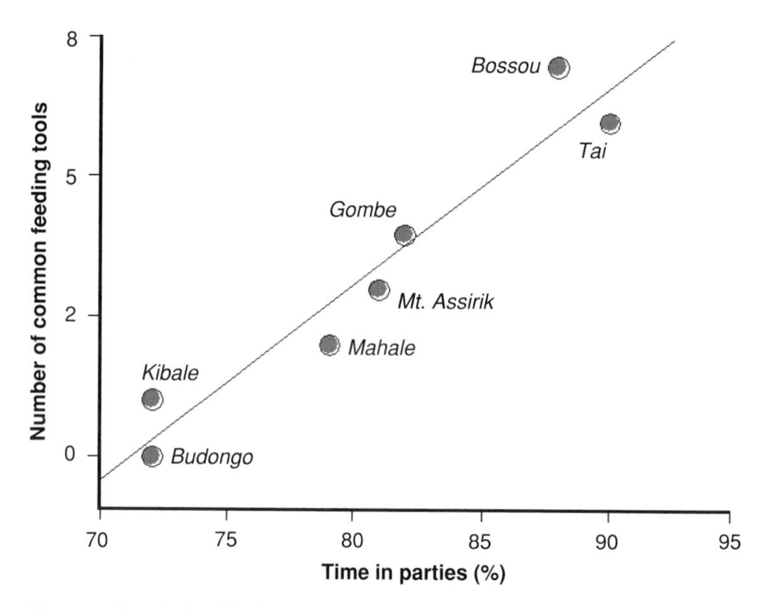

Fig. 11.5. The relationship between a measure for social tolerance (gregarious foraging as percentage time spent in parties) and the size of a community's tool kit among chimpanzee populations. (Based on the compilation of tool-use data from the literature by van Schaik *et al.* (1999).)

included all chimpanzee behaviors thought to be socially transmitted,[2] to provide a more rigorous test of the tolerant gregariousness hypothesis for chimpanzees. I repeated the analysis using all tool-assisted subsistence behaviors from their Table 1. The relationship remains quite strong ($r = +0.966; n = 6; p < 0.01$).

Indeed, one could argue that all socially transmitted skillful behaviors related to food acquisition (including drinking), rather than just those related to tool use, should be subject to the effect of social tolerance. If only the customary and habitual behaviors are included (a conservative approach assuming that the more rarely shown behaviors are individually invented), the correlation with social tolerance remains strong (Fig. 11.6a; $r = +0.943; n = 6; p < 0.01$). If all food-related socially transmitted skillful behaviors are included, the relationship becomes weaker but remains significant ($r = +0.885; n = 6; p < 0.02$).

[2]The data in their Table 1 were presented in four "bands," to eliminate alternative hypotheses to social transmission, but with social transmission firmly established there is no reason to believe that any of these bands are less likely to be transmitted socially.

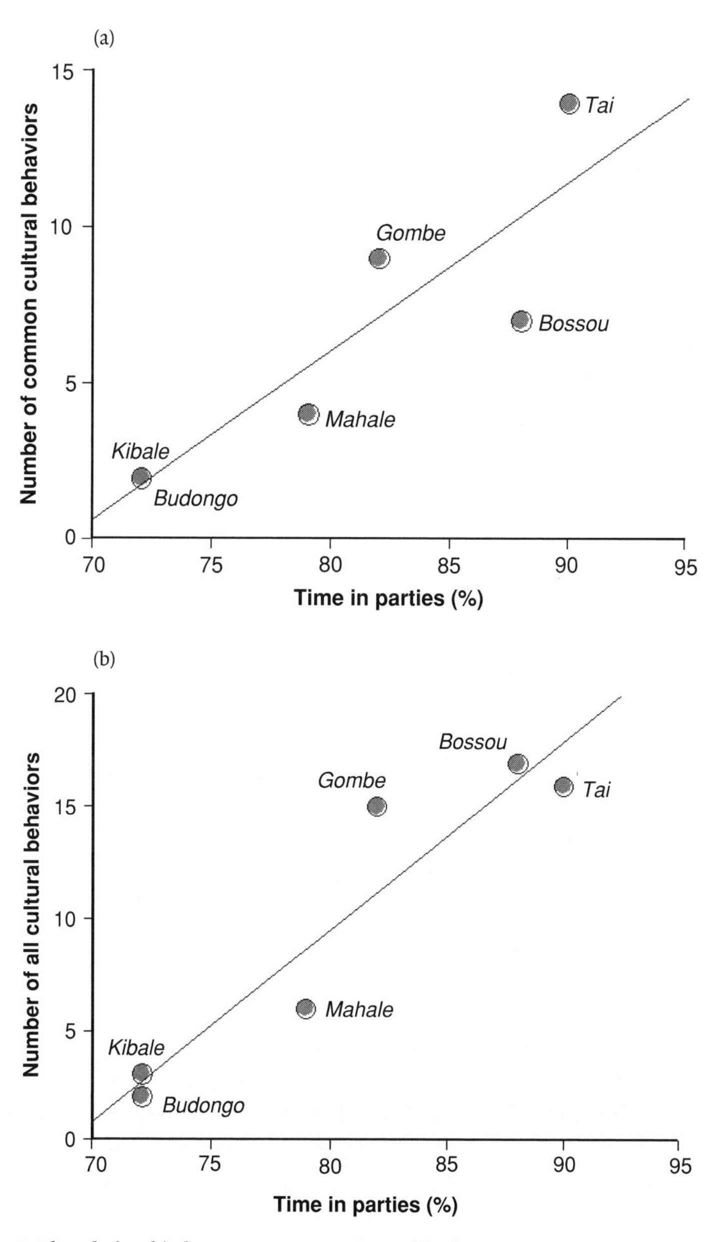

Fig. 11.6. The relationship between a measure for social tolerance (gregarious foraging (as percentage time spent in parties,) and (a) the number of customary and habitual food-related culturally transmitted behaviors and (b) all food-related socially transmitted behaviors. (Culture data taken from Whiten *et al.*, 1999.)

Feeding-related skills were considered here because they are likely to affect fitness and are, therefore, relevant for an investigation of the adaptive significance of tradition and culture. Other socially transmitted behaviors, for example communication signals, may be less critical to fitness (because their form is rather arbitrary) and may also depend less on social tolerance for their maintenance because they can be learned at greater distances than the feeding skills.

11.4 Discussion

The orangutan data presented and discussed in this chapter show the existence of local variants that are almost certainly traditions, just as those found among chimpanzees. The geographic distribution of tool-kit size, and perhaps other socially transmitted behavioral variants, in both species is explained well by variation in opportunities for social learning. Among orangutans, interindividual (within-group) variation in degree of tool-use specialization is also correlated with the frequency of exposure to other individuals in a socially relaxed foraging context. Finally, the greater variety of local traditions in chimpanzees than in orangutans can be attributed to a greater cumulative frequency of opportunities for social learning in the former.

Obviously, much more work is needed. Ideally, spread of novel behaviors is followed in different settings varying in social tolerance. More refined tests of the geographic pattern, especially in chimpanzees, require better measures of social tolerance than the *post hoc* one employed here, and especially data from additional sites. In addition, a strong test would be to relate variation in the size of tool kits or the efficiency of tool use among individual chimpanzees within a population to variation in the frequency of opportunities to learn the skills socially. Finally, alternative hypotheses should be formulated and tested. One possibility is that the number of possible models (i.e., community size or individual network (clique) size)) is a better predictor than the frequency of opportunities for social learning *per se* (M. Huffman, personal communication; but see Ch. 3).

The technique of between-site comparison that eliminates the ecological and genetic differences (Fig. 11.1) is a heuristic procedure to substantiate the presence of traditions. It cannot be applied to many local variants and hence cannot be used to estimate the tradition repertoire. However, acceptance of the opportunities for social learning hypothesis implies that other geographically invariant complex behaviors are also largely acquired through social transmission. A main task for the future is to design procedures that will allow us to estimate this number.

The hypothesis was developed for skilled behaviors such as tool use, but it should apply to all potentially socially transmitted behaviors. However, the tolerance requirement is likely to be especially critical for behaviors that need to be observed at close range before they can be acquired (i.e., usually highly complex behaviors) and that need to be practiced multiple times before animals become proficient. Therefore, the presence of variants of simple communicative gestures that are effective at longer distance should be less sensitive to social tolerance. In orangutans, analysis of geographic variation in communication signals supports this suggestion (van Schaik *et al.*, 2003a).

Among primates, local traditions have now been reported especially among chimpanzees and orangutans. Is this concentration among great apes simply an artefact of reporting biases? In general, holding ecological factors constant, I expect three factors to make independent and additive contributions to determining the extent of local traditions and the size of tradition repertoires: (a) gregariousness; (b) powers of social learning; and (c) social tolerance. Gregariousness allows social learning and should also determine the potential number of models for naïve individuals. It is clear that monkeys are more consistently gregarious than most great apes, but great apes do show evidence of possessing more efficient forms of observational learning (Russon, 1999; Whiten, 1998).

Limited efficiency of social learning can be compensated by high frequency of exposure (i.e., high social tolerance), which is less tightly linked to phylogenetic position than the first two factors. Only a subset of monkeys shows high social tolerance, including, perhaps not surprisingly, capuchin monkeys, as expressed in food sharing (Fragaszy *et al.*, 1997; Perry and Rose, 1994), grooming patterns (O'Brien, 1993), and close-range observation of skilled foragers (Ch. 13). Consequently one may expect that these three factors will create a bias toward traditions among great apes and the most tolerant of monkeys.

Great apes vary in their tendency toward social tolerance. It is interesting to explore the socioecology of social tolerance. Where predation underlies grouping, subordinates can be exploited to some extent by dominants because their options of leaving the groups are minimal. However, in large arboreal organisms such as orangutans, predation risk is negligible (if they stay in the trees), and solitary ranging is a very viable option. Consequently, associations are mainly voluntary, and the benefits of grouping are predominantly social (van Schaik, 1999). If dominants are to maintain these social benefits, they should limit their aggression. Hence, whenever the ecological conditions allow party formation, we should see that these parties show high social tolerance, instead of the tension that characterizes the groups of many other primates. Indeed, this is probably a more general great ape characteristic (van Schaik, Preuschoft, and Watts, 2003c) and contributes to the higher social tolerance of great apes.

At Suaq Balimbing, associations, including larger parties containing multiple adults, are common and clearly involve active coordination of travel among the participants. Hence, they reflect voluntary association in most cases (the exceptions being the usually rather brief associations accompanied by forced matings, which are relatively rare in Sumatra: Delgado and van Schaik, 2000). As to the benefits, infant socialization is an obvious function: much social play can be seen. But older offspring play less, and even adult females without offspring join these parties. I hypothesize that an important function of those parties is information gathering and active search for social learning. Many anecdotal observations of active interest in the foraging behavior of party members are consistent with this interpretation.

To many biologists, local traditions, or local tradition repertoires defined in this way, are equivalent to cultures (e.g., Bonner, 1980). The term culture is often restricted to the human situation involving a broader set of learning mechanisms, including imitation, and teaching, including language, thus allowing more faithful social learning and hence a more cumulative culture (Boesch and Tomasello, 1998; Galef, 1992; McGrew, 1998). In this book, we use the term traditions since we cannot usually make definitive statements about the nature of the social transmission process involved in the wild.

Nonetheless, arguments of continuity imply that human cultures almost certainly evolved from great ape traditions (*pace* Galef, 1992). Thus, the study of traditions should point to the factors that facilitated the enhancement of observational learning techniques and increased reliance on learned skilled behaviors that, during human evolution, led to the emergence of human culture. High social tolerance, gregariousness, and sophisticated copying techniques facilitate the development of extensive traditions. Yet, only one extant primate has evolved the capacity for culture, defined here as involving highly reliable social transmission of knowledge and skills based on imitation and language. What changed during hominid evolution that created the positive feedback loop between copying skills, invention abilities, and sharing and exchange (cf. van Schaik *et al.*, 1999)?

Technical skills are almost certainly adaptive for those who have them, as suggested by the nut cracking in chimpanzees and seed extraction in *Neesia* in orangutans. During hominid evolution, the dependence on technical skills acquired through social learning must have increased. This is consistent with increased interdependence and exchange of food, services, and artefacts between increasingly specialized individuals. The onset of the adoption of different roles and the reliance on foods produced by others, which culminated in the division of labor, may have provided strong selective pressures for the gradual improvement of the accuracy of the mechanisms underlying social learning and innovation. It is almost certain, however, that the neural substrates used in the various forms of learning, be they innovation, conditioning or other forms of individual learning, or socially biased learning, show extensive overlap. Therefore, all aspects of intelligence may have been favored simultaneously. The critical point is that the evolution of advanced technical skills relies on a combination of opportunities for innovation and social learning of these skills and fitness incentives for their use.

Finally, it is worth pointing out that a system that relies on strong social inputs during ontogeny is much more vulnerable to disturbances of the normal social learning processes. Consequently, it is tragically likely that habitat loss, disturbance, fragmentation, and hunting will negatively impact orangutan local traditions (van Schaik, 2002), potentially to the point of total disappearance of these traditions.

11.5 Conclusions

Among primates, extensive tool use for subsistence is found mainly in two great ape species, chimpanzees and orangutans, but in both of them the nature and variety of the tool use shows remarkable geographic variation. Careful comparisons in both chimpanzees and orangutans strongly point to the conclusion that the key to explaining the presence or absence of these behaviors (especially skillful ones, but perhaps many more) lies in aspects of the developmental process. Social learning is an important component of the process of acquisition of these behaviors. Variation in the frequency of opportunities for social learning can be invoked to explain intraspecific variation in the richness of the tradition repertoire in the two great apes. Especially in animals with such great abilities for social learning, social tolerance is a critical factor in the maintenance of learned skills in a population. Increased social tolerance may have been an important selective agent of technological evolution among hominids.

11.6 Acknowledgements

I thank the Indonesian Institute of Sciences (LIPI) and the Directorate General for Conservation (PKA) for permission to conduct research at Suaq Balimbing. I thank Beth Fox and Ian Singleton for overseeing much of the tool collecting at Suaq Balimbing; Arnold Sitompul and the late Idrusman for collecting tool data in Trumon-Singkil; Perry van Duynhoven for skillfully managing to get us in and out of many sites; Serge Wich for information on *Neesia* tool use elsewhere; the Leuser Management Unit coordinators for moral and logistic support and for permission to work in the Leuser Ecosystem; and Kim Bard, Rob Deaner, Doree Fragaszy, Jeff Galef, Michael Huffman, Michelle Merrill, Signe Preuschoft, Simon Reader, Maria van Noordwijk, and Elisabetta Visalberghi for valuable discussion and/or comments on the manuscript. The Wildlife Conservation Society provided significant long-term support of the project, and the L.S.B. Leakey Foundation additional support.

References

Anderson, J. R. 1985. Development of tool-use to obtain food in a captive group of *Macaca tonkeana*. *Journal of Human Evolution*, **14**, 637–645.

Aureli, F., Das, M., and Veenema, H. C. 1997. Differential kinship effect on reconciliation in three species of macaques (*Macaca fascicularis, M. fuscata, and M. sylvanus*). *Journal of Comparative Psychology*, **111**, 91–99.

Boesch, C. 1996a. Three approaches for assessing chimpanzee culture. In *Reaching into Thought: The Minds of the Great Apes*, ed. A. E. Russon, K. A. Bard, and S. T. Parker, pp. 404–429. Cambridge: Cambridge University Press.

Boesch, C. 1996b. Social grouping in Taï chimpanzees. In *Great Ape Societies*, ed. W. C. McGrew, L. F. Marchant, and T. Nishida, pp. 101–113. Cambridge: Cambridge University Press.

Boesch, C. and Boesch-Achermann, H. 2000. *The Chimpanzees of the Taï Forest: Behavioural Ecology and Evolution*. Oxford: Oxford University Press

Boesch, C. and Boesch, H. 1981. Sex differences in the use of natural hammers by wild chimpanzees: a preliminary report. *Journal of Human Evolution*, **10**, 585–593.

Boesch, C. and Boesch, H. 1990. Tool use and tool making in wild chimpanzees. *Folia Primatologica*, **54**, 86–99.

Boesch, C. and Tomasello, M. 1998. Chimpanzee and human cultures. *Current Anthropology*, **39**, 591–614.

Boesch, C. Marchesi, P., Marchesi, N., Fruth, B., and Joulian, F. 1994. Is nut cracking in wild chimpanzees a cultural behaviour? *Journal of Human Evolution*, **26**, 325–338.

Bonner, J. T. 1980.*The Evolution of Culture in Animals*. Princeton, NJ: Princeton University Press.

Chapman, C. A., Wrangham, R. W., and Chapman, L. J. 1995. Ecological constraints on group size: an analysis of spider monkey and chimpanzee subgroups. *Behavioral Ecolology and Sociobiology*, **36**, 59–70.

Courtenay, J., Groves, C., and Andrews, P. 1988. Inter- or intra-island variation? An assessment of the differences between Bornean and Sumatran orang-utans. In *Orangutan Biology*, ed. J. H. Schwartz, pp. 19–29. Oxford: Oxford University Press.

Coussi-Korbel, S. and Fragaszy, D. M. 1995. On the relation between social dynamics and social learning. *Animal Behaviour*, **50**, 1441–1453.

de Waal, F. B. M. and Luttrell, L. M. 1989. Toward a comparative socioecology of the genus *Macaca*: different dominance styles in rhesus and stumptail monkeys. *American Journal of Primatology*, **19**, 83–110.

Delgado, R., and van Schaik, C. P. 2000. The behavioral ecology and conservation of the orangutan (*Pongo pygmaeus*): a tale of two islands. *Evolutionary Anthropology*, **9**, 201–218.

Foster, S. A. and Endler, J. A. 1999. Thoughts on geographic variation in behavior. In *Geographic Variation in Behavior: Perspectives on Evolutionary Mechanisms*, ed. S. A. Foster and J. A. Endler, pp. 287–307. New York: Oxford University Press.

Fox, E. A., Sitompul, A. F., and van Schaik, C. P. 1999. Intelligent tool use in wild Sumatran orangutans. In *The Mentality of Gorillas and Orangutans*, ed. S. T. Parker, L. Miles, and R. Mitchell, pp. 99–116. Cambridge: Cambridge University Press.

Fragaszy, D. M. and Visalberghi, E. 1989. Social influences on the acquisition of tool-using behaviors in capuchin monkeys (*Cebus apella*). *Journal of Comparative Psychology*, **103**, 159–170.

Fragaszy, D. M. and Visalberghi, E. 1990. Social processes affecting the appearance of innovative behaviors in capuchin monkeys. *Folia Primatologica*, **54**, 155–165.

Fragaszy, D.M., Feuerstein, J.M., and Mitra, D. 1997. Transfers of food from adults to infants in tufted capuchins (*Cebus apella*). *Journal of Comparative Psychology*, **114**, 194–200.

Galef, B. G., Jr. 1976. Social transmission of acquired behavior: a discussion of tradition and social learning in vertebrates. In *Advances in the Study of Behavior*, ed. J. S. Rosenblatt, R. A. Hinde, E. Shaw, and C. Beer, pp. 77–100. New York: Academic Press.

Galef, B. G., Jr. 1992. The question of animal culture. *Human Nature*, **3**, 157–178.

Galef, B. G., Jr. 1998. Tradition and imitation in animals. In *Comparative Psychology: A Handbook*, ed. H. Greenberg and M. M. Haraway, pp. 614–622. New York: Garland.

Hirata, S., Watanabe, K., and Kawai, M. 2001. "Sweet-potato washing" revisited. In *Primate Origins of Human Cognition and Behavior*, ed. T. Matsuzawa, pp. 487–508. Tokyo: Springer-Verlag.

Huffman, M. A. 1984. Stone-play of *Macaca fuscata* in Arashiyama B troop: transmission of a non-adaptive behavior. *Journal of Human Evolution*, **13**, 725–735.

Huffman, M. A. and Quiatt, D. 1986. Stone handling by Japanese macaques (*Macaca fuscata*): implications for tool use of stone. *Primates,* **27**, 427–437.

Humle, T. and Matsuzawa, T. 2001. Behavioural diversity among the wild chimpanzee populations of Bossou and neighbouring areas, Guinea and Cote d'Ivoire, West Africa. *Folia Primatologica,* **72**, 57–68.

Itani, J. and Nishimura, A. 1973. The study of infrahuman culture in Japan. In *Precultural Behavior*, ed. J. E. Menzel, pp. 26–50. Basel: Karger.

Janson, C. H. 1986. The mating system as a determinant of social evolution in capuchin monkeys (*Cebus*). In *Proceedings of the Xth International Congress of Primatology*, ed. J. Else, and P. C. Lee, pp. 169–179. Cambridge: Cambridge University Press.

Kawai, M. 1965. Newly-acquired pre-cultural behavior of the natural troop of Japanese monkeys on Koshima Islet. *Primates,* **6**, 1–30.

Kummer, H. and Goodall, J. 1985. Conditions for innovative behaviour in primates. *Philosophical Transactions of the Royal Society London, Series B,* **308**, 203–214.

Laland, K. N. and Reader, S. M. 1999. Foraging innovation is inversely related to competitive ability in male but not in female guppies. *Behavioral Ecology,* **10**, 270–274.

Lee, P. C. 1994. Social structure and evolution. In *Behaviour and Evolution*, ed. P. J. B. Slater, and T. R. Halliday, pp. 266–303. Cambridge: Cambridge University Press.

Matsumura, S. 1999. Relaxed dominance relations among female Moor macaques (*Macaca maurus*) in their natural habitat, South Sulawesi, Indonesia. *Folia Primatologica,* **69**, 346–356.

McGrew, W. C. 1992.*Chimpanzee Material Culture*. Cambridge: Cambridge University Press.

McGrew, W. C. 1998. Culture in nonhuman primates? *Annual Review of Anthropology,* **27**, 310–328.

McGrew, W. C. and Tutin, C. E. G. 1978. Evidence for a social custom in wild chimpanzees? *Man,* **13**, 234–251.

McGrew, W. C., Ham, R. M., White, L. T. J., Tutin, C. E. G., and Fernandez, M. 1997. Why don't chimpanzees in Gabon crack nuts? *International Journal of Primatology,* **18**, 353–374.

Myowa-Yamakoshi, M. and Matsuzawa, T. 2000. Imitation of intentional manipulatory actions in chimpanzees (*Pan troglodytes*). *Journal of Comparative Psychology,* **114**, 381–391.

Nishida, T. 1987. Local traditions and cultural transmission. In *Primate Societies*, ed. B. B. Smuts, D. L. Cheney, R. M. Seyfarth, R. W. Wrangham, and T. T. Struhsaker, pp. 462–474. Chicago, IL: University of Chicago Press.

O'Brien, T. G. 1993. Allogrooming behaviour among adult female wedge-capped capuchin monkeys. *Animal Behaviour,* **46**, 499–510.

Ottoni, E. B. and Mannu, M. 2001. Semifree-ranging tufted capuchins (*Cebus apella*) spontaneously use tools to crack open nuts. *International Journal of Primatology,* **22**, 347–358.

Perry, S. and Rose, L. M. 1994. Begging and transfer of coati meat by white-faced capuchin monkeys, *Cebus capuchinus*. *Primates,* **35**, 409–415.

Peters, H. H. 2001. Tool use to modify calls by wild orang-utans. *Folia Primatologica,* **72**, 202–204.

Phillips, K. A. 1998. Tool use in wild capuchin monkeys (*Cebus albifrons trinitatis*). *American Journal of Primatology,* **46**, 259–261.

Preuschoft, S. 1995. "Laughter" and "Smiling" in Macaques: An Evolutionary Perspective. PhD Thesis, University of Utrecht.

Preuschoft, S. and van Schaik, C. P. 2000. Dominance and communication: conflict management in various social settings. In *Natural Conflict Resolution*, ed. F. Aureli, and F. B. M. de Waal, pp. 77–105. Berkeley, CA: University of California Press.

Reader, S.M. and Laland, K.N. 2001. Primate innovation: sex, age and social rank differences. *International Journal of Primatology*, **22**, 787–805.

Russon, A. E. 1999. Orangutans' imitation of tool use: a cognitive interpretation. In *Mentalities of Gorillas and Orangutans*, ed. S. T. Parker, R. W. Mitchell, and H. L. Miles, pp. 119–145. Cambridge: Cambridge University Press.

Russon, A. and Galdikas, B. 1995. Constraints on great apes' imitation: model and action selectivity in rehabilitant orangutan (*Pongo pygmaeus*) imitation. *Journal of Comparative Psychology*, **109**, 5–17.

Rijksen, H. D. and Meijaard, E. 1999.*Our Vanishing Relative: The Status of Wild Orangutans at the Close of the Twentieth Century*. Wageningen: Tropenbos.

Sterck, E. H. M., Watts, D. P., and van Schaik, C. P. 1997. The evolution of female social relationships in nonhuman primates. *Behavioral Ecology and Sociobiology*, **41**, 291–309.

Sugardjito, J., te Boekhorst, I. J. A., and van Hooff, J. A. R. A. M. 1987. Ecological constraints on the grouping of wild orang-utans (*Pongo pygmaeus*) in the Gunung Leuser National Park, Sumatra, Indonesia. *International Journal of Primatology*, **8**, 17–41.

Sugiyama, Y. 1997. Social traditions and the use of tool-composites by wild chimpanzees. *Evolutionary Anthropology*, **6**, 23–27.

Thierry, B. 1985. Patterns of agonistic interaction in three species of macaque (*Macaca mulatta, M. fascicularis, M. tonkeana*). *Aggressive Behavior*, **11**, 223–233.

Thierry, B. 1994. Social transmission, tradition and culture in primates: from the epiphenomenon to the phenomenon. *Techniques and Culture*, **23–24**, 91–119.

Uchida, A. 1998. Variation in tooth morphology of *Pongo pygmaeus*. *Journal of Human Evolution*, **34**, 71–79.

van Schaik, C. P. 1999. The socioecology of fission–fusion sociality in orangutans. *Primates*, **40**, 73–90.

van Schaik, C. P. 2002. Fragility of traditions: the disturbance hypothesis for the loss of local traditions in orangutans. *International Journal of Primatology*, **23**, 527–538.

van Schaik, C. P. and Knott, C. D. 2001. Geographic variation in tool use on *Neesia* fruits in orangutans. *American Journal of Physical Anthropology*, **114**, 331–342.

van Schaik, C. P. and van Noordwijk, M. A. 1986. The hidden costs of sociality: intra-group variation in feeding strategies in sumatran long-tailed macaques (*Macaca fascicularis*). *Behaviour*, **99**, 296–315.

van Schaik, C. P., Fox, E. A., and Sitompul, A. F. 1996. Manufacture and use of tools in wild Sumatran orangutans. *Nature*, **83**, 186–188.

van Schaik, C. P., Deaner, R. O., and Merrill, M. Y. 1999. The conditions for tool use in primates: implications for the evolution of material culture. *Journal of Human Evolution*, **36**, 719–741.

van Schaik. C. P., and Ancrenaz, M., Borgen, G., Galdikas, B., Knott, C. D., Singleton, I., Suzuki, A., Utami, S. S. and Merrill, M. 2003a. Orangutan cultures and the evolution of material culture. *Science*, **299**, 102–105.

van Schaik, C.P., Fox, E.A., and Fechtman, L.T. (2003b) Individual variation in the rate of use of tree-hole tools among wild orangutans: implications for hominid evolution. *Journal of Human Evolution*, in press.

van Schaik, C.P., Preuschoft, S., and Watts, D. 2003c. Great ape social systems. In *Evolutionary Origins of Great Ape Intelligence*, ed. A. Russon and M. Begun. Cambridge: Cambridge University Press, in press.

Westergaard, G. C. 1988. Lion-tailed macaques (*Macaca silenus*) manufacture and use tools. *Journal of Comparative Psychology*, **102**, 152–159.

Westergaard, G. C. and Fragaszy, D. M. 1987. The manufacture and use of tools by capuchin monkeys (*Cebus apella*). *Journal of Comparative Psychology*, **101**, 159–168.

Westergaard, G. C., Lundquist, A. L., Haynie, M. K., Kuhn, H. E., and Suomi, S. J. 1998. Why some capuchin monkeys (*Cebus apella*) use probing tools (and others do not). *Journal of Comparative Psychology*, **112**, 207–211.

Whiten, A. 1998. Imitation of the sequential structure of actions by chimpanzees (*Pan troglodytes*). *Journal of Comparative Psychology*, **112**, 270–281.

Whiten, A., Goodall, J., McGrew, W. C., Nishida, T., Reynolds, V., Sugiyama, Y., Tutin, C. E. G., Wrangham, R. W., and Boesch, C. 1999. Cultures in chimpanzees. *Nature*, **399**, 682–685.

Williams, G. C. 1966. *Adaptation and Natural Selection: A Critique of Some Current Evolutionary Thought*. Princeton, NJ: Princeton University Press.

Woodworth, R. S. 1958. *Dynamics of Behavior*. New York: Henry Holt.

Wrangham, R. W. 1977. Feeding behaviour of chimpanzees in Gombe National Park, Tanzania. In *Primate Ecology: Studies of Feeding, and Ranging Behaviour in Lemurs, Monkeys and Apes*, ed, T. H. Clutton-Brock, pp. 504–538. London: Academic Press.

Wrangham, R. W., de Waal, F. B. M., and McGrew, W. C. 1994. The challenge of behavioral diversity. In *Chimpanzee Cultures*, ed. R. W. Wrangham, W. C. McGrew, F. B. M. de Waal, P. G. Heltne, and L. A. Marquardt, pp. 1–18. Cambridge, MA: Harvard University Press.

Wrangham, R. W., Chapman, C. A., Clark-Arcadi, A. P., and Isabirye-Basuta, G. 1996. Social ecology of Kanyawara chimpanzees: implications for understanding the costs of great ape groups. In *Great Ape Societies*, ed. W. C. McGrew, L. F. Marchant, and T. Nishida, pp. 45–57. Cambridge: Cambridge University Press.

Xu, X. and Arnason, U. 1996. The mitochondrial DNA molecule of Sumatran orangutans and a molecular proposal for two (Bornean and Sumatran) species of orangutan. *Journal of Molecular Evolution*, **43**, 431–437.

Zuberbühler, K., Gygax, L., Harley, N., and Kummer, H. 1996. Stimulus enhancement and spread of a spontaneous tool use in a colony of long-tailed macaques. *Primates*, **37**, 1–12.

ANNE E. RUSSON

12

Developmental perspectives on great ape traditions

12.1 Introduction

Interest in nonhuman primate culture arose primarily because of the insights promised into human culture, given the likelihood that evolutionary continuities link the two. Concepts of human culture are not directly applicable to nonhuman primates, however, because nonhuman primates do not share all the capacities deemed intrinsic to human culture. Scholars interested in comparative evolutionary questions, therefore, set aside features considered beyond the reach of nonhuman primates in order to focus on what is taken as the core feature of culture: a collective system of shared, learned practices. The focal phenomena in studies of nonhuman primate culture are then its products, enduring behavioral traditions, and the processes that generate them, social influences on learning operating at the group level over long periods of time (e.g., Donald, 1991; Kummer, 1971; McGrew, 1998; Nishida, 1987). It would be surprising, in fact, if such traditions were not prominent in the lives of nonhuman primates. Nonhuman primates typically rely on intricate forms of sociality for survival (Humphrey, 1976; Jolly, 1966; Smuts *et al.*, 1987) and on lifelong learning for much of their expertise (Fobes and King, 1982; King, 1994; Parker and Gibson, 1990). How widely practices must be shared, how much they must owe to social influence, and how long they must endure to qualify as "traditions" remain matters of debate (see Ch. 1).

Great apes stand out in this enterprise because their traditions may be more complex than those of other nonhuman primates (Parker and Russon, 1996; Whiten *et al.*, 1999). In great apes, acquiring expertise is an especially protracted and complex process that can entail years of dedicated study and social support (Matsuzawa, 1996; McGrew, 1992). More so than other nonhuman primates, great ape learners' needs, their physical capabilities, and their abilities to absorb and apply new information change with development, as do their preferences for the experts to whom they turn (King, 1994; Parker and Gibson, 1990; Parker and McKinney, 1999; Russon and Galdikas, 1995). Individual and social influences likely intertwine in the acquisition process, as they do in humans (Boesch and Tomasello, 1998).

To explore the processes that generate great ape traditions, this chapter considers how developmental processes may contribute to the acquisition of shared expertise. My focus is food-processing expertise, a major type of behavioral tradition studied in great apes (McGrew, 1992; Whiten *et al.*, 1999) and orangutans, the species I study. Like other contributors to this book, I adopt a broad definition of traditions as shared practices that are relatively long lasting (i.e., enacted repeatedly over a period of time), shared among members of a group, and acquired in part through social influences on learning.

12.2 Food-processing expertise

Appreciating how great apes acquire food-processing expertise requires understanding of the problems their foods pose, because these set the specifications for the expertise. Foraging is considered to be the greatest ecological challenge for primates (Freeland and Janzen, 1974; Milton, 1984). It may pose special challenges for great apes because they are frugivorous by preference but their great size precludes a strictly frugivorous diet (Waterman, 1984). Partly for that reason, their diets span an exceptionally broad range of foods (over 200 species in some populations; Rodman, 2000, Russon, 2002) that includes fruits, other high-quality foods, and "difficult" foods; foods protected by antipredator defenses that make them hard to get.

Some consider difficult foods as the distinguishing feature of the diet of the great apes and obtaining them as their greatest cognitive challenge (Byrne, 1997; Parker and Gibson, 1979; Russon, 1998). Among their most difficult foods may be permanent foods, like barks and nest-building invertebrates; because they sustain survival through periods of food scarcity, avoiding them is not an option (Parker and Gibson, 1979; Russon, 1998; Yamakoshi, 1998). Embeddedness has been proposed as the major cognitive challenge they pose but other equally challenging defenses are also prevalent, including spines, inaccessible locations, companion protector species like ants, antipredator behavior in animal prey, irritant hairs, distasteful exudates, and toxins (Boesch and Boesch-Achermann, 2000; Byrne and Byrne, 1991; Fox, Sitompul, and van Schaik, 1999; Parker and Gibson, 1977; Russon, 1998).

To make it worse, the problems presented are often *multifaceted*. Multiple defenses often protect a single food, and social competition or cooperation may further complicate the job (Boesch and Boesch-Achermann, 2000; Stokes, 1999). Orangutans in East Borneo, for example, consume jelly from a wild coconut *(Borassodendron borneensis)* that is multiply protected. The jelly is embedded within a shell; the shell is embedded within a fibrous husk, and both become rock-hard with age. The coconuts grow in the palm's crown, up to 15 m above ground and surrounded by some 50 razor-edged leaf petioles. Companions regularly pester skilled foragers to share, and females may distract their male companion from his food-processing task. Foods themselves grow, so the challenges posed by any one food can multiply further. Young leaves are defended differently mature ones (Waterman, 1984), for example. Palm heart is to available in palms of all ages so, even in one species, very different techniques can be required to obtain heart from immature versus mature plants (e.g., slender rosettes at ground level versus 30 m trees with massive crowns).

The result is that great apes must acquire an exceptionally wide repertoire of expertise to cope with this broad and varied range of defenses. They must also be able to coordinate diverse forms of expertise to combat sets of multifaceted problems, not just mobilize a single form to combat one-dimensional problems. That chimpanzees acquire tool kits and use tool sets speaks to this challenge (e.g., Brewer and McGrew, 1990; McGrew, 1992), although fully appreciating the complexity requires considering expertise that is not tool based. This helps to explain why their techniques for difficult foods can be highly complex, involving flexible and lengthy tool-based and/or manipulative sequences that combine and recombine multiple forms of expertise in varied patterns (Byrne and Byrne, 1991; Inoue-Nakamura and Matsuzawa, 1997; Matsuzawa, 1996, 2001; McGrew, 1992; Russon, 1998, 2002; van Schaik, Fox, and Sitompul, 1996). It also helps to explain why acquiring food-processing techniques can take up to 10 years and why social input may be an important contributor. The diversity and multiplicity of defenses, social complications, risks of ingesting toxins, and low probabilities of independently discovering hidden, ephemeral, or unpredictable foods all favor socially mediated learning for its speed, safety, and power to cue novices to cryptic items and innovative techniques.

12.3 Social influences on the acquisition of food-processing expertise

12.3.1 Enculturation

The acquisition of food-processing expertise by great apes must be bound up with development because it occurs primarily during immaturity on the basis of experience (Parker and McKinney, 1999). The developmental construction of expertise in great apes has recently been framed in socio-cultural terms as a process of *enculturation*, based on Vygotskian views of human development as deeply socio-cultural (Parker, 1996; Tomasello and Call, 1997; Tomasello, Savage-Rumbaugh, and Kruger, 1993; Vygotsky, 1962). Exploring enculturation should contribute to understanding how the social context enables great apes to build expertise like that of other group members.

Enculturation, in its original anthropological sense, means immersing novices in a system of meaningful human relations, including language, behavior, beliefs, and material culture, so that they become active agents in the system and come to embody it in their own actions and understanding (Miles, 1978; Miles, Mitchell, and Harper, 1996). Researchers studying nonhuman primates first borrowed the term to refer to a comparable human process applied to great apes (e.g., Miles, 1978; Tomasello *et al.*, 1993) and, later, to refer to a comparable species-normal process in great apes (Parker and McKinney, 1999; Russon, 1999a).

Great ape enculturation, as a species-normal process, likely resembles apprenticeship (i.e., guided participation in activities of a shared nature: Rogoff, 1992) and contributes to expertise by supporting and perhaps extending the construction of natural repertoires (Matsuzawa *et al.*, 2001; Parker, 1996; Parker and McKinney, 1999; Russon, 2002, 2003; Suddendorf and Whiten, 2001). It may operate via an integrated complex of ecological and social abilities that is specific to great apes and hominids, including imitation, self-awareness, and demonstration teaching (Parker, 1996; Parker and McKinney, 1999). The process is less one of socially transmitting or implanting traditional expertise than one of guiding novices in reinvention of expertise by scaffolding, channelling, shaping, fleshing out, or honing their learning. Its contributions to the experiences upon which expertise is constructed are subject to developmentally channelled opportunities and constraints.

Many findings are consistent with this view. Development builds in opportunities for social influence in great apes, as it does in other mammals, via basic biologically designed tendencies in immatures, such as following and scrounging (Box and Gibson, 1999). Following introduces immatures to their community's ranging, navigation, and resource patterns. Where close following is tolerated, the acquisition of manipulative skills may gain from social input (Coussi-Korbel and Fragaszy, 1995). Intensive food scrounging characterizes immature great apes from infancy into the juvenile period and, at these ages, likely enhances their learning (Russon, 1997). Scrounging preselected, semiprocessed, or leftover foods, especially difficult ones, helps immatures to identify foods in the local repertoire, introduces them to items they cannot obtain on their own, and may show them something of how the foods are processed. Development also channels opportunities for social input. Mothers tolerate their immatures' scrounging initially but lose patience as they develop. Progressively waning tolerance may restrict the processing stage at which immatures can scrounge food, providing a step-by-step (backwards) guide to acquiring the expertise (Russon, 1997). Great apes do not appear to achieve imitative capabilities until juvenility; consequently, this powerful learning process is constrained with respect to when it is available and the range of expertise it can affect. Great apes also imitate selectively, favoring advanced facets of expertise that challenge their competencies over low-level motor action details, and their preferred models change with age (Byrne and Russon, 1998; Myowa-Yamakoshi and Matsuzawa, 1999; Russon, 1999b; Russon and Galdikas, 1995).

12.3.2 Niche construction

Group members may also influence the acquisition of expertise indirectly, through the physical traces they leave. Habitat is often altered by the usage, choices, and practices of community members, so communities set as well as solve their physical world problems (Laland *et al.*, 2000). For learners, habitat changes alter behavioral and learning opportunities and contingencies. Insofar as great apes alter their habitat, "niche construction" may be another community-generated influence on the acquisition of traditional expertise (Laland *et al.*, 2000; see Ch. 2).

Great apes substantially alter their physical habitat. They are extremely large frugivores and transport seeds long distances from parent plants, improving seed germination by passing seeds through their digestive systems (Lackman-Ancrenaz, 2001; Voysey *et al.*, 1999). They are also highly destructive. Orangutans are major seed predators (Rodman, 1977; Galdikas, 1988) and can kill their food plants (e.g., *Artocarpus sp.* or single-stemmed rattans: (E. Meyfarth, 1998, personal observation). Their use of foraging tools can expose otherwise inaccessible foods. If their impact occurs habitually and community wide, it may systematically alter their niche and the learning problems facing descendants. Gombe chimpanzees prey so heavily on red colobus that they seriously deplete the population (Stanford, 1996, 1998). Orangutans' habitual, community-wide tool use at Suaq Balimbing likely makes *Neesia* fruits accessible to age–sex classes otherwise unable to exploit them, and greater *Neesia* exploitation could lead to higher density and altered social life (van Schaik, Deaner, and Merrill, 1999; van Schaik and Knott, 2001).

Great apes may also construct their niche not just by altering physical pressures but also by altering information: their changes may affect learners by providing information that guides differential habitat use. In such a "conceptual" or "behavioral" niche, an object's meanings depend on how community members use it, not just on its physical characteristics. A leaf, for example, may be a drinking vessel, wiper, probe, grooming stimulator, courtship signaller, or medication (Huffman, 1997). Many species do alter habitat in ways that provide information, so our interest would lie primarily in species that can extract information from social traces, in traces that offer information *only*, or in traces made intentionally for communicative purposes.

Some evidence indicates that great apes can extract information from others' physical traces to guide their behavior (i.e., as indirect social input) even if other nonhuman primates cannot (Cheney and Seyfarth, 1990). Gombe chimpanzees display aggressively on encountering empty nests in a rival community's range (Goodall *et al.*, 1979); orangutans may follow hornbills to locate fruiting fig trees (MacKinnon, 1974); chimpanzees use diana monkey calls to detect red colobus prey (the two species travel together: Boesch and Boesch-Achermann, 2000), and a symbol-competent chimpanzee used indirect cues to locate hidden foods (Menzel, 2001).

Physical and informational contributions may be combined when great apes leave food or tool remains behind at feeding sites. The nut-cracking sites used by chimpanzees at Taï may offer the best example of the degree of support available (Boesch and Boesch-Achermann, 2000). At these sites, nuts and expert-selected tools conveniently co-occur, perhaps even correctly juxtaposed. Hammer stones may be brought from elsewhere in the forest, so these sites clearly involve the chimpanzees' construction. Some hammer or anvil stones bear signs of repeated use in the form of concave pits (Joulian, 1995). This may make them easier – or harder – to use than equally appropriate stones that are new, and it can inform users of the proper orientation for the stone.

Development is among the forces governing how niche construction contributes to shared practices. Mothers clearly guide their youngsters through the environment and its intricacies. If youngsters are not yet able to interpret informational changes to the physical world, their mothers are, and in so doing mothers usher their youngsters through the experiences that most clearly underpin interpretation. This may guide youngsters' learning by building consistent association patterns. Physically altered items may channel immatures' learning in conventional directions through the scaffolding they provide, especially when coupled with maternal choice and guidance. Chimpanzee youngsters, for instance, use their mothers' stone nut-cracking tools; as a result, they benefit from the tools themselves and their mothers' choice and demonstration (Boesch and Boesch-Achermann, 2000). Immature rehabilitant orangutans regularly scavenge fresh remains of others' foods, so their knowledge of food species grows in the direction of foods other community members have eaten, food locations others exploited, and the techniques others used to obtain the food (A. E. Russon, personal observations). On finding food remains, they sometimes scan the immediate area, apparently in search of the food's source, and if the source is visible they may travel to it and search for more.

The picture that emerges is that immature great apes experience a physical world that is selectively used, marked, and shaped by community members. This niche construction, in conjunction with apprenticeship, would channel the learning of immatures along conventional lines and favor some degree of conformity in behavioral practices.

12.4 "Life-history" perspectives on acquiring traditional expertise in great apes

Many factors affect the acquisition of expertise by immatures, especially where shared practices are concerned. Among the more important are probably their needs, physical capabilities, cognitive capabilities, and social roles. These factors and their interactions all change developmentally, in time with biologically framed schedules that operate roughly in concert. The patterning of parameters pacing the life cycle of individuals within a species, such as longevity, ontogenetic change and timing, reproduction, and philopatry, reflect the species' life history (De Rousseau, 1990; Fleagle, 1999). Species' life histories likely orchestrate ontogenetic change in the factors affecting expertise acquisition, so considering relevant life-history parameters may clarify how shared practices are generated in great apes and help to disentangle the role of social influence (Parker and Russon, 1996). Parker and Russon (1996) suggested the importance of two life-history parameters in generating traditional food-processing expertise in great apes: demography and social organization. Equally important may be ontogenetic parameters related to individual capabilities and sociality.

For great apes, these ontogenetic parameters are set in the context of life-history patterns in the primate order. As primates, the lives of the great apes are characterized by expertise acquired through experience-based learning, life within relatively permanent social groups organized along long-term interindividual relationship lines, and developmental change to physical abilities, cognition, and social position (Fleagle, 1999; Parker and McKinney, 1999). Great ape life histories are distinguished within nonhuman primates by disproportionately prolonged immaturity and extremely high adult body mass (Fleagle, 1999). Prolonged immaturity is important to foraging expertise because it extends intensive learning, physical immaturity, cognitive development, and parental dependency beyond infancy and through the juvenile period, to span the first seven to ten years of life (e.g., Parker, 1996).

Physical development assumes great importance in great apes because their great body mass as adults requires extreme ontogenetic change in size: adults are massively heavy and strong, but infants and even juveniles are not (Fleagle, 1999; Janson and van Schaik, 1993). Changes in body mass obviously affect feeding needs. They also affect feeding capabilities by changing strength and weight. Other changes associated with physical maturation, like dentition, digestion, puberty, and reproduction, likewise affect both foraging needs and physical capabilities.

Cognitively, great apes achieve their full potential after infancy. As juveniles, they achieve second-order (rudimentary symbolic) cognition to levels like those of human children 3 to 3.5 years of age, an order above the first-order (sensorimotor) levels they can muster in infancy (Parker and

McKinney, 1999). This alters their capacities for understanding experiences and generating expertise as a function of age.

Social development alters great apes' opportunities for socially mediated learning. Prolonged immaturity extends maternal support and tolerance, which can assist the acquisition of advanced facets of processing techniques (Parker, 1996). Other social encounters and companions also change with age, affecting opportunities for social input. The minimum condition for socially learning manipulative skills, tolerance of proximity, varies with the learner's age and sex status (van Schaik *et al.*, 1999). Tolerance patterns change markedly within the immaturity period alone, the period during which learning is concentrated in the great apes. For infants, parental tolerance is highest; for juveniles, tolerance from parents begins to wane and from same-sex peers to increase; and for adolescents, intolerance grows with same-sex adults and peers but tolerance increases with opposite-sex partners. By constraining the set of potential expert–learner pairs, developmental changes in tolerance constrain who is likely to share expertise, what accessible experts likely know, and learners' abilities to understand new experiences.

One consequence of the extensive ontogenetic change in the physical and cognitive capabilities of the great ape is that the multifaceted problems presented by difficult foods effectively change with the forager's age. For orangutans, for instance, obtaining hard-shelled arboreally located foods simplifies with age as strength and cognitive abilities improve; however, accessing the same foods becomes harder with weight increases. Great apes then face an unending cycle of *re-solving* the "same" food problems repeatedly as they progress through their life cycle. Ontogenetic parameters affecting dietary needs, strength, weight, cognition, and social opportunities probably pace the acquisition and repeated modification of food-processing techniques. If physical capabilities, cognitive abilities, and social tolerance affect great apes' acquisition of food-processing techniques, the patterns of social influence on acquisition should change with age. As new food difficulties and new capabilities arise, new social influences become available and old ones become inaccessible to learners.

12.5 Traditions through the life cycle in orangutans

To explore a life-history perspective on great ape traditions more closely, I focus on one species, the orangutan, because life-history parameters vary between species within adaptive arrays. Orangutans seem unlikely candidates for traditional expertise, having been type-cast as solitary. They show a distinct sociality, however, although it is dispersed and muted. They pursue long-term relationships beyond the mother–infant unit and may be members of loosely defined communities (van Schaik and van Hooff, 1996). They commonly have few companions but may associate with those for weeks on end. While adults tend to solitude, immatures are actively gregarious (Galdikas, 1995). Paradoxically, orangutans may be excellent subjects for studying the generation of traditions *because* their sociality is so spare. Their limited social contacts may show social influence routes more clearly than large interacting groups.

Ontogenetic life-history parameters likely to affect orangutans' food-related expertise are outlined in Table 12.1. Table 12.2 and Fig. 12.1 show corresponding social tolerance patterns. Overall, this suggests that individual physical capacities, cognitive abilities, ecological demands, and opportunities for social exchange all shift in tandem, in pace with global life stages. This has implications for the processes generating traditional expertise in orangutans, and probably other great apes.

First, in orangutans as in other primates, stage-related tolerance patterns constrain which social routes are open to each age/sex class within a community, but the contribution of social context to learning tends to vary with interindividual relationships among participants, primarily those based on kinship, dominance, mating, or affiliation (e.g., Coussi-Korbel and Fragaszy, 1995; Rijksen, 1978; Russon and Galdikas, 1995; van Schaik and van Hooff, 1996). Adolescent female orangutans are unlikely to share expertise if they remain strangers, for instance, although they tend to tolerate one another as a class. A major influence of stage-related social tolerance may concern facilitating or inhibiting the formation of particular interindividual relationships.

Second, stage-related tolerance patterns constrain the types of information likely to be shared. Infants and young juveniles are most highly tolerated by their mothers, for instance. They have immature capabilities and needs compared with their mothers' adult ones, however, so they probably absorb only the simple facets of her expertise commensurate with their own situation. Their achievements may be limited to identifying foods selected by their mothers and within their natal range, and basic food manipulations. Equally, they are unlikely to be influenced by adult males – not only because intolerance makes this association improbable but because adult male expertise is likely irrelevant to them and beyond their grasp.

Table 12.1. *Changes in orangutan food problems and problem-solving capacities*

Stage[a] (age[1])	Individual capacities		Food problem changes	
	Cognitive	Physical	Age related	Sex related
Infant	First order (sensorimotor)[2]	Maternal dependency, weak, light (2–15 kg),[1] immature dentition and poor motor control[3]	Nursing, semi-ready foods, scrounged remains, simple foods	
Juvenile	Second order (lower)[2]	Maternal help,[4] weak,[5] light (15–20 kg),[1,5] functional dentition and good motor control[3]	Weaning foods,[6] adult foods,[5] low volume [5]	M: disperse early[4]
Adolescent	Second order (higher)[3]	Strong[7], heavy (20–30 kg)[1,7] F: early sterility[8] M: mate guarding[1,9]	Adult competition, dispersal, high volume, arboreality	F: range small, disperse near[10] M: range large, disperse far[10]
Subadult		M: stronger, heavier (30–50 kg),[1] mate searching[10]		M: growth diet,[11] highly active[11]
Adult	Scaffold[2]	F: caregiving, heavier (30-50 kg)[1] M: strongest heaviest (50–90 kg),[1] mate competition[10]	Reproduction, higher volume, changed quality, arboreality, competition, association[10]	F: feed offspring, rich diet[11] M: great mass,[10] poorer diet,[11] range largest,[11] poor mobility[11]

[a] *Infant* (birth to 4/5 years), preweaned immature too young to survive independently (Pereira, 1993); *juvenile* (4/5–7/8 years), prepubertal immature who can survive losing adult caregivers (Pereira, 1993); *adolescent* (7/8–10/15 years), postpubertal individual not yet fertile (Pereira and Altmann, 1985); *subadult* ((9/11– ; years), male only), postadolescent lacking adult secondary sexual characteristics (SSCs); otherwise reproductively mature (van Schaik and van Hooff, 1996); *adult* (11/20–40+years), reproductively mature, females at first birth and males with SSCs and adult reproductive roles.

Source: Numbers indicate sources: 1, Rijksen, 1978; 2, Parker and McKinney, 1999; 3, Joffe, 1997; 4, Horr, 1977; 5, Janson and van Schaik, 1993; 6, Altmann, 1980; 7, Bogin, 1999; 8, Galdikas, 1995; Watts, 1985; 9, Galdikas, 1995; 10, van Schaik and van Hooff, 1996; 11, Galdikas and Teleki, 1981; Rijksen, 1978; Rodman, 1979; Utami, 2000.

Table 12.2. *Stage-related changes in orangutan social tolerance*

Actor (stage)	Social tolerance		
	Level	Dyads (actor: partner)[a]	Source[b]
Infant	Highly tolerant	All: I, AF	
	Tolerant	All: JF, AF IM: JM	Galdikas, 1995
	Neutral	IF: JM	
	Intolerant	All: AM, SM	
	Highly intolerant	All: AM	
Juvenile	Highly tolerant	All: AF, J (same sex)	Watt and Pusey, 1993
		JM: JF, AF	Rijksen, 1978; pers. obs.
		JF: AM	Galdikas, 1995; pers. obs.
	Tolerant	All: I, AM, S	Galdikas, 1995; Watt and Pusey, 1993; pers. obs.
	Neutral	JF: JM	Watt and Pusey; 1993; pers. obs.
	Intolerant	JM: AM	
Adolescent	Highly tolerant	AF: I, S, AM	Galdikas, 1995; Rijksen 1978; van Schaik and van Hooff, 1996
		AM: JF, AF, AM, F (estrus)	Watt and Pusey, 1993; pers. obs.
	Tolerant	AF: AF, AM, AF	Galdikas, 1995; pers. obs.
	Tolerant and intolerant	AM: JM, SM, AM	Rijksen, 1998; pers. obs.
	Neutral	AF: JF AM: I	
	Intolerant	AM: AM	
Subadult	Highly tolerant	SM: AF, AF	Rijksen 1978; van Schaik and van Hooff, 1996
Male	Tolerant	SM: JM, JF	Personal observation
	Tolerant and intolerant	SM: AM, SM, AM	Rijksen 1978; van Schaik and van Hooff, 1996
	Neutral	SM: I	
Adult	Highly tolerant	AF: I	
	High (tolerant–intolerant)	AM: AF	van Schaik and van Hoff, 1996
		AF: AM	van Schaik and van Hoff, 1996
	Tolerant	AF: J, AM, SM	Rijksen 1978; van Schaik and van Hooff, 1996
		AM: JM	
	Neutral	AM: AF, I	
	Tolerant and intolerant	AF: AF, AF	Galdikas, 1995; Rodman, 1973; Watt and Pusey, 1993
	Intolerant (to highly)	AM: SM, AM	Rijksen, 1978; Utami, 2000; van Schaik and van Hooff, 1996
	Highly intolerant	AM: AM	Mitani, 1985; Utami, 2000; van Schaik and van Hooff, 1996

[a]Age–sex class noted as XY, where X is age class (I, infant; J, juvenile; A, adolescent; S, subadult; A, adult and Y is sex class (M, male; F, female).
[b]Dyadic tolerance levels without sources cited are inferred by interpolating between tolerance levels for learners from adjacent age–sex classes.

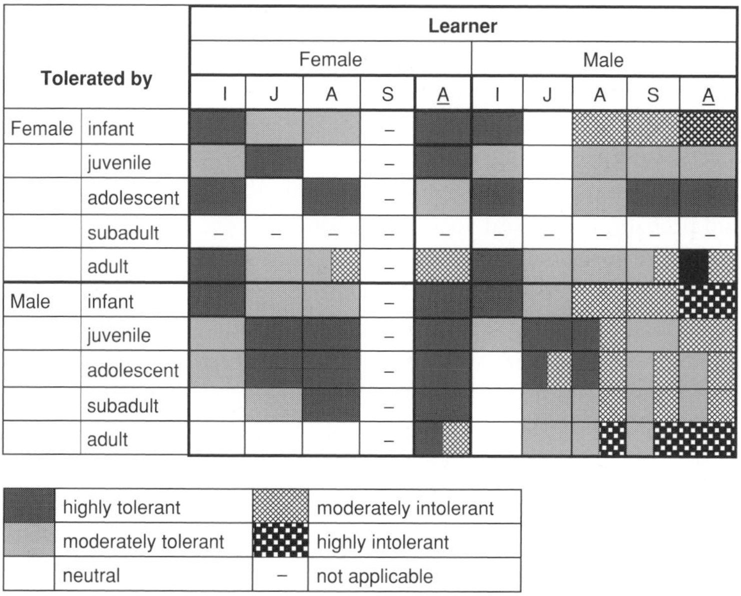

Tolerated by		Learner									
		Female					Male				
		I	J	A	S	A	I	J	A	S	A
Female	infant				–						
	juvenile				–						
	adolescent				–						
	subadult	–	–	–	–	–	–	–	–	–	–
	adult				–						
Male	infant				–						
	juvenile				–						
	adolescent				–						
	subadult				–						
	adult				–						

	highly tolerant		moderately intolerant
	moderately tolerant		highly intolerant
	neutral	–	not applicable

Fig. 12.1. Social tolerance across the lifespan in orangutans, by age–sex class: predicted tolerance level by group members (rows) for learners (columns; class indicated by initial letter only). Stage-related levels of dyadic social tolerance were established from Galdikas. (1995), Mitani (1985), Rijksen (1978), Rodman (1973), van Schaik *et al.* (2003), van Schaik and van Hooff (1996), Utami (2002), and Watts and Pusey (1993). In dyads for which no information was found, tolerance levels were inferred by interpolating between the tolerance levels for learners in adjacent age–sex classes. Tolerance levels are approximated in the five grades shown. Cells shown split, with two tolerance levels, represent oscillating or ambivalent responses.

Third, juveniles and adolescents stand out as important in sharing complex food expertise. Orangutans first extend their social contacts beyond the maternal unit as juveniles, mostly to peers, as maternal tolerance wanes and independence grows (Galdikas, 1995). Adolescents leave their natal range to establish independent home ranges, become highly social, and begin associating regularly outside their natal unit; they may associate with other adolescents or subadults for weeks, or with adults briefly (Galdikas, 1995; Rijksen, 1978). Learning needs probably intensify for adolescents with their range shift, which exposes them to novel foraging problems, and with their physical maturation, which alters their dietary needs. Their cognitive and physical capabilities have developed to more powerful levels; consequently, they can understand and attempt more difficult challenges. Conveniently, the sociability of both juveniles and adolescents enhances the social influence of peers. The importance of these two stages in generating traditions is underlined by arguments that good conditions for horizontal exchange are essential for community-wide diffusion of practices (van Schaik and Knott, 2001) and by evidence that orangutan learners prefer working with experts "like themselves", that is, those performing just beyond their own competence levels (Russon and Galdikas, 1995). At advanced levels, cultural processes in orangutans and probably other great apes may be tuned to peer more than to adult experts: to horizontal versus vertical or oblique social routes.

This makes it likely that the appropriate question about how traditional food-processing expertise is generated is not only *whether* one social route or another predominates but also *when* each is prominent. Over the lifespan, several social routes likely afford input. The relevance and difficulty of the various components of complex expertise along with changing contexts likely constrain when each component is acquired and so which social routes influence its acquisition. Probably important for the wide diffusion of complex food-processing expertise is patient vertical exchange in the early stages of learning, to support the slow process of acquiring the essential basics, *plus* extensive horizontal exchange in later stages, to spread complex elements widely.

12.5.1 Evidence on orangutan culture

The best evidence from the wild that orangutans use social learning and produce a collective culture concerns tool use at Suaq Balimbing, Sumatra (Fox *et al.*, 1999; van Schaik *et al.*, 1999; van Schaik and Knott, 2001; Ch. 11). This evidence does not provide experimental-level proof that acquiring this expertise owes something to the social context. It does, however, document the social and physical contexts in which that expertise occurs, intra- and intergroup variation in the expertise, and changes over time in individual performance (Ch. 1). With the wealth of contextual information, the relative contributions of these processes can be weighed to provide a well-informed best guess on how sociality influences acquisition.

Tool use in Suaq Balimbing orangutans is habitual and community wide, including using tools to extract *Neesia* seeds from hard, spiny shells. Suaq Balimbing is one of two isolated orangutan communities known to consume *Neesia* seeds as a major component of their diet, seasonally. The second is the Gunung Palung community in Western Borneo; its orangutans do not use tools to obtain *Neesia* seeds. Circumstances point to cultural processes as the basis for community-wide *Neesia* tool use at the former site (Ch. 11).

Neesia tool-use patterns are consistent with the suggestion that social tolerance is a precondition for obtaining social input when acquiring manipulative expertise. At Gunung Palung, adult males are the main *Neesia* consumers so even if they invented tool use, chances of their sharing it are slim because of the intolerance that surrounds them. Adult males are the most solitary class in the species. They invariably repulse other males. They consort with sexually receptive females, but as females have a seven to eight year interbirth interval and a dispersed range, these opportunities occur very rarely (Galdikas and Wood, 1990). Orangutan males typically find females by entering their ranges and when consorting they follow rather than lead the female, so males have little opportunity to initiate food choices and even if they do, females are unlikely to follow. At Suaq, Balimbing all age–sex classes extract and eat *Neesia* seeds using tools. Receptive females at Suaq Balimbing follow adult males, apparently for protective relief from unrelenting subadult harassment (van Schaik and van Hooff, 1996). This community favors community-level sharing of tool-use expertise not only because there are more tool users but also because tool users' age and sex classes offer broader tolerance. The evidence from Suaq Balimbing also suggests that ontogenetic patterns in social tolerance affect social exchange. During tool sessions at tree holes to obtain invertebrates by a mother–infant and an adult female–subadult male dyad, the partner watched the tool user closely and manipulated the tool user's tree hole and/or tool (Fox *et al.*, 1999). Both cases involved developmentally scheduled intimate dyads.

The mosaic, constructive process involved in acquiring cognitively complex expertise, where social influence is one of several contributors, is also evident in the findings at Suaq Balimbing. Fox *et al.* (1999) described some of a young female's tool-use acquisition. At three to four years of age, Andai often observed her mother, Ani, use tools to probe for invertebrates in tree holes. Five to six years of age, Andai was weaned and made her own tools and probed Ani's tree holes after Ani abandoned them, then subsequently initiated a tool session at her own tree hole independently. By seven to eight years of age, Andai was a frequent and competent tool user.

Additional field evidence of a mosaic constructive process involving collective-level social influences derives from my studies of foraging expertise in rehabilitant orangutans returned to free forest life by the Wanariset Orangutan Reintroduction Project (ORP) in East. Indonesian Borneo (Russon, 1998, 2002). Subjects ranged around two sites in Sungai

Wain Forest, K3 and K5, located 3–4 km apart. K3 and K5 orangutans did not meet one another for over two years, so they represented two isolated communities inhabiting ecologically similar habitat.

These rehabilitants offer evidence that acquisition of some facets of food-processing techniques is socially mediated (Russon, 1999a). One example illustrates the strength of social influences. Siti, a juvenile female about five years of age, appeared to acquire an ineffective technique for obtaining nest-building termites through Tono, a juvenile male about six years old. Siti was among the most forest naïve at release, having been captured and rescued under one year old then having lived with peers in ORP cages until she was old enough to release. In their first two months after being released at site K5, in 1996, in a group of 19, Siti and Tono often travelled and foraged together along with three or four other relatively naïve juveniles. Tono often tried to open lobed termite nests by banging them against hard objects (e.g., another orangutan's head); he persisted with this technique although it never worked. Knowledgeable Sungai Wain rehabilitants cracked these termite nests apart by hand then sucked termites from newly exposed cells. Siti began banging nest lobes after Tono did. She still used the technique a year later although it never worked for her either; at best, it chipped off small bits. Siti and Tono were the only individuals seen using this technique. Orangutans who knew effective techniques were not among Siti's companions; other than Tono, none of her companions even tried opening these nests.

Four cases are sketched that suggest how shared social practices may be generated. Three cases concern acquiring complex food-processing expertise in juveniles initially naïve to a difficult food. All concern multifaceted and variable problems, acquisition by repeatedly re-solving and modifying solutions, and social influences from several community members with age-appropriate tolerant relationships with learners. Two offer developmental views of part of the acquisition process and the third suggests how physical traces may aid acquisition. The fourth case suggests how information provided by niche modification may channel acquisition towards shared practices.

12.5.1.1 Hearts of palm

These rehabilitants regularly eat heart matter (apical meristem) of a tree palm, *Borassodendron borneensis*, locally called bandang. Techniques rely on one overall strategy: pull the newest leaf as a shoot emerging at the palm's tip then bite heart matter from the shoot's base. Techniques vary because the problem changes with palm growth. In immature bandang, the heart is in the centre of a small, slender rosette on the forest floor. In mature bandang, it is atop a 10–15 m trunk, embedded in a massive crown, and encircled by several ranks of huge leaves with robust, razor-edge petioles. Shoots range from slender and grass-like in small rosettes to stout and spear-like in trees, and rehabilitants select only those long enough to grasp and young enough to lack a petiole. Bandang heart then poses a set of multifaceted problems that are naturally graded in difficulty. Rehabilitants adjust for this variability by using two (sub) strategies. The basic strategy, for shoots in small to mid-sized rosettes, was to grab the whole shoot and pull it out all at once. The mature strategy, for shoots in large rosettes and trees, was to subdivide the shoot into sections of a few laminae each and pull out sections one by one.

I tracked Paul's acquisition of bandang heart expertise from a point when he was naïve until he had mastered the mature strategy. Observations started in June 1995, when he was about five years old and had resumed forest life for about six months at site K3. From this point, he took two years to acquire the mature strategy. The process involved at least five steps and the social influences of four other orangutans (Fig. 12.2).

Paul ranged near K3 along with two other juvenile males, Enggong and Bento, who were approximately five and six years old, respectively. Initially, all three behaved as if naïve to bandang heart: they made no move to obtain it even though bandang were plentiful and they often ate other bandang items. I first saw all three eat bandang heart when they scrounged remains discarded by Sariyem, a skilled five-year-old female, during her four day incursion to the K3 area. Only Paul tried to obtain bandang heart himself during her visit, and he used the basic strategy, incorrectly. He pulled a shoot from a small rosette but ate its tip instead of its base. That he tried at all probably owed to Sariyem's tolerance. She tolerated Paul but not the other two males in proximity while she worked, so Paul alone could observe her technique. Sariyem used mature as well as basic strategies, so Paul's successes and his errors probably owed to his limited understanding, perhaps related to his young age. Charlie, an adolescent male proficient in obtaining bandang heart, also periodically visited K3. The three K3 males typically fled at Charlie's sight, however, because he was already aggressively intolerant of other males, so Charlie's presence was not conducive to their social learning.

Six months later (March 1996) Paul had acquired the basic strategy plus two idiosyncratic tactical enhancements, a two-step pull (pull the shoot

	1995 age–sex	1995	1996	1997
Paul	J–M (5)	N S B–	B B+	M
Enggong	J–M (5)	N S	M	M
Bento	J–M (6)	N S	M	–
Charlie	A–M (9)	M	M	M
Sariyem	J–F (5)	B M	–	–

(dark shading)	tolerant
(light shading)	intolerant
(no shading)	not available

Fig. 12.2. Paul's acquisition of bandang heart techniques and the presence of social tolerance. Age–sex classes noted as X–Y (Z) where X is the age class (J, juvenile; A, adolescent), Y is the sex (M, male; F, female), and Z is the approximate, age (years). The technique variations are: N, naïve; S, scrounge; B–, basic strategy-failed; B, basic strategy; B+, basic strategy enhanced; M, mature strategy. Social tolerance is at two levels; tolerant and intolerant as indicated by shading.

down through the side of the crown, then pull it out) and bracing against a nearby tree while pulling. His enhanced technique succeeded for shoots from mid-sized but not large rosettes; he did not even try trees. Given his size, the limitation owed to a strategic error: he tried to strengthen himself rather than weaken the palm. Only adult males may have the strength to pull whole large spears (P. Rodman, personal communication). In the same period, Enggong and Bento acquired basic and mature strategies. Social intolerance may explain Paul's lagging progress. All three K3 males had similar chances to invent techniques, Paul's cognitive capacities should have resembled Enggong's, and Paul had a head start. Paul suffered Bento's intolerance, however, perhaps because Paul was a relative newcomer. Bento rebuffed Paul's approaches and kept Paul away from Enggong, even though Enggong tolerated Paul. Accordingly, Paul had little opportunity to track their progress. This, plus the idiosyncratic tactics, suggests that Paul's enhancements owed little to direct social exchange. Paul may have gleaned some information from their food remains. He often lurked nearby while they foraged and on several occasions was observed to enter and rework their foraging site after they left.

By mid-1997, Paul had acquired the mature strategy. Changing social tolerance as well as indirect social influences likely contributed to his progress. Bento, probably with approaching adolescence, grew intolerant of Enggong in 1996. This allowed Paul and Enggong to associate and they became steady foraging companions. This gave Paul better opportunities to learn directly with Enggong. Paul also scrounged Enggong's bandang heart remains often, including those from mature palm trees. Twice, after eating remains from the ground below the tree and waiting until the tree was vacated, Paul climbed up and retrieved remains lodged in the crown. Once, after eating Enggong's remains from the ground below the tree, Paul left and foraged in other areas. Two hours later he returned to the same tree, travelling directly to it through 150 m of dense forest, climbed directly into its crown, and retrieved remains lodged there. He must have used some physical cues to relocate the palm, and fresh shoot remains are the obvious candidates. Scrounging of this sort exposed Paul to large bandang shoots subdivided into sections and to their source, the tops of bandang trees. It was after several months of scrounging and foraging with Enggong that Paul began to subdivide large shoots himself and obtain them from bandang trees. The palm's own physical qualities may also have contributed to Paul's progress. Bandang shoot laminae are separate at the tip and sometimes a few slipped free accidentally while Paul was pulling the whole shoot. Noticing the accidental subdividing then reproducing it deliberately could generate the mature strategy. The year's development may have brought this within Paul's cognitive reach, however he discovered it.

This illustrates part of the acquisition process for bandang heart techniques but it shows only one male's juvenile-level achievements. Some later patterns have been seen in other adolescents and subadults. Two notable changes emerge with puberty: increasing size and strength linked to a growth spurt in males, and increasing peer contact in the form of sexually motivated pairings. Greater strength likely simplifies the task of extracting bandang heart differentially for males. Gregariousness affords horizontal sharing of expertise. Four adolescent females travelling with an adolescent or subadult male were observed scrounging the male's bandang heart remains on a regular basis. In two instances, the male shared a newly pulled and uneaten section of his shoot with his female companion, unsolicited (Grundmann *et al.* 2000; A. E. Russon, personal observation).

All four females were less adept than the males with the mature technique – in particular, less efficient and less able to extract extremely stout spears – probably because they lacked male strength. This scrounging and sharing contributes to females' foraging beyond their independent means and could also contribute to their own skill acquisition.

12.5.1.2 Palm pith

Rehabilitants also eat pith (parenchyma) from bandang petioles. The common strategy is to bite then tear the petiole open lengthwise, pull strips of pith away from the sheath, and chew the pith for juice. Plant growth alters the task by changing the size, toughness, and location of petioles, so mature techniques differ from basic ones in the tactics for handling more difficult defenses.

I tracked Siti's pith technique over 18 months, from her release into the forest at site K5 in May 1996. Over that time, she had opportunities for social exchange with several rehabilitants who ranged near K5 (Fig. 12.3). Immediately after release, Siti associated closely with Kiki and Ida, two like-aged juvenile females she already knew. They all behaved as if naïve to bandang pith until Kiki discovered it after a month and invented a basic technique for obtaining it from tiny rosettes at ground level. Immediately, Siti and Ida scrounged from Kiki – they chewed her petiole, another petiole on the same plant, or her leftovers – and within four weeks could obtain pith independently from small rosettes. Within another week, Siti tackled bandang trees, adding tactics to handle the petiole's arboreal location (arboreal feeding postures) and robust size (making the first bite into a petiole, U-shaped in cross-section, over one arm of the "U"). She probably invented the latter tactic because other orangutans in the forest made their first bite over the rounded bottom of the "U". The tactic may reflect an age-related constraint; with Siti's small size and teeth, she may have lacked the strength to bite the petiole open the common way. How she advanced to arboreal petioles is less clear: her closest companions worked pith only from the ground but some occasional companions obtained it from trees.

A year later, in 1997, Siti associated with three other orangutans who still ranged near K5; others had all moved elsewhere. Her steady companion was Judi, a near-adult female, from whom she obtained parent- or sibling-like support. Of the four, only Siti and Judi were observed eating bandang pith. They ate it frequently, often working the same petiole together and shared two idiosyncratic tactics. First, both

	1996 age–sex	1996	1997
Siti	J–F (5)	N S B M_1	M_2 M_3 M_4
Kiki	J–F (4.5)	N S B	–
Ida	J–F (4)	N S B	–
Tono	J–M (6)	N	–
Mojo	J–F (6.5)	M	M
Judi	A–F (13)	N	M_2 M_3
Dan	J–M (4)	M	–
Aming	A–M (9)	M	–
Panjul	A–M (9)	M	M

	highly tolerant
	moderately tolerant
	neutral
–	not available

Fig. 12.3. Siti's acquisition of palm pith techniques and social tolerance. Shown are only those community members with whom Siti had regular contact. Age–sex classes noted as X–Y (Z) where X is the age class (J, juvenile; A, adolescent), Y is the sex (M, male; F, female), and Z is the approximate age (years). Technique variations noted as N, naïve; S, scrounge; B, basic strategy; M_1–M_4, mature strategy + tactics 1, 2, 3, and 4, respectively (see text). Social tolerance at three levels: tolerant, moderately tolerant, and neutral, as indicated by shading.

always chose the second newest petiole, never newest or mature ones. K3 rehabilitants, only 4 km away in similar habitat, always chose mature petioles. No contacts had been reported between K5 and K3 rehabilitants, bandang are abundant, and each palm has about 50 mature petioles to one second-newest one, so neither ecological differences nor competition explain their choice. Second, both typically made their first bite into the petiole close to the leaf then tore open the petiole towards the crown. All K3 rehabilitants typically made their first bite a third of the way down from the leaf then opened the petiole in both directions. Siti used a third tactic that was not evident in Judi. She chose bandang with a liana or branch running diagonally through the crown below her chosen petiole. Her first bite typically cracked the petiole and it flopped over the liana/branch. The liana/branch acted as a hanger that probably helped to anchor the petiole while she tore it open. I did not notice this pattern until late in my observations so I lack reliable data on its occurrence. Siti used it on at least five occasions, however, so the layout, the place of biting, and the hanger were likely deliberate tactics. No others were observed using these three tactics so Siti probably acquired the first two in tandem with Judi and invented the third herself.

Both cases illustrate the multifaceted nature of orangutans' food problems as well as the lengthy acquisition process of piecing together effective strategies and tactics for solving them. Acquisition is necessarily piecemeal and protracted because some particular defenses and some forms of a food problem may be beyond youngsters' physical or cognitive capabilities. Progress took the form of multiple small advances, many of them tactical but some of them strategic or program-level. Advances appear to owe to a mix of independent and social influences, some of which involve participating in problem solving with a knowledgeable partner. Direct social input was enabled by relationship-based tolerance, probably enabled itself by species-typical tolerance patterns linked with age–sex classes. Social influences drew from multiple companions at several different points in the acquisition process. Finally, these cases illustrate how developmental changes to cognition, physical capacities, problems, and social tolerance all affect acquisition in concert. Other than cases where novices' first attempts directly followed scrounging and/or observation, this evidence speaks only to co-occurrence and as such is only suggestive. Nonetheless, the emerging picture is consistent. Regularly associating community members share elements of their practices.

12.5.1.3 Palm fruit

Paul appeared to learn to eat young bandang coconuts in 1995 by scavenging food remains then backtracking to their source. He initially behaved as if naive to this food. I next observed him eating discarded young coconut remains from Bento and Enggong. Both ate only small, immature coconuts soft enough to tear apart; they bit the fruit's leathery shell at its base, tore it apart lengthwise, bit off and chewed the fibrous material beneath, then spat it out. Paul scavenged their remains from the ground below the palm, partially processed fruits and chewed fibre alike. Sometimes he waited below for discards while they worked in the palm's crown; sometimes he searched for old remains on his own. Subsequently, I observed Paul climb bandang and pick his own young coconuts. In his early attempts at processing, he correctly bit into the fruit's base but gnawed off bits of leathery shell then fibre instead of tearing it lengthwise. For that reason, he obtained only a fraction of its edible material. He struggled with each coconut for over 10 minutes (compared with Enggong's 2–5 minutes) and seemed to have difficulty selecting fruit of an appropriate size (he often discarded fruit he had picked after a few ineffective bites). These observations suggest that Paul relied on social assistance to learn about this food and how to process it, but at least as much from others' physical traces as from their behavior.

12.5.1.4 Bandang traces

Orangutans' social life is so dispersed that they may be more sensitive to the ghosts of orangutans past, the enduring traces of social activity, than group-living apes who enjoy almost constant direct social input. One candidate trace is the damage orangutans inflict on vegetation in feeding. Damage caused in pulling out new bandang shoots for palm heart is one example. It can remain visible for months and could provide information to others in the area. Pulling breaks off the shoot's tip but leaves its basal section and petiole intact. Fully grown, the leaf looks as if neatly trimmed with scissors. The damage is highly distinctive and visible, and it lasts for months if not years. No other species has been seen obtaining this food in this fashion, nor do I know of any with the manipulative capabilities and strength to do so. Trimmed palm leaves then record the history of orangutan foraging and travel through an area and their distribution shows something of usage patterns.

We systematically recorded rehabilitants' damage to bandang along trails and in focal individual follows. Regularly, we find bandang-rich areas where some individual bandang palms have many trimmed leaves (up to 12) but others just a few meters away have none. This implies that rehabilitants selectively choose and revisit individual palms and reuse travel routes. Several individuals have been observed feeding from the same palm and travelling along the same route at different points in time. This suggests that rehabilitants themselves may be reading trimmed leaves to guide their choice of palms and travel routes.

12.6 Life-history perspectives on traditions in other great apes

Some of these patterns likely characterize other great apes, who share many life-history parameters with orangutans. The ontogenetic patterns discussed in the apprenticeship hypothesis are likely candidates.

Chimpanzees' acquisition of stone nut-cracking illustrates similarities (Boesch and Boesch-Achermann, 2000; Inoue-Nakamura and Matsuzawa, 1997; Matsuzawa *et al.*, 2001). The task is multifaceted: *Panda* nuts have three kernels independently embedded within a hard wooden shell; old nuts pose different problems to new ones; some but not all nuts can be cracked with wood tools; stone hammers and anvils vary in their qualities; and social concerns may affect the nut-cracking abilities of males. Acquiring nut-cracking skills is constrained by cognitive, physical, and social development. Cognitively, infants can master basic operations using nut-cracking items in their first two years but cannot combine them appropriately before three years of age. Physically, lack of strength limits infants' early successes in nut cracking and too much strength may limit success in maturing males. Socially, infants' opportunities for social learning are mainly with their mothers, although with age they increasingly scrounge from and observe other group members. Infants' scrounging is tolerated but juveniles may be chased away. Sex-related changes in tolerance may affect individuals approaching maturity: for example subadults may progress slowly because of difficulty accessing good hammers, and sex differences emerging in late adolescence may owe to differential maternal support for male versus female offspring. The apprenticeship process is clear: mothers provide substantial support tuned to their offspring's skills into adolescence (e.g., food and tool sharing, teaching), and offspring actively solicit maternal input and assistance (e.g., observe, scrounge, borrow tools); exchanges seem to facilitate acquisition. Full mastery involves a mix of independent and social experience; controlling strength, for example, requires direct practice and cannot be understood by watching. The role of niche construction is especially clear in Taï nut-cracking sites.

Taï chimpanzees illustrate several additional similarities. At Taï, expertise is distributed through the community. The most efficient nut crackers are adult females; hunters are mostly adult males, and no one male holds or enacts all the skills needed to hunt successfully. This affects skill acquisition: cracking nuts with stones is learned primarily with mothers, hunting with adult males, and hunting skills must be acquired via multiple experts. The extent to which skills advance through relationships is evident in an orphan male's acquisition of hunting skills; he was adopted when five years old by Brutus, the community's best hunter. The orphan followed Brutus everywhere and began hunting apprenticeship earlier than normal; the head start and privileged access to Brutus probably allowed him to progress earlier and farther than normal. Hunting also suggests niche construction. Ambushers may force a target colobus to flee downwards, into the lower canopy; they have better chances of catching it in the continuous tree cover of the lower canopy because there they can run faster than colobus. Staying high in the canopy reduces chances of capture because red colobus, weighing about 13 kg, can access branches that will not support chimpanzee adult males, weighing 40–50 kg. From a learning perspective, forcing prey downwards favors a specific segment of the habitat that facilitates capture. For males acquiring the capture role, this would facilitate or even enable their learning. Forcing prey downwards would probably occur only to ambushers already aware of the needs of capturers.

Great ape species differ in their social systems; consequently, life-history parameters that concern sociality should generate different tradition patterns. Parker and Russon (1996) suggested differences associated with interbirth intervals, subgrouping patterns, philopatry, and demography. Further differences may be linked with age- and sex-linked social tolerance patterns.

A smattering of evidence allows exploration how acquiring shared practices may vary across great apes relative to species-specific social tolerance patterns. Three cases have been reported of infants whose mothers lacked a food technique shared by most other group members: two in chimpanzees (Inoue-Nakamura and Matsuzawa, 1997) and one in mountain gorillas (Byrne and Byrne, 1993). Both chimpanzee youngsters but not the gorilla acquired the shared practice. As with *Neesia*, the likely explanation is that novices must be introduced to this operation socially, as infants. Mothers, the normal guides, could offer no assistance for this expertise. Nonrelatives do not tolerate infants in proximity during foraging in mountain gorillas (Byrne and Russon, 1998) but they do in chimpanzees (Inoue-Nakamura and Matsuzawa, 1997). Different outcomes are consistent with species differences in social tolerance patterns during development.

A second example involves differences between chimpanzees and orangutans in adult male tolerance. Adult male orangutans show extreme mutual intolerance and avoidance; close encounters are invariably agonistic and readily escalate to fights and injuries (van Schaik and van Hooff, 1996). Adult male chimpanzees are mutually tolerant, mutually affiliative, and associate in parties (Nishida, 1979; Wrangham, 1979). Based on tolerance, the potential for social learning between adult males should differ between these species. Calls used by adult males in long-distance communication, orangutan long calls and chimpanzee pant hoots, are consistent with this prediction. Long calls show no evidence of learned similarities between adult males; what stands out is their individuality (Galdikas, 1985). Pant hoots suggest learned similarities between adult males. Adult males mutually alter their pant hoots when chorusing with other males, to converge with one another (Mitani and Gros-Louis, 1998). Adult male pant hoots differ systematically between communities, so convergence may reach collective levels. Socially mediated vocal learning is the favored explanation (Mitani, Hunley, and Murdoch, 1999). Species differences in age- and sex-linked tolerance offer a plausible explanation for these behavioral differences.

The inevitable comparison with humans reveals a very similar process. In humans, traditional complex skills are not acquired as whole packages. These skills are composites designed to handle multifaceted tasks, comprising multiple components that are combined and recombined to make the whole (Gosselain, 2000). In a variety of human societies, traditional craft skills may be acquired from various different sources, and social learning routes vary with the particular skills and social structures involved (Shennan and Steele, 1999). The influence of community can, at times, be traced in the multiple inputs to one individual. In humans, as in orangutans, it is possible to identify an individual's teacher by performance details or style (i.e., tactics) as well as by overall pattern (i.e., strategy). It is even possible to identify multiple teachers in one individual's performance. This emphasizes that humans, too, imprint community-wide influence in their own expertise.

12.7 Discussion

My orangutan cases may not qualify as traditions – I can only show sharing within small cliques, for example – but they illustrate how great apes acquire complex foraging expertise. The foraging challenges they face are multifaceted and changeable. Accordingly, acquiring the relevant expertise is a mosaic, constructive process that can consume vast amounts of time: in some cases the whole of immaturity and occasionally intruding into adulthood. Many food-processing techniques are not tool based. Chimpanzees are the only habitual tool users among the great apes, so greater attention to manipulative techniques may be one key to better understanding of how traditional foraging expertise is generated. The acquisition process entails first building basic elements and, subsequently, combining and recombining these into integrated programs, or strategies, that are multilayered in organization. Social influence is an integral part of the acquisition process, directly via apprentice-like relationships and probably indirectly via niche construction. Because of the time span involved, direct social influence, indirect social influence through habitat modification, and individual inventiveness may all contribute to the same expertise: each on many occasions, with several knowledgeable conspecifics, and in varied physical conditions.

Entre autre, this means that explaining how social influences affect the acquisition of expertise requires concepts beyond socially mediated learning. Socially mediated learning does not capture the identity or the multiplicity of social influences, the distributed nature of the expertise, indirect social influences via niche modification, or the basis for the tolerance affording direct social influence. It also fails to capture the role that the community must play in generating such complex expertise.

Social influences on acquisition must involve the community. First, great apes master their most sophisticated expertise very late in life and, in some cases, not at all. This implies that some expertise strains their highest physical or cognitive capacities and/or requires special social supports (Russon *et al.*, 1998; Parker, 1996). It is, therefore, likely that great apes can profit *only* from social influences that are very directly and closely related to their own current competencies. In other words, to be effective, social input must come from conspecifics that experience the same feeding problems and practices. Second, sharing is only possible with conspecifics that tolerate proximity, and tolerance is strongly tied to interindividual relationships in primates. Relationships develop primarily within communities; in chimpanzees and gorillas, intercommunity relations are actively hostile. Therefore, complex expertise in great apes must be generated with assistance from many expert members *from the same community*, who share the target food problems and food-processing practices.

If this is how great apes normally acquire expertise, then practices are virtually always shared within communities because acquisition relies on extensive social support within communities. The question of traditions may be reduced to, "What makes expertise spread *widely*?" One prerequisite is probably expertise with the potential for broad usage, at least within some definable subgroup. Extensive horizontal and oblique routes of social influence are also probably critical (van Schaik and Knott, 2001).

Four facets of great ape sociality may afford the wide spreading of expertise. First, relatively egalitarian, fission–fusion societies, which may characterize all great apes (e.g., Fuentes, 2000; van Schaik *et al.*, 2003), may privilege horizontal and oblique sharing of expertise as well as broad and varying networks of social exchange. Working out what traditions are likely to occur would entail analyzing the membership, interaction, and activity patterns of the temporary subgroups that form. Second, horizontal and oblique routes may be favored in the period between puberty and adulthood when dispersal and extensive socio-sexual affiliation occur. Both changes likely affect social routes; female orangutans and chimpanzees, for example, join groups primarily for reproductive purposes or in large fruit patches (van Schaik *et al.*, 2003; Yamagiwa, 2003). Dispersal may open important opportunities for diffusion through sexual liaisons, because immigrants typically gain admission to new groups on the basis of sexual attractiveness and sexually fuelled alliances with group members. In orangutans, dispersal, high sexually based gregariousness, and advanced individual capabilities co-occur during adolescent and subadult periods. That these social changes coincide with advanced cognitive and physical capabilities may not be accidental, because more powerful capabilities support more rapid sharing of complex expertise. Extending expertise is critical during this period because of the new needs created by changing roles and ranges. Third, the great apes engage in considerable food sharing; even gorillas supplant feeding spots that others occupy (van Schaik *et al.*, 2003; Yamagiwa, 2003). Sharing food would facilitate sharing food-related expertise, along routes defined by relationships in which food is commonly shared. Finally, shared expertise may be accentuated in rehabilitants (Rogers and Kaplan, 1994; Russon, 2003). Maternal bonds having been destroyed, peer- and sibling-like bonds take on greater importance. Horizontal and oblique routes may then play an especially important role, and earlier in life.

If great ape expertise is structured this way, some confusion may surround the concept of traditions because the multilevel structures involved have not been taken into account (Joulian, 1995). If great ape expertise can consist of multilevel programs that integrate multiple behavioral elements, then practices may be shared at any of the levels or elements involved. Some traditions have been identified in terms of whole cloth, or strategy, as in using versus not using tools to obtain *Neesia* seeds. Other traditions represent specific tactics within one form of expertise, as in kiss-squeaking with versus without leaves (Peters, 2001), or ant dipping with a long stick and two hands versus a short stick and one hand (McGrew, 1998). If traditions, like the devil, are in the details, then the question of how great apes generate traditions is likely to resist resolution until the multiple levels at which their expertise is organized, at which acquisition occurs, and at which social influences operate have been systematically factored into conceptual and methodological equations.

This approach also suggests reconsidering methodological assumptions. First, these longitudinal data on acquisition illustrate that it is not necessary to rely on group-differences logic to show that social influences operating at the community level affect the acquisition of shared practices. The influence of the community in generating shared practices can be seen within single individuals, by tracing the multiple sources of social influence that contribute to that expertise over time. It is possible to collect information on the process of acquisition in a social context, and this information can be used to support the argument that what is seen in individuals represents "traditions". Second, ironic as it may seem, clear evidence of traditions in great apes may come from the "least social" species among them: the orangutan. It is largely because their sociality is so spare that the sharing of practices is so clear: it is not embedded in a buzzing confusion of overlapping and perhaps contradictory influences. In having only one companion at a time, who has been dallying with whom is patently clear. This suggests that high-density sociality is not the only condition favoring traditions, and some of the more sparsely social species may offer especially clear perspectives on the processes that generate them.

Several more general points issue from this life-history perspective. First, great ape traditions are constructed, developmentally, in line with interacting biological, psychological, and social parameters. Affordances and constraints associated with all three sets of parameter alter learners' needs, capabilities, and social positions in predictable fashion throughout their lives. Together, these parameters channel what facets of the community's expertise are available to any given learner at any given time. If cultural processes in great apes are this closely tied to life-history parameters, then they operate not as a separate module patched onto an individual learning system nor as a distinct process that intertwines with individual processes in indeterminate fashion. Great ape traditions then appear to depend on emergent processes in which physical, cognitive, and social factors interweave in a developmentally organized fashion. This is in accord with views of development as a function of co-evolving biological and cultural systems and as an open system with an architecture that is structured, incompletely, by biological, ecological, and cultural parameters (Baltes, Staudinger, and Lindenberger, 1999; Durham, 1991). It also accords with views of socially mediated learning as a normal facet of the behavioral biology of many species and not simply a lead-up to, or incomplete version of, human processes (Giraldeau, 1997; Laland *et al.*, 2000; Box and Gibson, 1999).

In this view, what distinguishes individuals acting in social settings is their heightened propensity to generate behaviors that are similar to one another. Under certain circumstances, this propensity translates into traditions. This analysis concurs with the common view that similar social influences on learning likely generate traditions in other primates and in other orders as well, and that differences likely concern the power of supporting cognitive processes (Byrne, 1995; Donald, 1991; Parker and McKinney, 1999; Parker and Russon, 1996; Russon *et al.*, 1996). Possibly, in addition, fission–fusion, egalitarian social structures are important in sharing practices widely. Given the parameters that appear to have significance, traditions may also characterize nonprimate species that are longlived, that live social lives within flexible social structures, and that rely on learning for the bulk of their expertise. Parrots, corvids (crows and ravens), cetaceans, and elephants are likely candidates. The list is hardly novel. What this perspective offers is another explanation for its consistency. Species often differ in their social structures as well as in their cognition, and both factors afford and constrain different avenues of sharing practices.

12.8 Acknowledgements

Studies of the ex-captive orangutans were sponsored by the Research Institute of the Indonesian Ministry of Forestry and Estate Crops, Samarinda, East Kalimantan, Indonesia, the TROPENBOS Foundation of the Netherlands, and the Wanariset Orangutan Reintroduction Project. Funding was provided by Glendon College and York University, Toronto, and the Natural Sciences and Engineering Research Council of Canada. My thanks to the Wenner Gren Foundation, which provided the impetus to develop some of the ideas central to this paper through its *Culture and the Cultural* conference. Thanks also to one anonymous reviewer and both volume editors, whose comments contributed substantially to the chapter's quality.

References

Altmann, J. 1980. *Baboon Mothers and Infants*. Cambridge, MA: Harvard University Press.

Baltes, P. P., Staudinger, U. M., and Lindenberger, U. 1999. Lifespan psychology: theory and application to intellectual functioning. *Annual Review of Psychology*, **50**, 471–507.

Boesch, C. and Boesch-Achermann, H. 2000. *The Chimpanzees of the Taï Forest*. Oxford: Oxford University Press.

Boesch, C. and Tomasello, M. 1998. Chimpanzee and human cultures. *Current Anthropology*, **39**, 591–614.

Bogin, B. 1999. Evolutionary perspective on human growth. *Annual Review of Anthropology*, **28**, 109–153.

Box, H. O. and Gibson, K. R. (ed.) 1999. *Mammalian Social Learning*. Cambridge: Cambridge University Press.

Brewer, S. M. and McGrew, W. C. 1990. Chimpanzee use of a tool-set to get honey. *Folia Primatologica*, **54**, 100–104.

Byrne, R. W. 1995. *The Thinking Ape*. Oxford: Oxford University Press.

Byrne, R. W. 1997. The technical intelligence hypothesis: an alternative evolutionary stimulus to intelligence? In *Machiavellian Intelligence II*, ed. A. Whiten and R. W. Byrne, pp. 289–311, Cambridge, UK: Cambridge University Press.

Byrne, R. W. and Byrne, J. M. E. 1991. Hand preferences in the skilled gathering tasks of mountain gorillas (*Gorilla gorilla berengei*). *Cortex*, **27**, 521–546.

Byrne, R. W. and Byrne, J. M. E. 1993. Complex leaf-gathering skills of mountain gorillas (*Gorilla g. beringei*): variability and standardization. *American Journal of Primatology*, **31**, 241–261.

Byrne, R. W. and Russon, A. E. 1998. Learning by imitation: a hierarchical approach. *Behavioural and Brain Sciences*, **21**, 667–721.

Cheney, D. and Seyfarth, R. 1990. *How Monkeys See the World*. Chicago, IL: University of Chicago Press.

Coussi-Korbel, S. and Fragaszy, D. M. 1995. On the relation between social dynamics and social learning. *Animal Behaviour*, **50**, 1441–1453.

De Rousseau, C. J. (ed.). 1990. *Monographs in Primatology*, Vol, 14, *Primate Life History and Evolution*. New York: Wiley-Liss.

Donald, M. 1991. *Origins of the Modern Mind: Three Stages in the Evolution of Culture and Cognition*. Cambridge, MA: Harvard University Press.

Durham, W. H. 1991. *Coevolution: Genes, Culture and Human Diversity*. Stanford, CA: Stanford University Press.

Fleagle, J. G. 1999. *Primate Adaptation and Evolution*, 2nd edn. San Diego, CA: Academic Press.

Fobes, J. L. and King, J. E. 1982. *Primate Behaviour*. New York: Academic Press.

Fox, E. A. Sitompul, A. F., and van Schaik, C. P. 1999. Intelligent tool use in wild Sumatran orangutans. In *The Mentalities of Gorillas and Orangutans: Comparative Perspectives*, ed. S. T. Parker, R. W. Mitchell, and H. L. Miles, pp. 99–116. Cambridge: Cambridge University Press.

Freeland, W. J. and Janzen, D. H. 1974. Strategies in herbivory by mammals: the role of plant secondary compounds. *American Naturalist*, **108**, 269–289.

Fuentes, A. 2000. Hylobatid communities: changing views of pair bonding and social organization in hominoids. *Yearbook of Physical Anthropology*, **43** (Supplement 31 to the *American Journal of Physical Anthropology*), 33–60.

Galdikas, B. M. F. 1985. Adult male sociality and reproductive tactics among orangutans at Tanjung Puting. *Folia Primatologica*, **45**, 9–24.

Galdikas, B. M. F. 1988. Orangutan diet, range, and activity at Tanjung Puting, Central Borneo. *International Journal of Primatology*, **9**, 1–35.

Galdikas, B. M. F. 1995. Social and reproductive behaviour of wild adolescent female orangutans. In *The Neglected Ape*, ed. R. D. Nadler, B. F. M. Galdikas, L. K. Sheeran, and N. Rosen, pp. 163–182. New York: Plenum Press.

Galdikas, B. M. F. and Teleki, G. 1981. Variations in subsistence activities of male and female pongids: new perspectives on the origins of hominid labour division. *Current Anthropology*, **22**, 241–256.

Galdikas, B. M. F. and Wood, J. 1990. Great ape and human birth intervals. *American Journal of Physical Anthropology*, **83**, 185–192.

Giraldeau, L.-A. 1997. The ecology of information use. In *Handbook of Behavioral Ecology*, ed. N. Davies and J. Krebs, pp. 42–68. Oxford: Blackwell Scientific.

Goodall, J., Bandora, A., Bergmann, E., Busse, C., Matama, H., Mpongo, E., Peirce, A., and Riss, D. 1979. Intercommunity interactions in the chimpanzee population of the Gombe National Park. In *The Great Apes*, ed. D. Hamburg and E. R. McCown, pp. 15–33. Menlo Park, CA: Benjamin/Cummings.

Gosselain, O. P. 2000. Materializing identities: an African perspective. *Journal of Archaeological Method and Theory*, **7**, 187–217.

Grundmann, E., Lestel, D., Boestani, A. N., and Bomsel, M.-C. 2000. Learning to survive in the forest: what every orangutan should know. Presented at *The Apes: Challenges for the 21st Century*, Brookfield Zoo, Chicago, 10–13 May.

Horr, D. A. 1977. Orang-utan maturation: growing up in a female world. In *Primate Bio-social Development: Biological, Social, and Ecological Determinants*, ed. S. Chevalier-Skolnikoff and F. E. Poirier, pp. 289-321. New York: Garland.

Huffman, M. A. 1997. Current evidence for self-medication in primates: a (Supplement 28 to the *American Journal of Physical Anthropology*), multidisciplinary perspective. *Yearbook of Physical Anthropology*, **40** 171–200.

Humphrey, N. K. 1976. The social function of intellect. In *Growing Points in Ethology*, ed. P. P. G. Bateson and R. A. Hinde, pp. 303–317. Cambridge: Cambridge University Press.

Inoue-Nakamura, N. and Matsuzawa, T. 1997. Development of stone tool use by wild chimpanzees (*Pan troglodytes*). *Journal of Comparative Psychology*, **111**, 159–173.

Janson, C. H. and van Schaik, C. P. 1993. Ecological risk aversion in juvenile primates: slow and steady wins the race. In *Juvenile Primates: Life History, Development, and Behaviour*, ed. M. E. Pereira and L. A. Fairbanks, pp. 57–76. New York: Oxford University Press.

Joffe, T. H. 1997. Social pressures have selected for an extended juvenile period in primates. *Journal of Human Evolution*, **32**, 593–605.

Jolly, A. 1966. Lemur social behaviour and primate intelligence. *Science*, **153**, 501–506.

Joulian F. (1995). Mise en evidence de differences traditionnelles dans le cassage des noix chez les chimpanzés (*Pan troglodytes*) de la Côte d'Ivoire, implications paléoanthropologiques. *Journal des Africanistes*, **65**, 57–77.

King, B. J. 1994. *The Information Continuum*. Santa Fe, NM: School of American Research.

Kummer, H. 1971. *Primate Societies: Group Techniques of Ecological Adaptation*. Chicago, IL: Aldine.

Lackman-Ancrenaz, I. 2001. Report on the Kinabatangan Orangutan Conservation Project. Presentation to the *Orangutan Reintroduction and Protection Workshop*, Balikpapan, Indonesia, 15–18 June.

Laland, K., Odling-Smee, J., and Feldman, M. 2000. Niche construction, biological evolution, and cultural change. *Behavioral and Brain Sciences*, **23**, 131–146.

MacKinnon, J. R. 1974. The behaviour and ecology of wild orang-utans. *Animal Behaviour*, **22**, 3–74.

Matsuzawa, T. 2001. Primate foundations of human intelligence: a view of tool use in nonhuman primates and fossil hominids. In *Primate Origins of Human Cognition and Behavior*, ed. T. Matsuzawa, pp. 3–25. Tokyo: Springer-Verlag.

Matsuzawa, T. 1996. Chimpanzee intelligence in nature and in captivity: isomorphism of symbol use and tool use. In *Great Ape Societies*, ed. W.C. McGrew, L.F. Marchant, and T. Nishida, pp. 196–209. Cambridge: Cambridge University Press.

Matsuzawa, T., Biro, D., Humle, T., Inoue-Nakamura, N., Tonooka, R., and Yamakoshi, G. 2001. Emergence of culture in wild chimpanzees: education by master–apprenticeship. In *Primate Origins of Human Cognition and Behavior*, ed. T. Matsuzawa, pp. 557–574. Tokyo: Springer-Verlag.

McGrew, W. C. 1992. *Chimpanzee Material Culture: Implications for Human Evolution*. Cambridge, UK: Cambridge University Press.

McGrew, W. C. 1998. Culture in nonhuman primates? *Annual Review of Anthropology*, **27**, 301–328.

Menzel, C. 2001. Umprompted recall and reporting of hidden objects by a chimpanzee (*Pan troglodytes*) after extended delays. *Journal of Comparative Psychology*, **113**, 426–434.

Meyfarth, E. 1998. Biological conservation: orangutan-rattan relationships in Indonesian Borneo. Master's Thesis, York University, Toronto.

Miles, H.L. 1978. Conversations with apes: the use of sign language with two chimpanzees. PhD Thesis, Yale University New Haven, CT.

Miles, H. L., Mitchell, R. W., and Harper, S. 1996. Simon says: the development of imitation in an enculturated orangutan. In *Reaching into Thought: The Minds of the Great Apes*, ed. A. E. Russon, K. A. Bard, and S. T. Parker. Cambridge: Cambridge University Press.

Milton, K. 1984. The role of food processing factors in primate food choice. In *Adaptations for Foraging in Nonhuman Primates: Contributions to an Organismal Biology of Prosimians, Monkeys and Apes*, ed. P. S. Rodman and J. G. H. Cant, pp. 249–279. New York: Columbia University Press.

Mitani, J. C. 1985. Mating behaviour of male orangutans in the Kutai Game Reserve, Indonesia. *Animal Behaviour*, **33**, 392–402.

Mitani, J. C. and Gros-Louis, J. 1998. Chorusing and call convergence in chimpanzees: test of three hypotheses. *Behaviour*, **135**, 1041–1064.

Mitani, J. C., Hunley, K. L., and Murdoch, M.E. 1999. Geographic variation in the calls of wild chimpanzees: a reassessment. *American Journal of Physical Anthropology*, **47**, 133–151.

Myowa-Yamakoshi, M. and Matsuzawa, T. 1999. Factors influencing imitation of manipulatory actions in chimpanzees (*Pan troglodytes*). *Journal of Comparative Psychology*, **113**, 128–136.

Nishida, T. 1987. Local traditions and cultural transmission. In *Primate Societies*, ed. B. B. Smuts, D. L. Cheney, R. M. Seyfarth, R. W. Wrangham, and T. T. Struhsaker, pp. 462–474. Chicago IL: University Chicago Press.

Nishida, T. 1979. The social structure of chimpanzees of the Mahale Mountains. In *The Great Apes*, ed. D. A. Hamburg and E. R. McCown. pp. 73–121. Menlo Park, CA: Benjamin/Cummings.

Parker, S. T. 1996. Apprenticeship in tool-mediated extractive foraging: the origins of imitation, teaching and self-awareness in great apes. In *Reaching into Thought: The Minds of the Great Apes*, ed. A. E. Russon, K. A. Bard, and S. T. Parker, pp. 348–370. Cambridge: Cambridge University Press.

Parker, S. T. and Gibson, K. R. 1977. Object manipulation, tool use and sensorimotor intelligence as feeding adaptations in *Cebus* monkeys and great apes. *Journal of Human Evolution*, **6**, 623–641.

Parker, S. T. and Gibson, K. R. 1979. A model of the evolution of language and intelligence in early hominids. *Behavioural and Brain Sciences*, **2**, 367–407.

Parker, S. T., and Gibson, K. R. (ed.) 1990. *"Language" and Intelligence in Monkeys and Apes: Comparative Developmental Perspectives*. New York: Cambridge University Press.

Parker, S. T. and McKinney, M. L. 1999.*Origins of Intelligence: The Evolution of Cognitive Development in Monkeys, Apes, and Humans*. Baltimore, MD: Johns Hopkins.

Parker, S. T., and Russon, A. E. 1996. On the wild side of culture and cognition in the great apes. In *Reaching into Thought: The Minds of the Great Apes*, ed. A. E. Russon, K. A. Bard, and S. T. Parker, pp. 430–450. Cambridge: Cambridge University Press.

Pereira, M. E. 1993. Juvenility in animals. In *Juvenile Primates: Life History, Development, and Behaviour*, ed. M. E. Pereira and L. A. Fairbanks, pp. 7–27. New York: Oxford University Press.

Pereira, M. E. and Altmann, J. 1985. Development of social behaviour in free-living nonhuman primates. In *Nonhuman Primate Models for Human Growth and Development*, ed. E. S. Watts, pp. 217–309. New York: Liss.

Peters, H. 2001. Tool use to modify vocalisations by wild orang-utans. *Folia Primatologica*, **72**, 242–244.

Rijksen, H. D. 1978. *A Field Study on Sumatran Orangutans (*Pongo pygmaeus *Abelii Lesson 1827): Ecology, Behaviour, and Conservation*. Wageningen: H. Veenman and Zonen B.V.

Rodman, P. 1973. Population composition and adaptive organization among orangutans of the Kutai Reserve. In *Comparative Ecology and Behaviour of Primates*, ed. J. H. Crook and R. P. Michael, pp. 171–209. London: Academic Press.

Rodman, P. S. 1979. Individual activity patterns and the solitary nature of orangutans. In *The Great Apes*, ed. D. A. Hamburg and E. R. McCown, pp. 235–255. Menlo Park, CA: Benjamin-Cummings.

Rodman, P. S. 2000. Great ape models for the evolution of human diet. www.cast.uark.edu/local/icaes/conferences/wburg/posters/psrodman/GAMHD.htm, Dec.

Rodman, P. W. 1977. Feeding behaviour of orang-utans of the Kutai Nature Reserve, East Kalimantan. In *Primate Ecology*, ed. T. H. Clutton-Brock, pp. 383–413. New York: Academic Press.

Rogers, L. and Kaplan, G. 1994. A new form of tool-using by orangutans in Sabah, East Malaysia. *Folia Primatologica*, **63**, 50–52.

Rogoff, B. 1992. *Apprenticeship in Thinking*. New York: Oxford University Press.

Russon, A. E. 1997. Exploiting the expertise of others. In *Machiavellian Intelligence II: Evaluations and Extensions*, ed. A. Whiten and R. W. Byrne, pp. 174–206. Cambridge: Cambridge University Press.

Russon, A. E. 1998. The nature and evolution of orangutan intelligence. *Primates*, **9**, 485–503.

Russon, A. E. 1999a. Naturalistic approaches to orangutan intelligence and the question of enculturation. *International Journal of Comparative Psychology*, **12**, 181–202.

Russon, A. E. 1999b. Orangutans' imitation of tool use: a cognitive analysis. In *Mentalities of Gorillas and Orangutans*, ed. S. T. Parker, R. W. Mitchell, and H. L. Miles, pp. 117–146. Cambridge: Cambridge University Press.

Russon, A. E. 2002. Return of the native: Cognition and site specific expertise in orangutan rehabilitation. *International Journal of Primatology*, **23**, 461–478.

Russon, A. E. 2003. Comparative developmental perspectives on culture: the great apes. In *Between Biology and Culture: Perspectives on Ontogenetic Development*, ed. H. Keller, Y. H. Poortinga, and A. Schoelmerich. Cambridge: Cambridge University Press, in press.

Russon, A. E., and Galdikas, B. M. F. 1995. Constraints on great apes' imitation: model and action selectivity in rehabilitant orangutan imitation (*Pongo pygmaeus*). *Journal of Comparative Psychology*, **109**, 5–17.

Russon, A. E., Bard, K. A., and Parker, S. T. (eds.) 1996. *Reaching into Thought: The Minds of the Great Apes*. Cambridge: Cambridge University Press.

Russon, A. E., Mitchell, R. W., Lefebvre, L., and Abravanel, E. 1998. The comparative evolution of imitation. In *Piaget, Evolution, and Development*, ed. J. Langer and M. Killen, pp. 103–143. Hillsdale, NJ: Erlbaum.

Shennan, S. J. and Steele, J. 1999. Cultural learning in hominids: a behavioural ecological approach. In *Mammalian Social Learning: Comparative and Ecological Perspectives*, ed. H. O. Box and K. R. Gibson, pp. 367–388. Cambridge: Cambridge University Press.

Smuts, B. B., Cheney, D. L., Seyfarth, R. M., Wrangham, R. W, and Struhsaker, T. T. (ed.) 1987. *Primate Societies*. Chicago, IL: University of Chicago Press.

Stanford, C. B. 1996. The hunting ecology of wild chimpanzees: implications for the evolutionary ecology of Pliocene hominids. *American Anthropologist*, **98**, 96–113.

Stanford, C. B. 1998. *Chimpanzee and Red Colobus: The Ecology of Predator and Prey*. Cambridge, MA: Harvard University Press.

Stokes, E. J. 1999. Feeding skills and the effect of injury on wild chimpanzees. PhD Thesis, University of St. Andrews, St. Andrews, Scotland.

Suddendorf, T. and Whiten, A. 2001. Mental evolution and development: evidence for secondary representation in children, great apes, and other animals. *Psychological Bulletin*, **127**, 629–650.

Tomasello, M. and Call, J. 1997. *Primate Cognition*. New York: Oxford University Press.

Tomasello, M., Savage-Rumbaugh, E. S., and Kruger, A. 1993. Imitative learning of actions on objects by children, chimpanzees, and enculturated chimpanzees. *Child Development*, **64**, 1688–1705.

Utami, S. S. 2000. *Bimaturism in Orang-Utan Males: Reproductive and Ecological Strategies*. Utrecht: Central Reproductie FSB Universiteit Utrecht.

van Schaik, C.P. and Knott, C. 2001. Geographic variation in tool use in *Neesia* fruits in orangutans. *American Journal of Physical Anthropology*, **114**, 331–342.

van Schaik, C. P., and van Hooff, J. A. R. A. M. 1996. Toward an understanding of the orangutan's social system. In *Great Ape Societies*, ed. W. C. McGrew, L. F. Marchant, and T. Nishida, pp. 3–15. Cambridge: Cambridge University Press.

van Schaik, C. P., Fox, E. A., and Sitompul, A. F. 1996. Manufacture and use of tools in wild Sumatran orangutans. *Naturwissenschaften*, **83**, 186–188.

van Schaik, C. P., Deaner, R. O., and Merrill, M. Y. 1999. The conditions for tool use in primates: implications for the evolution of material culture. *Journal of Human Evolution*, **36**, 719–741.

van Schaik, C. P., Preuschoft, S., and Watts, D. P. 2003. Great ape social systems. In *The Evolutionary Origins of Great Ape Intelligence*, ed. A. E. Russon and D. Begun. Cambridge: Cambridge University Press, in press.

Voysey, B. C., McDonald, K. E., Rogers, M. E., Tutin, C. E. G., and Parnell, R. J. 1999. Gorillas and seed dispersal in the Lope Reserve, Gabon. II: survival and growth of seedlings. *Journal of Tropical Ecology*, **15**, 39–60.

Vygotsky, L. 1962 *Thought and Language*, Cambridge, MA: MIT Press.

Waterman, P. G. 1984. Food acquisition and processing as a function of plant chemistry. In *Food Acquisition and Processing in Primates*, ed. D. J. Chivers, B. A. Wood, and A. Bilsborough, pp. 177–211. New York: Plenum Press.

Watts, D. P. and Pusey, A. E. 1993. Behaviour of juvenile and adolescent great apes. In *Juvenile Primates: Life History, Development, and Behaviour*, ed. M.E. Pereira and L. A. Fairbanks, pp. 148-171. New York: Oxford University Press.

Watts, E. 1985. Adolescent growth and development in monkeys, apes, and humans. In *Nonhuman Primate Models for Human Growth and Development*, ed. E. Watts, pp. 41–65. New York: Alan R. Liss.

Whiten, A., Goodall, J., McGrew, W. C., Nishida, T., Reynolds, V., Sugiyama, Y., Tutin, C. E. G., Wrangham, R. W., and Boesch, C. 1999. Culture in chimpanzees. *Nature*, **399**, 682–685.

Wrangham, R. W. 1979. On the evolution of ape social systems. *Social Science Information*, **18**, 334–368.

Yamagiwa, J. 2003. Diet and foraging of the great apes: ecological constraints on their social organizations and implications for their divergence. In *The Evolutionary Origins of Great Ape Intelligence*, ed. A. E. Russon and D. Begun. Cambridge: Cambridge University Press, in press.

Yamakoshi, G. 1998. Dietary responses to fruit scarcity of wild chimpanzees at Bossou, Guinea: possible implications for ecological importance of tool use. *American Journal of Physical Anthropology*, **106**, 283-295.

SUE BOINSKI, ROBERT P. QUATRONE, KAREN SUGHRUE,
LARA SELVAGGI, MALINDA HENRY, CLAUDIA M. STICKLER,
AND LISA M. ROSE

13

Do brown capuchins socially learn foraging skills?

13.1 Introduction

Tool use and complex object manipulation skills are of intense interest to many disciplines. Yet the number of nonhuman primate taxa exploited in these comparative studies is usually limited to the great apes, and especially the chimpanzee, *Pan troglodytes*. The focus on chimpanzees is understandable. In the wild, chimpanzees greatly exceed all other apes in the frequency and complexity of tool manufacture and object and tool use (Sugiyama, 1997; Whiten *et al.*, 1999). In captivity, however, tool use and complex object manipulation is common and can be readily elicited from all great ape species (Visalberghi *et al.*, 1995).

In recent years, primatologists and comparative psychologists have paid increasing attention to the manipulative skills of capuchins, the New World primate genus *Cebus*. Not only does the proclivity of capuchins to use tools surpass that of all other monkeys either in the Old or the New World, but in many respects the spontaneous manipulative activities and dexterity of capuchins and chimpanzees share many characteristics (Anderson, 1996; Antinucci and Visalberghi, 1986; Panger, 1998; Parker and Gibson, 1977). Capuchins are well known for strenuous arthropod-extraction techniques and complex manipulation of difficult to process fruits (Fragaszy and Boinski, 1995; Janson and Boinski, 1992). Pounding and rubbing of fruits, invertebrates, and other food items against hard substrates is another food-processing technique exhibited by all four capuchin species (*C. apella*, brown capuchin, in Colombia and Peru: Izawa and Mizuno, 1977; Struhsaker and Leland, 1977; Terborgh, 1983; *C. albifrons*, white-fronted capuchin in Peru: Terborgh, 1983; *C. capucinus*, white-faced capuchin, in Costa Rica: Panger, 1998; Rose, 2001; and

C. olivaceus, wedge-capped capuchin, in Venezuela: Fragaszy and Boinski, 1995; Robinson, 1986). Tool use by wild capuchins is rare but occurs in foraging contexts. A notable example is Fernandes' (1991) observation of a *C. apella* in a mangrove swamp in Brazil using a chunk of oyster bed to break open a closed oyster shell so that the oyster meat could be ingested. Boinski (1988) also reports a *C. capucinus* using a large branch to club a venomous snake repeatedly, eventually killing it after more than 50 blows. Perhaps the most fascinating reports to-date of tool use among capuchins not confined in cages are those recent reports of stones employed by *C. apella* to crack open the nuts of the palm *Syragrus romanzoffiana* (Arecaceae) in the Brazilian Caatinga (Langguth and Alonso, 1997) and a reforested, semi-free ranging area in a Brazilian park (Ottoni and Mannu, 2001). As is the case with great apes, capuchins in captivity display complex manipulative skills and high rates of tool use, greatly exceeding that documented for wild capuchins (Costello and Fragaszy, 1988; Klüver, 1933; Vevers and Weiner, 1963; Visalberghi, 1997). Nearly all capuchins studied in captive situations have been *C. apella*, whereas *C. capucinus* has most often been the subject of field investigations.

The exceptional manipulative abilities of capuchins merit attention not merely because they provide diverting natural history anecdotes. Instead the burgeoning number of investigations addressing tool and object use by capuchins is part of the current intense scrutiny, experimentation, and debate as to what is primate intelligence and what cognitive abilities are reflected in the tool use of nonhuman primates (Bard and Vauclair, 1989; Tomasello and Call, 1997; Visalberghi, 1993). Do individual primates, be they capuchin, great ape, or human, arrive at tool use and other goal-directed manual activities through the same cognitive activities and learning processes? Current evidence from laboratory studies leads many researchers to believe that capuchins rely primarily on phylogenetically common mechanisms of associative learning and procedural knowledge (i.e., knowing how or what to do; Shettleworth, 1998) when they modify their exploratory behavior to capitalize on useful outcomes. The discovery of tool use can be considered a fortuitous outcome of combinatorial activity in this view (Fragaszy and Adams-Curtis, 1991; Visalberghi and Limongelli, 1994). It may be that capuchins go beyond associative learning to some broader understanding of a problem, so that they can, for example, select the correct tool for a particular task (e.g., Anderson and Henneman, 1994). Capuchins' successes in various tool-using situations lead some to think that the abilities of apes and capuchins are not very far apart in the instrumental domain (i.e., using objects to achieve a purpose; Anderson 1996; Tomasello and Call, 1997). Some researchers believe, however, that chimpanzees can arrive at a deeper comprehension of a problem and its solution via the use of a tool than do capuchins (Custance, Whiten, and Fredman, 1999; Lavallee, 1999; Visalberghi and Limongelli, 1996) In any case, laboratory reports from both sides of the controversy agree that individual *C. apella* vary considerably in interest and aptitude in solving manipulative problems, can change tactics frequently, and use tactics that are sometimes awkward to the point of appearing random and patently destined to fail.

Our primary purpose in this chapter is to consider the evidence and potential for "traditions" (i.e., relatively long-lasting behavioral practices shared among group members in part via social learning: see Ch. 1) as contributing to the deft manipulative abilities of brown capuchins documented in our recent and ongoing observations of wild *C. apella*. Brown capuchin at Raleighvallen, an undisturbed site in the interior of Suriname, employ persistent pounding and even a remarkable instance of tool use (striking with stout branch as a club) to fracture large, thick-husked fruits containing edible, highly nutritious seeds and pulp. These instances of complex manipulation occur at a frequency and level of dexterity not previously documented in wild or captive capuchins. In many respects, the manipulative abilities of capuchins at Raleighvallen are comparable to those described for wild chimpanzees (McGrew and Marchant, 1997). The inefficiencies and random components of object manipulation described in captive *C. apella* (i.e., Visalberghi, 1993, 1997) are found usually only in the youngest, least practiced individuals at this Surinamese site. If these foraging techniques truly represent traditions, the wild population of brown capuchins at Raleighvallen should be a particularly propitious situation to document and study this phenomenon because of the diversity of apparently specialized manipulative skills and the high frequency at which such skills are exhibited.

We first detail the ecological and biogeographic foundations of the remarkable and previously unappreciated abilities in this population of *C. apella*. Next we recount our observations of manipulative skill and tool use and then compare these with previous reports based on *C. apella* in the wild and captivity. Testable hypotheses are offered to explain the apparently rapid learning of the necessary skills and a strong male-bias in the exploitation of substrate-use techniques. We also describe a series of manipulative protocols and purely observational studies to be implemented at Raleighvallen in coming years. Many of these proposed observations are not unique to us but are also suggested by other chapters in this book. A major implication of our findings is that social interaction and close observation by naïve or relatively naïve immatures of the complex manipulative activities of accomplished group members is so ubiquitous that it would be difficult to deny at least a catalyst role for social facilitation in the generation and maintenance of manipulative skills among Raleighvallen brown capuchins. Feasible field research planned for the future, moreover, should be able to provide more concrete conclusions.

13.2 Site description and field methods

13.2.1 Site description

Observations of *C. apella* in Suriname come primarily from the Raleighvallen site (4° 0′ N, 56° 30′ W) within the Central Suriname Nature Preserve, a virtually undisturbed primary tropical forest that receives an annual rainfall averaging 2300 mm (Reichart, 1993). Preliminary fieldwork began there in June and July 1996 and October 1997. A long-term, investigation of squirrel monkeys (*Saimiri sciureus*) and *C. apella* began in Raleighvallen in January 1998 and is continuing. Squirrel monkeys were initially the primary focus of this field study, but as this species frequently forms mixed-species groups with *C. apella* for approximately 50% of daylight hours (Mittermeier and van Roosmalen, 1981), extensive opportunities exist for detailed observation of the latter species. Some of the observations reported below (those involving Brazil nuts) come from July 1995 when S. Boinski studied these two monkey species in swamp forest in the coastal province of Saramacca, as well as from the two month long preliminary study at Raleighvallen in 1996.

13.2.2 Study animals

At the time when the results presented here were collated (July 2000), one *C. apella* study group (ST) in Raleighvallen had been tolerant of observation by members of our study team for about two years and had been well habituated for at least 18 months. Three other troops were encountered less often in our study site and were not well habituated until approximately mid-2000. By September 2000, individual recognition was established for about one half of the 100 individuals comprising these four brown capuchin troops. Although some members of the study groups were individually recognized beginning in 1998, particularly adults, in the data reported here we typically distinguished immature capuchins at the level of age–sex classes and estimated ages. Visibility of social, foraging, and manipulative behaviors of these habituated animals is often excellent as *C. apella* is usually active in the understory or lower portion of the canopy (Mittermeier and van Roosmalen, 1981).

13.2.3 Definitions of behaviors reflecting complex manipulation

Two behavioral categories encompass the range of behaviors we deemed complex manipulative skill: substrate use and tool use. Our definition of substrate use, the transformation of an object by its (usually forceful) application to a stable, usually hard or resistant substrate, is similar to what Parker and Gibson (1977) term "proto-tool use" and what Panger (1998) terms "object use". Nearly all instances of substrate use we describe in this report involved accessing food encased in an object with hard protective covering (i.e., a fruit with a thick, durable husk) by fracturing the protective covering through pounding or hitting the object against a sturdy branch. This activity is what Panger characterizes as "pound". The exception was when brown capuchins pounded the fruit of *Jacaratia spinosa* (Caricaceae). The peel of this large, soft berry (7 cm × 5 cm; closely related and morphologically similar to the domesticated papaya) contains copious amounts of viscous, noxious latex, which capuchins appear extremely hesitant to bite through. We also observed frequent instances of substrate use that were the "rub" and "fulcrum" subcategories of Panger's (1998) object use. "Rub" and "fulcrum" were such extremely common processing techniques that we did not consider them noteworthy. We conservatively estimate that they were observed on at least a near daily basis. For tool use we concur with many previous workers and employ Beck's (1980, p. 10) definition: "... tool use is the external employment of an unattached environmental object to alter more efficiently the form, position, or condition of another object, another organism, or the user itself when the user holds or carries the tool during or just prior to use and is responsible for the proper and effective orientation of the tool". We fully agree with Panger (1998) that branch dropping, waving, and carrying are best not included within tool use, as these behaviors were ubiquitous and contextually ambiguous in our study animals.

13.2.4 Behavior sampling

Here we do not attempt a robust quantitative presentation of our observations using our admittedly opportunistic sampling. Instead we aim to alert other researchers to the singular characteristics of complex object manipulation relevant to traditions that are now documented in a wild capuchin population. In June 2000, a dedicated study of the social behavior and ecology of our four *C. apella* study troops was initiated. A major objective of this new field investigation was to obtain detailed data on individual differences in substrate and tool use by *C. apella*, especially regarding the acquisition of object manipulation skills by immatures and the potential role of social learning and traditions.

From January 1998 through July 2000, we or other members of our research group contacted at least one of the four *C. apella* study troops at Raleighvallen on a minimum of 600 days. When *C. apella* study troops were in mixed-species groups with squirrel monkeys, scan samples at 15 minute intervals (Fragaszy, Boinski, and Whipple, 1992) were taken of their location. More than 1100 hours of this mixed-species association were documented in this period. Note was also often taken of *C. apella* troops encountered elsewhere apart from a squirrel monkey troop.

Documentation of substrate and tool use and prominent social interactions and other foraging activities by capuchins during scan samples and the intervening periods were *ad libitum*. In total, there is written and tape-recorded documentation of at least 120 instances in this 31-month-long period during which one or more individuals in a *C. apella* troop exhibited complex object manipulation in foraging or there was apparent attempts to do so by immature capuchins (Table 13.1). This figure is best interpreted as a gross underestimate of the true frequency observed. In general our detail and thoroughness in documenting instances of complex manipulation improved over the course of the study as our appreciation of the distinctiveness of these behaviors in Raleighvallen relative to other capuchin populations increased. Nevertheless, many pertinent episodes of substrate use at Raleighvallen were never described in field notes; these instances were perceived as so commonplace that they often received a low priority in the panoply of *ad libitum* behavioral data we strove to harvest. For example, mention might be made in field notes that a *C. apella* troop processed a species of hard-husked fruit with substrate use that day. Even if this minimal account was entered, description of the hours for which the troop was occupied in this foraging activity and the number of troop members and their individual success and techniques rarely was included.

Table 13.1. *A summary from February 1998 through June 2000 of the instances of different types of substrate-use categories by* C. apella *at Raleighvallen on which this report is based*[a]

Substrate use category	Documented	Recalled
Pounding *Astrocaryum* spp. (Arecaceae)	6	6
Pounding *Carapa procera* (Meliaceae)	5	4
Pounding *Clusia grandiflora* (Clusiaceae)	5	4
Pounding *Couratori* spp. (Lecythidaceae)	3	4
Pounding *Cynometra* spp. (Caesalpinioideae)	0	7
Pounding *Escheweilera* spp. (Lecythidaceae)	4	20
Pounding *Gustavia* spp. (Lecythidaceae)	0	2
Pounding *Jacaratia spinosa* (Caricaceae)	20	2
Pounding *Lecythis davisii* (Lecythidaceae)	0	6
Pounding *Hymenaeae courbaril* (Caesalpinioideae)	1	5
Pounding *Pachira aquatica* (Bombacaeae)	2	5
Pounding *Phenakospermum guyannense* (Strelitziaceae)	34	70
Pounding *Strychnos mitscherlichii* (Loganiaceae)	6	0
Pounding *Vouacapouca americana* (Caesalpinaceae)	0	8
Pounding unknown species of Lecythidaceae fruit	6	0
Pounding unknown or other husked fruit species	20	20
Immatures closely observing substrate use by others	15	43
Immatures *unsuccessfully* attempting to open husked fruit	7	30
Immatures *successfully* opening husked fruit	7	8
Females exhibiting substrate use	2	5
Adult female closely observing substrate use by an adult male	1	0

[a]The sum total of those instances documented in field notes is indicated in column 2, while the third column represents a qualitative (but conservative) estimate of additional instances based on the recollection of field observers. Within each column, the category frequencies do not sum to the cumulative total of instances observed. This is because some episodes of substrate use may be relevant to and reported under more than category.

Therefore, in reporting our qualitative observations of specific categories of manipulative activities in Table 13.1 we also include the estimated minimum number of instances this was observed based, in part, on our and other team members' recollections.

13.3 Field observations and pertinent ecological background

13.3.1 Fruit resources harvested with extensive manipulation

Lecythidaceae, the Brazil nut family, tree species are common in Suriname with six genera and many species (van Roosmalen, 1985). These bear abundant crops of large, nut-like, thick-husked fruits, and the seeds are

Fig. 13.1. Examples of five species of husked fruits harvested by *Cebus apella* in Raleighvallen (clockwise from top far left): *Lecythis poiteaui* (Lecythidaceae), *Capparis maroniensis* (Capparaceae), *Carapa procera* (Meliaceae), *Lecythis corrugata* (Lecythidaceae), and *Couratari stellata* (Lecythidaceae). (Photograph courtesy of Marc van Roosmalen.)

an important food resource for *C. apella* at this site. The seeds and adherent pulp of other well-protected fruits, with fruit walls too hard and large to be processed with the capuchin's powerful jaws, also comprise a significant component of the diet: *Pachira* (Bombaceae), *Cynometra*, *Vouacapoua*, and *Hymenaea* (Caesalpinaceae), *Carapa* (Meliaceae), *Phenakospermum* (Strelitziaceae), and *Strychnos* (Loganiaceae) spp. at Raleighvallen (Figs. 13.1 and 13.2; also see the illustrations and descriptions of these fruits in van Roosmalen, Mittermeier, and Fleagle, 1988). We roughly estimate that 5% of time spent foraging fruits by adult and subadult male *C. apella* in Raleighvallen is allocated to fruit species with edible seeds and pulp that are reliably extracted from intact fruits usually with dexterous object manipulation involving forceful blows. This estimate is based on extrapolation from observations of *C. apella* foraging in varied circumstances and our current estimates of this species' allocation of time across these circumstances. Certainly, the proportion of time spent foraging husked fruits fluctuates seasonally. In Suriname, and elsewhere in South America, fruits from these tree taxa are mostly available in the wet season, usually January through July, with flowering in the dry season (Mori and Lepsch-Cunha, 1995; Mori and Prance, 1987; S. Boinski, unpublished

Fig. 13.2. An adult brown capuchin, *Cebus apella*, in Raleighvallen employing substrate use to break open a *Couratari oblongifolia* (Lecythidaceae) pyxidium. This photograph has been digitally enhanced to retrieve additional information from key image components. (Photograph by R. Quatrone.)

data). Some species, however, fruit in the dry season (Lepsch-Cunha and Mori, 1999; Oliveira-Filho and Galetti, 1996). In Raleighvallen, some representatives of this category of fruits are generally available throughout the year. By contrast, large tough-husked fruits providing seeds and pulp are less abundant and diverse and contribute relatively minor components of the diet for *C. apella* in Peru and Colombia (Izawa, 1979; Izawa and Mizuno, 1977; Janson, Stiles, and White, 1986; Terborgh, 1983), *C. capucinus* in Costa Rica (Chapman, 1987; Chapman and Fedigan, 1990), and *C. olivaceus* in Venezuela (Robinson, 1986). No Lecythidaceae is listed as a food source in Peru, Costa Rica, or Venezuela, and only two species of this family, *Grias haughtii* and *Gustavia superba,* are intermittently exploited in Colombia. In Panama, however, *G. superba* is an important food source for *C. capucinus* (Mitchell, 1989).

Here we refer to the general category of fruits with difficult to penetrate mechanical protection of nutritious fruit contents (i.e., seeds and pulp) as "husked fruits". When referring to the fruit of a particular species, however, the specific botanical term for that fruit is employed. The fruit of Lecythidaceae species, for example, is termed a "pyxidium", a woody cup-like capsule with a lid or operculum at one end (Prance and Mori, 1978). Fruits of Bombacaceae, Meliaceae, and Musaceae species mentioned here are woody "capsules": dry fruits consisting of more than one carpel and each carpel with more than one seed.

So why are husked fruits, especially the Lecythidaceae, common and speciose in Suriname and uncommon or absent at other sites where extended behavioral field studies of capuchins have been undertaken? The answer lies ultimately in the mosaic of soil types now found in the neotropics, which has a diverse geological history (Terbough and Andreson, 1998). Soils of Eastern Amazonia and the Guianas (including Suriname) are nutrient poor and highly weathered as they are derived from ancient Precambrian shields. Central Amazonia, where soils originate from strongly weathered tertiary marine deposits, also has exceptionally poor soils. In contrast, the soils of Western Amazonia (including Peru, Colombia, and Venezuela) are much younger and more fertile than those of Central or the Eastern Amazonia as they were produced during the Andean orogeny during the Miocene. The plant composition of Amazonian forests, in turn, covaries with the underlying soil types and geological history. Lecythidaceae and Chrysobalanaceae, both tree families with exceptionally large and nutrient-laden seeds, are among the predominant plant families in the nutrient-poor soils of the Guianas and Eastern and Central Amazonia (Millikin, 1998; Mori *et al.*, 2001; Terborgh and Andreson, 1998).

The rewards obtained by *C. apella* for harvesting the contents of these husked fruits may be great. To cite a typical instance, Brazil nuts (*Bertholletia excelsa*) can exceed 72% of the dry seed weight in fat content (Peres, 1991). Seven (approximately 28 g dry weight) of these nuts, the same ones that occupy the nut bowls of the Northern hemisphere during winter months, represents 186 kilocalories, 4g protein 4g carbohydrate, 19g fat, and a panoply of vitamins and minerals (Whitney and Rolles, 1993). Each Brazil nut pyxidium may hold 12 to 22 such nuts (van Roosmalen, 1985). Furthermore, the potential nutritional value of any individual husked fruit to capuchins in the Guianas should probably be weighted upward compared with its value for capuchins in Western Amazonia and Central America. For relative to the latter two regions, the Guianas have a markedly lower primary productivity and a reduced biomass of primary consumers, including that of primates (Kay *et al.*, 1997; Peres, 1999; Stevenson, 2001). In effect, a capuchin from the Guianas is likely more motivated to invest the time and energy needed to harvest, successfully process, and ingest a specific husked fruit than would a capuchin from other, more fertile and productive neotropical regions.

13.3.2 General description: persistence and precision of object manipulation by *C. apella*

Aside from the youngest cohort of independent foragers (see below), substrate manipulation by *C. apella* of husked fruits usually is exceedingly directed, goal-oriented, adroit, and persistent. The individual capuchin appears to have decided upon one tactic to process the fruit prior to initiation. Rapid alternation between diverse tactics until one finally succeeds within a few moments, the common strategy in captive situations when a capuchin is given a novel task, is not observed in Suriname. Inept, sloppy, and uncertain movements or seemingly random motion components are uncommon among mature animals in this wild population. Instead, the substrate and tool use of these capuchins has the smooth, rapid flow that in humans is associated with tasks so well practiced that the motor actions themselves are no longer consciously attended.

The persistence and precision typical of husked-fruit processing is exemplified by the three adult *C. apella* observed in Saramacca, each methodically pounding a Brazil nut pyxidium on a thick branch. Each capuchin employed the identical technique. Seated on top of a thick branch, the capuchin held the large round and heavy pyxidium (9–12 cm in diameter) in both hands, raised the fruit to about head height, and then hit the fruit forcefully (and loudly) on the branch for four to eight blows in succession before stopping, carefully inspecting, and occasionally gnawing at the damaged spot. The capuchin then resumed another bout of hitting the pyxidium against the branch and again scrutinizing the damage incurred. All of the capuchins continued processing in this manner for periods of a minimum 10 minutes in duration and 90 blows. At no time did these capuchins appear to fumble in manipulation of the capsule or did a blow of the pyxidium miss hitting the branch. We estimate, based on more distant and interrupted observations of and the resounding sounds produced by *C. apella* processing Brazil nuts, that successful efforts to access the nutritious seeds commonly requires more than 30 minutes of continuous efforts. A two-year-old juvenile female was observed to process the pyxidia of a green *Eschweilera congestifolia* fruit in much the same manner, using calm and deliberate blows, and rotating the pyxidium in a slow, smooth, careful manner during inspection, while immature animals looked on at close range.

13.3.3 Variety of techniques employed

Most mature *C. apella* are not readily distinguished in the motor actions and techniques used to harvest the contents of each species of husked fruit. Much greater variation was apparent in how different species of husked fruit are processed than in the individual variation exhibited within processing any single species. We fully expect that the within-individual variation in substrate-use techniques will expand further as the size and level of detailed analyses of our sample increases.

The capsules of *Phenakospermum guyannense* are well fortified against seed predators: they are 11–13 cm by × 5–7 cm in dimension and so woody that persistent effort is needed by a human to hack through an unopened capsule with a machete (S. Boinski, personal observation). In Raleighvallen, capuchins commonly harvest the contents of undehisced capsules, an abundant and predictable resource at this site during the wet season, thereby presumably reducing competition from seed predators that can only open dehisced capsules.

The technique employed by Raleighvallen capuchins to open the undehisced capsule, moreover, entails a particular sequence of steps and is effectively invariant among the numerous subadult and adult capuchins observed to use it. The capsule has three longitudinal valves. Two of the intervening capsule facets are flat, and the third facet has a pronounced outward or convex curvature. By hitting the convex surface at its apex, the entire downward force is applied to a fulcrum point rather than being distributed over the entire surface. All mature capuchins specifically pound only the apex of the facet with the convex curvature against a tree surface; the two flat surfaces are never processed. After several strikes of the convex surface against the tree, the capuchin usually rolls the capsule over and examines its progress in breaching the capsule's walls. At this point the capuchin might also gnaw with its teeth and tear with its fingers at the opening so far created before resuming the strikes against the tree surface. These blows are specifically directed at one and only one target area. Eventually a "window" about 2–3 cm^2 is created in the convex facet and fingers and teeth are used to extract the seeds from that carpel of the fruit. Efforts to open the two carpels protected with flat facets, either through internal or external walls, have never been observed. Instead the capsule is dropped after the capuchin has apparently extracted all the seeds and arils that can be accessed through the "window". On at least three occasions, including that described above, a capuchin abandoned efforts to open an undehisced capsule after extensive pounding on the branch.

13.3.4 Opportunities for social learning by immatures

13.3.4.1 Intense visual monitoring by immatures

A prominent concomitant to many, but not all, instances of substrate use by capuchins is that the actor's (whether it is an adult or immature) every movement is closely and persistently monitored by an audience of immatures. From one to as many as four immatures (ranging in estimated ages from about one to two years old) are immediately adjacent to the actor (substrate user) such that the mouths and eyes of the immatures are within 10 to 20 cm of the husked fruit. No active food sharing (*sensu* Fragaszy, Feuerstein, and Mitra, 1997) or cooperation between two individuals in processing a husked fruit has yet been seen, but tolerated scrounging occurs (*sensu* de Waal, 1989; de Waal and Berger, 2000). Immature capuchins pick up food fragments that have fallen onto the tree limb, surrounding vegetation, and the ground in the course of food processing by the mature troop member. One adult male, however, after removing the skin of a gecko by rubbing and pounding it against a stout branch, handed the detached tail of the gecko to an immature capuchin that had been closely watching his activities.

13.3.4.2 Practicing skills

Of the more than 50 instances in which immature capuchins were observed practicing substrate-use skills or otherwise performing the motor skill component of the maneuver, at least 35 occurred at approximately the same time or immediately following more mature troop members using substrate-use techniques to process the same species of fruit. In fact, for only one instance of immature substrate use can we claim that an immature *probably* engaged in substrate use without an adult troop member also having done so with the same fruit species in the preceding 30 minutes.

Patently inept attempts at substrate use by young capuchins are not common. The infants and juveniles appear to acquire rapidly the link between the movement of arms while holding the husked fruit and the eventual penetration of the husk. Instead, immatures are most likely to fail because they select an inappropriate anvil. An immature capuchin, for example, one to two years of age, unsuccessfully attempted to harvest seeds from intact *Couratari oblongifolia* fruit by breaking the fruit open against the ground. The technique employed was reminiscent of the usual two-handed blow employed by adults. The immature capuchin held the pyxidium in both hands and made the same downward motion with its arms that larger and more mature animals use. However, this juvenile was on the ground. While making the smashing motion it would jump simultaneously in the air. At the end of its swing, the juvenile tossed the fruit in the air while throwing itself on the ground. Despite the intense effort of the juvenile, the pyxidium did not contact any hard substrate. The small capuchin repeated this procedure at least three times before apparently noticing how close the human observer was and scurrying up the nearest tree without the fruit. The fruit had teeth marks and scratches on the exterior but remained intact. Other juveniles have been observed on at least two occasions trying to break open Lecythidaceae pyxidia upon the ground with a similar lack of success. No adult capuchin has ever been seen trying to break a husked fruit upon the ground.

Another illustration of an immature pairing correct actions with an inappropriate pounding substrate was an infant male, approximately one year old, who harvested an undehisced *P. guyannense* capsule. After harvesting the capsule, the infant male did not leave this nonwoody plant to seek a hard, sturdy substrate. Instead he proceeded to strike the capsule repeatedly (> 10 blows) against the elongate (and extremely resilient) petiole at the base of the broad leaf blade as well as the broad leaf blade itself. The arm motions and body stances exhibited by the infant were indistinguishable in form and adroitness from those commonly employed by mature capuchins processing *P. guyannense* capsules. The infant, however, did not exclusively present the convex facet of the *P. guyannense* capsule to the substrate in his strikes. Suture edges and the two flat facets also received a minority of the strikes.

13.4 A link between seasonal availability of husked fruits and timing of skill (tradition) acquisition by immatures?

We have a series of hypotheses to explain our observations so far concerning, first, the variable attendance by immatures at substrate-use episodes (and none at the sole instance of documented tool use) and, second, the relatively few instances of patently unskilled processing of husked fruits by immatures. Our qualitative observations of the capuchins, together with ongoing phenological studies of fruit availability at Raleighvallen, suggest that successful learning of substrate use is rapid and focused to a considerable extent at the level of husked fruit species, not merely broad categories of fruit. For each species of husked fruit separately, the instances of intense visual monitoring of mature animals successfully processing husked fruits and the more limited "practicing" by immature capuchins appear to coincide with the initiation of that species' seasonal availability. In other words, the annual and supra-annual cycles of availability of these diverse husked fruit species effectively present self-feeding, but immature, capuchins with a multiyear series of completely novel husked fruit species or of species that were last encountered after a time lag approaching a minimum of a year. After a period of familiarity with a husked fruit species, attentiveness by immatures to others processing that fruit species appear to decline. These qualitative observations, of course, must be verified by future fieldwork.

Timing of skill acquisition does not identify the mechanisms of skill acquisition but is an essential first phase of study. As yet undetermined forms of the many potential social and associative (trial and error) learning processes are involved (Anderson, 1996; Tomasello and Call, 1997). However, even at this early stage of our studies of brown capuchins at Raleighvallen, we are confident that this phenomenon falls within the compass of a social tradition as defined by the organizers of this volume (Ch. 1): "... behavior patterns shared among members of a group that depend to a measurable degree on social contributions to individual learning, resulting in shared practices among members of a group". The linkage between proficient adult model processing of husked fruits and the direction of the immatures' attention to those husked fruits as objects of interest is strong. Our observations of intense attention by immatures to episodes of proficient object manipulation and tool use by adults is extremely similar to the observations of immature *C. apella* closely monitoring the use of stones to crack open palm nuts by adult group members (Ottoni and Mannu, 2001) and the intense interest of captive *C. apella* in others cracking and eating pecans (Fragaszy *et al.*, 1997). It is also quite plausible at this early stage of research that the adult models in these situations provide the additional scaffolding necessary for the generation of substrate use in immatures by indicating that the "bashing" motions of the arms are somehow involved in a successful outcome (i.e., accessing the nutritious foods protected by the husks). Even the youngest immatures in Raleighvallen appear to employ a stringently limited range of arm motions and related strategies to break open husked fruits when processing with teeth and hands fail. The possibility is great that the social context influences more than just selection of nuts and promotion of pounding them. Social facilitation may be important in sustaining the tradition of substrate use among brown capuchins at Raleighvallen.

Our observations of inept "practicing" of substrate use by immature *C. apella* in Raleighvallen are extremely suggestive of the development of stone-tool use by wild chimpanzees at Bossou, Guinea (Inoue-Nakamura and Matsuzawa, 1997). In both wild primate populations, the appropriate motor actions were swiftly attained. Integration of the motor actions into a successful multiaction sequence, however, took far longer, particularly the selection of appropriate objects (substrates) and the requisite functional relations. The cognitive processes (e.g., emulation, imitation) underlying the acquisition of these skills by our study subjects remain unclear.

The social tolerance model presented by van Schaik (Ch. 11) is relevant to the question of why foraging traditions may be well developed in wild *C. apella*. He argued that the amount of social tolerance group members express toward subordinates in foraging situations is a critical mechanism in the social transmission of complex foraging skills. Because the appearance of new foraging techniques is uncommon in wild populations, social tolerance is essential for the skills to be propagated among group members and eventually, via dispersal, to a larger population of conspecifics. This insight complements that of Kummer and Goodall (1985), who emphasized that the opportunity for technical innovation by individuals is enhanced when they are unfettered by social constraints and often forage alone or in small groups. One of the reasons that between-population variation in complex foraging techniques is so apparent among captive and wild brown capuchins might well be that brown capuchin social structure represents a propitious balance between these facilitating factors. Brown capuchin social organization and the consequent within-group competitive regimes for food are distinctive among many primates in that they are characterized both by despotic hierarchies among adults for access to desirable food patches and by adults tolerating the presence of immatures at these same food patches (Di Bitetti and Janson, 2001; Janson, 1990; our observations reported here). Therefore, a dynamic cycle is effectively created among brown capuchin group members in that immatures enjoy great social tolerance at feeding sites and, consequently, abundant opportunities for social learning. As young adults, however, these same immatures are exiled to the group periphery where, until they attain higher social rank, they are able and motivated to improvise, practice, and develop new foraging techniques on less-preferred and less-efficiently harvested foodstuffs (i.e., the hard husked fruits in the Guianas). Finally, as mature and socially dominant animals (presumably attainment of high status is more likely among brown capuchins who are adept and successful foragers), they return to the social and spatial center where a new crop of immatures await transmission of the honed complex foraging skills of these elders.

13.5 How brown capuchins in Suriname are distinctive from other wild populations

We stress that the difference in complex manipulative skills between Raleighvallen and other wild populations is best considered quantitative, not qualitative. Nevertheless, we conclude that the substrate use brown capuchins exhibit at the Raleighvallen study site to process hard-husked fruits is singularly skilled, persistent, common, and technique specific relative to reports from other extended field studies of capuchins in Peru, Colombia, Venezuela, and Costa Rica. Evidently the abundance, high quality, and diverse morphological structures of husked fruits in Raleighvallen provide the opportunity and incentive to brown capuchins for development of object manipulation skills. The difficulty of accessing these foods increases the probability that social influences contribute to the maintenance of the practices (i.e., that they are traditions). We believe these skills are so unusual that they are more likely to reflect social traditions than the less-elaborate skills that characterize capuchin foraging at other sites. For example, in relatively few instances, which involved immature *C. apella*, at Raleighvallen did we note a capuchin display manipulation of a husked fruit that was not deft, efficient, and smoothly effected. In contrast, the descriptions of motor actions commonly exhibited in substrate and tool use at other sites seldom convey this impression (but see the description of *Luehea candida* (Tiliaceae) capsule processing by *C. capucinus* in Panger (1998)). Unfortunately, this motor skill disparity we describe would undoubtedly be better conveyed and quantified with videotapes. In the absence of that medium, we provide examples from the literature and the primary author's observations of substrate and tool use in wild *C. capucinus*. Among adult *C. capucinus* in Corcovado, Costa Rica, the only instances of substrate use documented were several involving the large (football-sized), gourd-like fruits of *Enallagama latifolia* (Bignoniaceae) and the slightly smaller gourd-like fruits of feral cacao trees (*Theobroma cacao,* Sterculiaceae) (S. Boinski, unpublished data). For both fruit species, adult capuchins would smash the fruits against thick branches to break the fruit walls enclosing tasty pulp. Juvenile *C. capucinus* attempts at processing these same fruits were inept, and no successful efforts were observed. Instead of the swift, sure actions of the mature brown capuchins in Raleighvallen, these mature white-faced capuchins were clumsy, repeatedly fumbling and dropping the fruits, and frequently missing the branch completely when striking a blow with the fruit. At no time was a capuchin seen to attempt more than 10 blows, and most far fewer. Likewise, the *C. capucinus* adult male in Manuel Antonio, Costa Rica, which used a branch as a club to attack a venomous snake, often fumbled and missed striking the snake and frequently dropped the branch (Boinski, 1988). Izawa and Mizuno (1977) also note that *C. apella* juveniles at their Colombian site were less skilled and successful than adults, but Terborgh (1983) noted no comparable age difference in substrate-use skills among Peruvian *C. apella*.

Our observations in Raleighvallen are probably representative of the object manipulation skills of *C. apella* throughout the Guianas and Central and Eastern Amazonia, the biogeographic region where Lecythidaceae are common. Consistent with our observations, brief anecdotes of *C. apella* processing or ingesting husked fruits are found in publications focused closely on the ecology (as opposed to foraging behavior) of *C. apella* and other primates and Lecythidaceae plant species (Galetti and Pedroni, 1994; Peres, 1991; Prance and Mori, 1978). Guillotin, Dubost, and Sabatier (1994), for example, noted that in French Guiana *C. apella,* but not the black spider monkey or the red howler, eats seeds from Lecythidaceae fruits, although these workers do not detail the foraging technique *C. apella* employed. Perhaps the most convincing corroboration comes from Marc van Roosmalen's ongoing primate fieldwork in Central Amazonia (personal communication). He describes *Cariniana micrantha* (Lecythidaceae) pyxidia as a "keystone resource" for *C. apella* in this region. Although available in at least small quantities throughout the annual cycle, *C. micrantha* is in peak abundance in the late dry season, when alternative fruit sources for *C. apella* are at the annual nadir of availability. During the late dry season, van Roosmalen characterized the Central Amazonian forest as being filled with the resonant sounds of *C. apella* vigorously and persistently striking *C. micrantha* pyxidia against tree branches. No tool use has yet been noted by van Roosmalen (personal communication) and his observations of the substrate-use techniques are not sufficiently detailed to compare closely with ours from Suriname.

13.6 Locality-specific conditions affect skills maintained as traditions

Local conditions are crucial in the expression of object manipulation and tool use in primates. This is hardly a new insight, but a refrain repeated in the literature for more than 20 years. Geographic variation in the presence, absence, and seasonality of desirable food resources has long been suggested to account for variable expression in frequency of substrate and tool use within and between species (Boinski, Quatrone, and Swarts, 2000; de Waal, 1997; Ingmanson, 1996; Izawa and Mizuno, 1977; Parker and Gibson, 1977). Local disparities in proclivities to manipulate objects are thought to be further amplified by the nutritional quality of accessibility to alternative food resources (Boesch and Boesch, 1993; McGrew, 1992). Our report now provides a concrete example of contextual variation between sites, namely the biogeography of Lecythidaceae and Lecythidaceae-like husked fruits, in promoting the expression of complex object manipulation in capuchins. The wide between-site differences in the expression of skilled object manipulation now documented for wild and captive capuchins are fully consistent with what has long been accepted for chimpanzees. Researchers seeking to understand complex manipulation skills in primates must, therefore, not only incorporate species differences (i.e., van Schaik, Deaner, and Merril, 1999) but also consider the within-species heterogeneity associated with local conditions.

13.7 Quest for useful data

From our anecdotal dataset, we have extracted nearly all the useful and reasonably robust insights regarding possible social traditions in substrate use among brown capuchins in Raleighvallen. Some colleagues might even suggest that we have overinterpreted our data. Nevertheless, our preliminary observations clearly warrant a more structured research program into this phenomenon. Here we outline our future research tactics. The basic strategy is simple. The challenge will rest nearly completely on collecting the desired data. First, we seek to identify disparities in the expression of complex substrate-processing techniques at levels ranging from between population to within groups. Once disparities are found, then hypothesis testing on the mechanisms underlying the expression of these differences will proceed.

Cultural differences in object and tool use and manufacture have been proposed for chimpanzees (McGrew, 1992; Whiten *et al.*, 1999). Given that *C. apella* in Suriname appears to use processing techniques finely tuned to different taxa (i.e., structural categories) of husked fruit, we suggest that detailed studies of the processing techniques *C. apella* employs at other sites in the Guianas and Central and Eastern Amazonia for these husked fruit taxa can be useful. If differences in husked-fruit processing techniques by *C. apella* are found that are not explained by fruit morphology or abundance, then the likelihood of social traditions in *C. apella* is strengthened. A simple, relatively easy-to-implement approach that encompasses all substrate uses and complex foraging behaviors (as well as the social traditions described in Ch. 14) would be to compare substrate-use methods between brown capuchin troops on opposite banks of major rivers where the rates of river meandering are low and the banks distant. In this situation, the environmental similarities would usually be quite comparable, but the opportunities for social transmission between the two populations would be at best indirect and infrequent. This is the group comparsion method (see Ch. 1; identical to the regional contrast approach described in Ch. 5), which can suggest candidates for investigations of the diversity of skills.

Another situation that plausibly engenders traditions in husked-fruit processing follows from the uneven distribution of trees species bearing husked fruit among ranging areas of *C. apella* troops in the same locality. Mori, Becker, and Kincaid, (2001) documented extreme heterogeneity in the distribution of the 38 species of Lecythidaceae found in the 100 ha sample grid at their Amazonian site. Although this site in Amazonia has the greatest species diversity and absolute abundance of Lecythidaceae trees reported for any location in the world, only one or a few individuals of some Lecythicidae species were found anywhere in this sample area. Consequently, the opportunity for members of capuchin troops to learn, practice, and model specialized processing techniques for such husked fruits could markedly vary across the foraging landscapes at the scale of 1 or 2 km distances.

In particular, we expect discernable contrasts in locality-specific techniques used to extract seeds from the well-protected Brazil nut (*B. excelsa*) pyxidium. Natural populations of Brazil nut trees are distinctively distributed within their range in lowland Amazonia and the Guianan Shield. Typically in Amazonia, these emergent trees are found clumped in widely spaced "groves" of 50–300 adult trees in 20–50ha areas with few intervening trees (Mori, 1992; Mori *et al.*, 2001; Peres, Schiesari, and DiasLeme, 1997). Consequently, it is quite plausible that nearby brown capuchin troops with non-overlapping ranges could differ dramatically in their opportunities to process Brazil nut fruits. If social traditions had a significant effect, we could make several predictions. First, capuchin troops with ranges that encompass different Brazil nut groves, and thus have non-overlapping radii for transmission of traditions, would be expected to have divergent techniques. Also males emigrating into groups might be expected to exploit a more diverse set of techniques than the natal females.

Manipulative protocols to explore the acquisition of substrate use appear feasible in Raleighvallen. A diverse set of hard-husked fruit species (largely gourds and pumpkins) not native to Suriname are grown by rural and indigenous peoples and are well documented not to survive as feral or escaped plants. These foodstuffs, and perhaps simulated fruits (e.g., tasty, odoriferous foodstuffs encapsulated in resilient, difficult to penetrate materials) could be presented to troops, subgroups, or peripheral troop members to create inequity among individuals in their experience with processing these food items. Social contributions to skill development would be substantiated if individuals with strong social affinities to those individuals introduced to exotic foodstuffs diverged from those group members without experience in how they processed these foods. The development of shared within-group idiosyncratic methods to process novel foodstuffs would also support social traditions. Of course, social traditions hypotheses in substrate use would not be supported if all individuals across troops and subgroups quickly converged on a set of common methods despite variation among individual capuchins in their opportunities for social learning of skills. Our interpretation in this instance would be that inherent species-specific processing actions in conjunction with experience were adequate to generate the same substrate-use techniques in all individuals.

13.8 Acknowledgements

Fieldwork was supported by the National Science Foundation (SBR-9722840; BCS-0078967) and the L. S. B. Leakey Foundation. Also essential to this study are the continuing kindness and support of STINASU (Suriname Nature Conservation Foundation), the Nature Conservation Division of the Suriname Forest Service, and Stuart Vervuurt and his gracious family. We thank the editors and the reviewers, Michael Huffman and Eduardo Ottoni, for their considered and helpful comments. This chapter is based on a previously published article: Boinski *et al.*, 2000. *American Anthropologist* **102**, 741–761.

References

Anderson, J. R. 1996. Chimpanzees and capuchin monkeys: Comparative cognition. In *Reaching into Thought: The Minds of the Great Apes*, ed. A. Russon, K. Bard, and S. Parker, pp. 23–56. Cambridge, MA: Cambridge University Press.

Anderson, J. R. and Henneman, M. C. 1994. Solutions to a tool use problem in a pair of *Cebus apella*. *Mammalia*, **58**, 351–361.

Antinucci, F. and Visalberghi, E. 1986. Tool use in *Cebus apella*: a case study. *International Journal of Primatology*, **54**, 138–145.

Bard, K. A. and Vauclair, J. 1989. What's the tool and where is the goal? *Behavioral and Brain Sciences*, **12**, 590–591.

Beck, B. 1980. *Animal Tool Behavior: The Use and Manufacture of Tools by Animals*. New York: Garland.

Boesch, C. and Boesch, H. 1984. Mental map in chimpanzees: an analysis of hammer transports for nut cracking. *Primates*, **25**, 160–170.

Boesch, C. and Boesch, H. 1993. Diversity of tool use and tool-making in wild chimpanzees. In *The Use of Tools by Human and Nonhuman Primates*, ed. A. Berthelet and J. Chavaillon, pp. 158–168. Oxford: Oxford University Press.

Boinski, S. 1988. Use of a club by a white-faced capuchin (*Cebus capucinus*) to attack a venomous snake (*Bothrops asper*). *American Journal of Primatology*, **14**, 177–179.

Boinski, S., Quatrone, R., and Swarts, H. 2000. Substrate and tool use by brown capuchins in Suriname: ecological contexts and cognitive bases. *American Anthropologist*, **102**, 741–761.

Chapman, C. A. 1987. Flexibility in diets of three species of Costa Rican primates. *Folia Primatologica*, **29**, 90–105.

Chapman, C. A. and Fedigan, L. M. 1990. Dietary differences between neighboring *Cebus capucinus* groups: local traditions, food availability, or response to food profitability? *Folia Primatologica*, **54**, 177–186.

Costello, M. B. and Fragaszy, D. M. 1988. Prehension in *Cebus* and *Saimiri*: I. Grip type and hand preference. *American Journal of Primatology*, **15**, 235–245.

Custance, D., Whiten, A.,and Fredman, T. 1999. Social learning of an artificial fruit task in capuchin monkeys (*Cebus apella*). *Journal of Comparative Psychology*, **113**, 13–23.

de Waal, F. B. M. 1989. Food sharing and reciprocal obligations among chimpanzees. *Journal of Human Evolution*, **18**, 433–459.

de Waal, F. B. M. 1997. *Bonobo: The Forgotten Ape*. Berkeley, CA: University of Berkeley Press.

de Waal, F. B. M. and Bergei, M. L. 2000. Payment for labour in monkeys. *Nature*, **404**, 563.

Di Bitetti, M. S. and Janson, C. H. 2001. Social foraging and the finder's share in capuchin monkeys, *Cebus apella*. *Animal Behaviour*, **62**, 47–56.

Fernandes, M. E. B. 1991. Tool use and predation of oysters (*Crassostrea rhizophorea*) by the tufted capuchin, *Cebus apella apella*, in brackish water mangrove swamp. *Primates*, **3**, 529–531.

Fragaszy, D. M. and Adams-Curtis, L. E. 1991. Environmental challenges in groups of capuchins. In *Primate Responses to Environmental Change*, ed. H. O. Box, pp. 239–264. New York: Chapman & Hall.

Fragaszy, D. and Boinski, S. 1995. Patterns of individual choice and efficiency of foraging and diet in the wedge-capped capuchin monkey, *Cebus olivaceus*. *Journal of Comparative Psychology*, **109**, 339–34.

Fragaszy, D., Boinski, S., and Whipple, J. 1992. Behavioral sampling in the field: comparison of individual and group sampling methods. *American Journal of Primatology*, **26**, 259–275.

Fragaszy, D., Feuerstein, J., and Mitra, D. 1997. Transfers of food from adults to infants in tufted capuchins (*Cebus apella*). *Journal of Comparative Psychology*, **111**, 194–200.

Galetti, M. and Pedroni, F. 1994. Seasonal diet of capuchin monkeys (*Cebus apella*) in a semi-deciduous forest in south-east Brazil. *Journal of Tropical Ecology*, **10**, 27–39.

Guillotin M, Dubost G, Sabatier D. 1994. Food choice and food competition among the three major primate species of French Guiana. *Journal of Zoology*, **233**, 551–579.

Ingmanson, E. J. 1996. Tool-using behavior in wild *Pan paniscus*: social and ecological considerations. In *Reaching into Thought: The Minds of the Great Apes*, ed. A. E. Russon, K. A. Bard, and S. T. Parker, pp. 190–210. Cambridge: Cambridge University Press.

Inoue-Nakamaura, N. and Matsuzawa, T. 1997. Development of stone tool use by wild chimpanzees (*Pan troglodytes*). *Journal of Comparative Psychology*, **111**, 159–173.

Izawa, K. 1979. Foods and feeding behavior of wild black-capped capuchins (*Cebus apella*). *Primates*, **18**, 773–792.

Izawa, K. and Mizuno, A. 1977. Palm-fruit cracking behavior of wild black-capped capuchin (*C. apella*). *Primates*, **18**, 773–792.

Janson, C. H. 1990. Social correlates of individual spatial choice in foraging groups of brown capuchins. *Animal Behaviour*, **40**, 910–921.

Janson, C.H. and Boinski, S. 1992. Morphological and behavioral adaptations for foraging in generalist primates: the case of the Cebines. *American Journal of Primatology*, **88**, 483–498.

Janson, C.H., Stiles, E. W., and White, D. W. 1986. Selection on plant fruiting traits by brown capuchin monkeys: a multivariate approach. In *Frugivores and Seed Dispersal*, ed. A. Estrada and T. H. Fleming, pp. 83–92. Dordrecht: W. Junk.

Kay, R. F., Madden, R. H., van Schaik, C., and Higdon, D. 1997. Primate species richness is determined by plant productivity: implications for conservation. *Proceedings of the National Academy of Sciences, USA*, **94**, 13023–13027.

Klüver, H. 1933. *Behavior Mechanisms by Monkeys*. Chicago, IL: University of Chicago Press.

Kummer, H. and Goodall, J. 1985. Conditions of innovative behavior in primates. *Philosophical Transactions of the Royal Society of London, Series B*, **308**, 203–214.

Langguth, A. and Alonso, C. 1997. Capuchin monkeys in the Caatinga: tool use and food habits during drought. *Neotropical Primates*, **5**, 77–78.

Lavallee, A. C. 1999. Capuchin (*C. apella*) tool use in a captive naturalistic environment. *International Journal of Primatology*, **20**, 399–414.

Lepsch-Cunha, N. and Mori, S. A. 1999. Reproductive phenology and mating potential in a low density tree population of *Couratari multiflora* (Lecythidaceae) in central Amazonia. *Journal of Tropical Ecology*, **15**, 97–121.

McGrew, W. C. 1992. *Chimpanzee Material Culture: Implications for Human Evolution.* Cambridge: Cambridge University Press.

McGrew, W. C. and Marchant, L. F. 1997. Using the tools and hand: manual laterality and elementary technology in *Cebus* spp. and *Pan* spp. *International Journal of Primatology*, **18**, 787–810.

Mitchell, B. 1989. Resources, group behavior, and infant development in white-faced capuchin monkeys, *Cebus capucinus*. Dissertation, University of California, Berkeley.

Mittermeier, R. A. and van Roosmalen, M. G. M. 1981. Preliminary observations on habitat utilization and diet in eight Surinam monkeys. *Folia Primatologica*, **36**, 1–39.

Mori, S. A. 1992. The Brazil nut industry – past, present, and future. In *Sustainable Harvest and Marketing of Rain Forest Products*, ed. M. Plotkin and L. Famarole, pp. 241–251. Washington, DC: Island Press.

Mori, S. A. and Lepsch-Cunha, N. 1995. Lecythidaceae of a central Amazonian moist forest. *Memoirs of the New York Botanical Garden*, **75**, 1–55.

Mori, S. A. and Prance, G. T. 1987. A guide to collecting Lecythidaceae. *Annals of the Missouri Botanical Garden*, **74**, 321–330.

Mori, S. A., Becker, P., and Kincaid, D. 2001. Lecythidaceae of a central Amazonian lowland forest: implications for conservation. In *Lessons from Amazonia*, ed. R. O. Bierregard Jr., C. Gascon, T. E. Lovejoy, and R. Mesquita. New Haven, CT: Yale University Press.

Oliveira-Filho, A. T. and Galetti, M. 1996. Seed predation of *Cariniana estrellensis* (Lecythidaceae) by black howler monkeys, *Alouatta carya*. *Primates*, **37**, 87–90.

Ottoni, E. B. and Mannu, M. 2001. Semifree-ranging tufted capuchins (*Cebus apella*) spontaneously use tools to crack open nuts. *International Journal of Primatology*, **22**, 347–358.

Panger, M. A. 1998. Object-use in free-ranging white-faced capuchins (*Cebus capucinus*) in Costa Rica. *American Journal of Physical Anthropology*, **106**, 311–321.

Parker, S. T. and Gibson, K. R. 1977. Object manipulation, tool use and sensorimotor intelligence as feeding adaptations in Cebus monkeys and great apes. *Journal of Human Evolution*, **6**, 623–641.

Peres, C. A. 1991. Seed predation on *Cariniana micrantha* (Lecythidaceae) by brown capuchin monkeys in Central Amazonia. *Biotropica*, **23**, 262–270.

Peres, C. A. 1999. Effects of subsistence hunting and forest types on the structure of Amazonian primate communities. In *Primate Communities*, ed. J. G. Fleagle, C. Janson, and K. E. Reed, pp. 268–283. Cambridge: Cambridge University Press.

Peres, C. A., Schiesari, L. C., and DiasLeme, C. L. 1997. Vertebrate predation of Brazil-nuts (*Bertholletia excelsa*, Lecythidaceae), an agouti-dispersed Amazonian seed crop: a test of the escape hypothesis. *Journal of Tropical Ecology*, **13**, 69–79.

Prance, G. T. and Mori, S. A. 1978. Observations on the fruit and seeds of neotropical Lecythidaceae. *Brittonia*, **30**, 21–33.

Reichart, H. 1993. *Management Plan for Raleighvallen Nature Preserve, 1993–1998*. Paramaribo, Suriname: Suriname Nature Conservation Foundation.

Robinson, J. G. 1986. Seasonal variation in the use of time and space by the wedge-capped capuchin, *Cebus olivaceus*. *Smithsonian Contributions to Zoology*, **431**, 1–60.

Rose, L. M. 2001. Meat and the early human diet: insights from Neotropical primate studies. In *Meat-eating and Human Evolution*, ed. C. B. Stanford and H. T. Bunn, pp. 141–158. Oxford: Oxford University Press.

Shettleworth, S. 1998. *Cognition, Evolution, and Behavior*. Oxford: Oxford University Press.

Stevenson, P. R. 2001. The relationship between fruit production and primate abundance in Neotropical communities. *Biological Journal of the Linnean Society*, **72**, 161–178.

Struhsaker, T. T. and Leland, L. 1977. Palm-nut smashing by *Cebus a. apella* in Colombia. *Biotropica*, **9**, 124–126.

Sugiyama, Y. 1997. Social tradition and the use of tool-composites by wild chimpanzees. *Evolutionary Anthropology*, **6**, 23–27.

Terborgh, J. 1983. *Five New World Primates: A Study in Comparative Ecology*. Princeton, NJ: Princeton University Press.

Terborgh, J. and Andreson, E. 1998. The composition of Amazonian forests: patterns at local and regional scales. *Journal of Tropical Ecology*, **14**, 645–664.

Tomasello, M. and Call, J. 1997. *Primate Cognition*. Oxford: Oxford University Press.

van Roosmalen, M. G. M. 1985. *Fruits of the Guyanan Flora*. Utrecht: Institute of Systematic Botany.

van Roosmalen, M. G. M., Mittermeier, R. A., and Fleagle, J. G. 1988. Diet of the northern bearded saki (*Chiropotes satanas chiropotes*): a neotropical seed predator. *American Journal of Primatology*, **14**, 11–35.

van Schaik, C. P., Deaner, R.O., and Merrill, M. Y. 1999. The conditions of tool use in primates: implications for the evolution of material culture. *Journal of Human Evolution*, **36**, 719–741.

Vevers, G. M. and Weiner, J. S. 1963. Use of a tool by a captive capuchin monkey. In *The Primates, Symposia of the Zoological Society of London*, ed. J. Napier and N. A. Barnicot, pp. 115–117. London: Zoological Society of London.

Visalberghi, E. 1993. Tool use in a South American monkey species: an overview of the characteristics and limitations of tool use in *Cebus apella*. In *The Use of Tools by Human and Non-Human Primates*, ed. A. Berthelet and J. Chavaillon, pp. 118–131. Oxford: Clarendon Press.

Visalberghi, E. 1997. Success and understanding in cognitive tasks: a comparison between *Cebus apella* and *Pan troglodytes*. *International Journal of Primatology*, **18**, 811–830.

Visalberghi, E. and Limongelli, L. 1994. Lack of comprehension of cause–effect relations in tool-using Capuchin monkeys (*Cebus apella*). *Journal of Comparative Psychology*, **108**, 15–22.

Visalberghi, E. and Limongelli, L. 1996. Acting and understanding: tool use revisited through the minds of capuchin monkeys. In *Reaching into Thought: The Minds of the Great Apes*, ed. A. Russon, K. Bard, and S. Parker, pp. 57–79. Cambridge: Cambridge University Press.

Visalberghi, E., Fragaszy, D. M., and Savage-Rumbaugh, S. 1995. Performance in a tool-using task by common chimpanzees (*Pan troglodytes*), bonobos (*Pan paniscus*), an orangutan (*Pongo pygmaeus*), and capuchin monkeys (*Cebus apella*). *Journal of Comparative Psychology*, **109**, 52–60.

Whiten, A., Goodall, J., McGrew, W. C., Nishida, T., Reynolds, V., Sugiyama, Y., Tutin, C. E. G., Wrangham, R. W., and Boesch, C. 1999. Chimpanzee cultures. *Nature*, **399**, 682–685.

Whitney, E. N. and Rolles, S. R. 1993. *Understanding Nutrition*. St. Paul, MN: West Publishing.

SUSAN PERRY, MELISSA PANGER, LISA M. ROSE, MARY BAKER, JULIE GROS-LOUIS, KATHERINE JACK, KATHERINE C. MACKINNON, JOSEPH MANSON, LINDA FEDIGAN, AND KENDRA PYLE

14

Traditions in wild white-faced capuchin monkeys

14.1 Introduction

Primatologists have long recognized that social learning could play an important role in food choice and food processing in primates, since the discovery (by Itani in 1958) of innovative food-processing techniques disseminated among Japanese macaques (see Ch. 10 for a review of subsequent findings). It is somewhat surprising that, after the initial discovery of the importance of social learning in Japanese macaques, practically all subsequent research on social learning in wild nonhuman primates has been on apes (e.g. Boesch, 1996a,b; Boesch and Boesch-Achermann, 2000; Boesch and Tomasello, 1998; McGrew, 1992, 1998; van Schaik, Deaner, and Merrill, 1999; Whiten *et al.*, 1999; see Chs. 10 and 11). To remedy the gap in what we know about social learning in natural settings in other primates, and because a truly comparative framework is necessary to understand the biological underpinnings of social learning (see Ch. 1), we began a comprehensive study of social learning in wild capuchin monkeys (Cebus spp.). Our study investigates the probable role of social learning in a number of behavioral domains.

Capuchins seem particularly likely to exhibit extensive reliance on learning, and social learning in particular, for the following reasons (Fragaszy, Visalberghi, and Fedigan, 2003). Several aspects of capuchin ecology promote behavioral flexibility. First, the genus *Cebus* occupies a wider geographic area than any other New World genus apart from *Alouatta* (Emmons, 1997), and it uses many different habitat types. Therefore, capuchins face a wide variety of environmental challenges. Second, capuchins include a wide range of plants and animals in their diets (Freese, 1976; Terborgh, 1983), and diets vary even between adjacent groups at the same site (Chapman and Fedigan, 1990). Third, capuchins are capable of producing a great variety of motor movements, enabling them to have more "building blocks" in their behavioral repertoire, which can be used in the production of new behaviors (see Ch. 10). For example, capuchins, like chimpanzees, spontaneously exhibit many types of tool use in laboratory settings (e.g., using objects as hammers, probes, levers, containers, etc.: Fragaszy *et al.*, 2003; Westergaard, 1994; Westergaard and Suomi, 1995). Occasionally they use objects as tools in the wild as well (see Ch. 13 for a review). Capuchins' propensity for tool use in captivity would seem to make them likely candidates for "material culture" (*sensu* McGrew, 1992; van Schaik *et al.*, 1999) in the wild.

In addition to the above-mentioned factors which are expected to favor innovation and advanced generalized learning capacities in capuchins, there are several factors that would seem to favor social learning in particular, in a variety of behavioral domains. First, because capuchins are extraordinarily tolerant of the close proximity of others (particularly immatures) while they are foraging (Perry and Rose, 1994), there is ample opportunity for group members to observe food choice and processing. Documentation of learning opportunity does not, of course, necessarily demonstrate that social learning is actually occurring (see Ch. 1.). Second, interactions with members of other species are typically also social activities: they involve multiple capuchins mobbing a predator, chasing a prey item, or harassing an ecologically neutral species (Rose *et al.*, 2003). Therefore, there is ample opportunity for young animals to observe adults' mode of interaction with other species. Third, capuchins rely on one another's cooperation in a number of important behavioral domains, for example for protection from predators, for cooperation in within-group aggression, for expulsion of would-be (and potentially infanticidal) immigrants (Perry, 1996a,b, 1997, 1998a,b, 2003; Rose, 1994). Consequently, they have devised many means of negotiating aspects of their social relationships. Some of these communication signals are fairly stereotypical, but others appear to be more flexible, and hence prime candidates for traditions (S. Perry, unpublished data).

In addition to these reasons why we expect capuchins to show an unusual degree of social learning (beyond most primates), we also expect them to show typical degrees of social learning propensities in the domain of vocal communication (in which social learning has been documented for vocal usage and comprehension: Cheney and Seyfarth, 1990; Ch. 8).

In this chapter, we review how capuchin monkeys at four sites in Costa Rica vary in social connections, in behavior toward other species, and in feeding techniques. This work adopts some of the logic of the group contrast approach to identify candidate traditions (see Ch. 1). We also seek evidence of traditions within groups in a joint analysis of patterns of acquisition by individuals and their patterns of social affiliation, adopting the process model of traditions laid out by Fragaszy and Perry in Ch. 1. We are clearly at the beginning of this project. This chapter constitutes a preliminary report, not a definitive statement. More important, this chapter serves as an example of how researchers can move from using a group contrast approach (which prompted our initial inquiries) to using the process model to guide the study of potential traditions in nonhuman animals living in natural conditions.

14.2 Methods

14.2.1 The study sites

There have been multiple long-term studies of groups of *Cebus capucinus* (the white-faced capuchin monkey). The sites of these studies are closely spaced geographically, thus increasing the number of animals available for these analyses and thereby minimizing the likelihood of substantive ecological or genetic differences between study populations (addressing the concerns of those adopting a group contrast approach). Figure 14.1 shows the locations of the study sites. Two of the study sites, Palo Verde (PV) and Lomas Barbudal (LB), are connected by a thin forest corridor, and wider corridors were available until quite recently. Hence, it is safe to assume that there has been genetic intermingling of these two populations at least until the past generation, and probably continuing into the present. Santa Rosa (SR) is about 50 km from LB. It is not known exactly when deforestation would have separated these two populations, but it probably occurred sometime within the past 30–50 years. All three of these sites consist largely of tropical dry forests and have broadly overlapping plant species lists. The fourth site, Curú, is least similar to the others; it is a coastal forest, including most of the dry forest plants but also some species not present at the other three sites. It is not known when Curú became geographically isolated from the other sites, but it probably happened within the past 50 years.

Detailed descriptions of the sites are available in other publications (LB: Frankie *et al.*, 1988; SR: Fedigan, Rose, and Avila, 1996; Hartshorn,

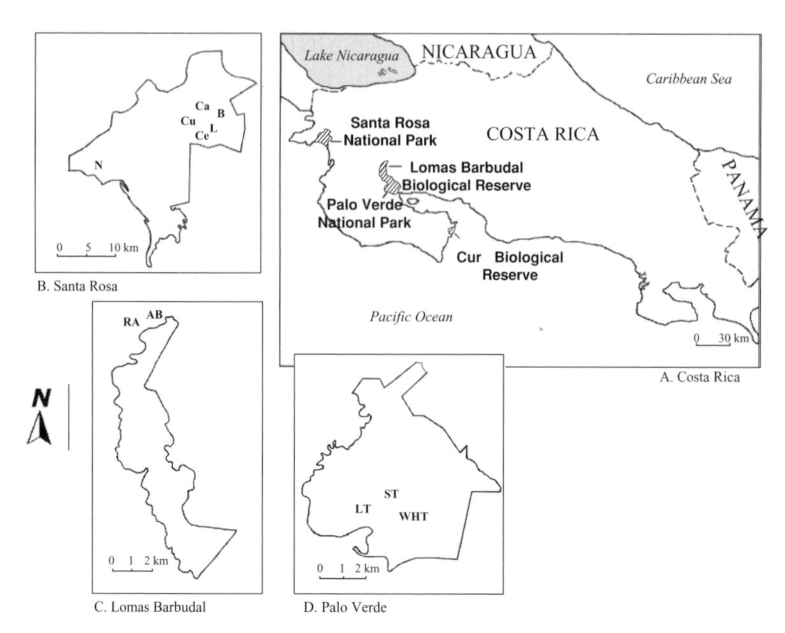

Fig. 14.1. Map of the study sites. Letters represent the core areas of different *Cebus capucinus* study troops. (B) Santa Rosa: N, Nancite; Ca, Cafetal; B, Bosque Humido; Cu, Cuajiniquil; L, Los Valles; Ce, Cerco de Piedra. (C) Lomas Barbudal: R, Rambo's group; A, Abby's group. (D) Palo Verde: ST, station troop; LT, lagoon troop; WHT, water hole troop.

1983; PV: Panger, 1997, 1998). Detailed descriptions of habitat are unavailable for Curú. Secondary dry forest is the most common habitat type at all sites. The ranges of the monkeys at LB include far more riparian forest than is typical at the other sites. Curú and Nancite group at SR both have some coastal forest, including mangroves, which is lacking for other study groups. PV monkeys have access to a large seasonal marsh, though they rarely utilize it while foraging. There are many domestic fruit trees at Curú, which the monkeys frequent.

14.2.2 The datasets

The data discussed in this chapter come from 10 different researchers studying 13 social groups of monkeys at four sites. In most cases, the data sets were collected to answer quite different questions than those addressed in this paper, so the methods used vary from study to study, and not all data sets can be used to address all topics in this chapter. The full data set of 20786 contact hours is shown in Table 14.1. Approximately 19000 hours were used for analysis of social conventions; smaller subsets

Table 14.1. *Periods of data collection on social behavior at each study site*

Site and study group	Time period (months/year)	Observation (h)	Principal investigators during each time period	Uses of data[a]
Lomas Barbudal	5–8/90	337	SP	S[b], F, I
Abby's group	5–12/91	619	SP	S[b], F, I[b]
	1–12/92	1850	SP	S[b], F, I[b]
	1–5/93	1234	SP	S[b], F, I[b]
	2/94	72	SP, JM	S[b], F, I
	7–8/95	282	SP, JM	S[b], F, I
	12/96	48	SP, JM	S[b], F, I
	1–8/97	914	SP, JM, JGL	S[b], F, I
	2–5/98	381	JGL	S[b], F, I
	1–7/99	356	SP, JM, JGL	S[b], F, I
	1–6/00	372	JGL	S[b], F, I
	1–6/01	784	SP, JM	S[b], F[b], I
Rambo's group	1–8/97	964	SP, JGL	S[b], F, I
	1–5/98	315	JGL	S[b], F, I
	1–7/99	759	SP, JGL, JM	S[b], F, I
	1–5/00	542	JGL	S[b], F, I
	1–6/01	655	SP, JM	S[b], F[b], I
Santa Rosa	1–6/86	69	LMF	S[b]
Sendero	5–9/92	10	KM	S[b], F, I
	1–4/93	35	KM	S[b], F, I
Cerco de piedra	1–6/86	123	LMF	S[b]
	1–7/91	285	LR	S[b], F, I[b]
	5–9/92	150	KM	S[b], F, I
	1–4/93	120	KM	S[b], F, I
	1–9/95	405	LR	S[b], F, I[b]
	12/95–8/96	327	LR	S[b], F, I[b]
	1–12/98	770	KM, KJ	S[b], F, I[b]
	1–4/99	168	KJ	S[b], I[b]
Los Valles	1–7/91	260	LR	S[b], F, I[b]
	5–9/92	170	KM	S[b], F, I
	1–4/93	200	KM	S[b], F, I
	1–9/95	656	LR	S[b], F, I[b]
	12/95–8/96	341	LR	S[b], F, I[b]
	1–12/98	1332	KM, KJ	S[b], F, I[b]
	1–4/99	204	KJ	S[b], F, I[b]
Nancite	12/95–6/9	408	LR	S[b], F, I[b]
Cuajiniquil	2–12/98	264	KJ	S[b], I[b]
	1–2/99	1	KJ	S[b], I[b]
Cafetal	3–4/99	56	KJ	S[b], I
Bosque Humido	2–12/98	588	KJ	S[b], I[b]
(BH)	1–4/99	240	KJ	S[b], I[b]
Palo Verde	4–12/95	852	MP	S[b], F[b], I[b]
Station troop	1/96	36	MP	S[b], F[b], I[b]
Water hole troop	3/95	84	MP	F, I[b]
Lagoon Troop	3–7/95	228	MP	F, I[b]

Table 14.1. (*cont.*)

Site and study group	Time period (months/year)	Observation (h)	Principal investigators during each time period	Uses of data[a]
Curú	8–9/91	189	MB	S[b], F
Bette's group	1–6/93	692.5	MB	S[b], F
	7–8/94	147.5	MB	S[b], F
	1–4,6–9/95	665	MB	S[b], F
	7–9/96	226	MB	S[b],F

[a] Uses of data: S, social conventions; F, food processing; I, interspecific interactions
[b] Used for quantitative analyses presented or summarized in this chapter; otherwise, these data were used as a source of descriptions and anecdotes only.

of the data were used for analysis of food processing and interspecific interactions, as described below.

The LB researchers (Perry, Manson, Gros-Louis, and Pyle) all studied social behavior. Their methods consisted of focal animal follows (hereafter, follows), during which all-occurrence sampling of social behavior involving a single focal animal was noted along with 2.5-minute scan samples of activity (including foraging behaviors) and proximity of other group members to the focal animal. The length of focal samples was typically 10 minutes, though the standard sampling protocol was supplemented by 4-hour dyad follows in 1997 (Perry) and all-day focal animal follows in 2001 (Perry and Manson). Extensive *ad libitum* data were collected in all years of the study. Adults as well as juveniles were focal subjects at LB during most years. All of the data from PV were collected by Panger, who focused on object manipulation and handedness in all age–sex classes and collected her data primarily in the form of 10-minute follows and *ad libitum* data. Baker studied fur rubbing in members of all age–sex classes and provided all of the observations for Curú. Most of the data included from her study come from *ad libitum* observations. The following researchers all collected data on some aspect of social behavior at SR, though they varied as to the age–sex classes studied: Fedigan (adults), Rose (adults and subadults), Jack (adult and juvenile males), and MacKinnon (primarily infants and juveniles, but also some data on adults). Most data were in the form of 10- or 15-minute follows supplemented by *ad libitum* data and scans. (Further details about the data set are provided in Perry *et al.*, 2003.)

14.3 The study animal

White-faced capuchin monkeys live in relatively stable, female philopatric social groups. The closest bonds (measured by proximity and grooming frequencies) are typically among female–female dyads (Perry, 1996a; Rose 1998). The alpha male is highly central and has much closer relationships with females than do subordinate males (Perry, 1997), but subordinate males do regularly associate with other group members, particularly with juveniles (Perry, 1998b). Capuchins are exceedingly tolerant of close-range observation and begging during foraging (Perry and Rose, 1994). When resting or foraging on fruit, it is fairly common for the group to be compact enough that most group members can be seen from a single vantage point, at least during the dry season. However, during foraging for insects or travelling, the group is often widely dispersed such that only a few monkeys can be seen at any one time.

Group size is about 18 animals at SR and 19 at PV, though group size for the two study groups at LB has ranged from 20 to 37 (Fedigan *et al.*, 1996; Panger, 1997; S. Perry, unpublished data). Sex ratio at SR and PV is 1 male to 1.3 females and 1.2 females, respectively (Fedigan *et al.*, 1996; Panger, 1997). At LB, sex ratio is closer to one male to two females, and immatures constitute about 55% of the population (S. Perry, unpublished data). Consequently, although group sizes were a little larger and groups contained more females in LB than at SR and PV, the slight differences in demography between the three main sites seem insufficient to explain the intersite behavioral differences noted below.

14.4 Social conventions

14.4.1 Operational definitions

Social conventions are dyadic social behaviors of a communicative nature that are shared among members of particular social networks. Although much work has been done on vocal traditions in birds and marine mammals (see Ch. 8), and many of the geographically distinct communication patterns in birds might well be termed "social conventions", there is surprisingly little in the primate literature about social conventions. Some noteworthy exceptions include unique grooming styles in Japanese macaques (Tanaka, 1995, 1998) and chimpanzees (Boesch, 1996a,b; de Waal and Seres, 1997; McGrew and Tutin, 1978; Nakamura *et al.*, 2000; Whiten *et al.*, 1999), such as social scratching, hand-clasp grooming, and leaf grooming, which are found only in particular social networks or sites. Some social conventions, such as leaf clipping (Boesch, 1996a,b), are exhibited at multiple sites in identical form but are used to convey different meanings at different sites. Other social conventions are different in their form yet are apparently used to convey the same meaning (e.g. leaf clipping and knuckle knocking are both used by chimpanzees at different sites in the context of courtship (Boesch, 1996b)).

Most communicative signals in primates are standard elements of the species-typical behavioral repertoire. Although a certain amount of social influence may be necessary to facilitate a juvenile primate's proper contextual usage of, and response to, particular signals, the production of these signals is relatively inflexible developmentally (Seyfarth and Cheney, 1997; see also Ch. 8). We were interested not so much in documenting the ontogeny of species-typical signals but in documenting the innovation and subsequent acquisition by new practitioners of signals that are not part of the species-typical repertoire. As explained in Ch. 1, we defined a behavioral tradition as a behavioral practice that is (a) relatively long lasting, (b) shared among members of a group, and (c) aided to a measurable degree by social context for the generation of the practice in new individuals. We imposed some additional criteria so as to be conservative in our assessments of the likelihood that these behaviors are traditions (see Perry *et al.* (2003) for discussion of the rationale for these criteria).

1. The trait in question must exhibit some intergroup variation: that is, be present in some groups and absent in others. To qualify as unequivocally present in a particular group, the behavior has to have been seen at a rate of at least once per 100 hours of observation, and it must have been performed by at least three different individuals. To qualify as absent in a particular group, the behavior must never have been seen, and the observer must have logged at least 250 hours of observation.
2. The trait in question must exhibit some within-group variation, and there must be an increase over time in the number of performers of the behavior. Whenever possible, we tried to document more than two links in a social transmission chain (i.e., when B acquires a behavior "from" A, that is one link; when C acquires the same behavior "from" B, that is a second link), but gaps in observation did not permit the reliable construction of social transmission chains at all sites.
3. The behavior must endure, spanning at least six months within a particular group.

Using an electronic network, we invited capuchin researchers to submit behaviors that they considered to be likely candidates for a behavioral tradition. Many behaviors were quickly dismissed from our analysis because they were observed in a single individual or because they were exhibited universally (thus making it difficult to determine social contribution to their acquisition). The following behaviors remained as likely candidates for behavioral traditions: hand sniffing, sucking of body parts, and "games" (see below for definitions, and Perry *et al.* (2003) for further details). Each potential tradition is discussed in turn, and further details for all of these behavioral patterns are provided in Perry *et al.* (2003).

14.4.2 The behaviors

14.4.2.1 Hand sniffing

In "hand sniffing", one monkey inserts his/her fingers up the other's nose or cups his/her hand over the nose and mouth of the other monkey. This behavior is often performed mutually, with each monkey inserting his/her fingers in or over the mouth of the other. The behavior can be initiated either by placing one's own hand on the partner's face, or by seizing the partner's hand and placing it on one's own nose. Hand sniffing can last for several minutes at a time, and the participants have a trance-like expression on their faces while performing it. Hand sniffing qualified as a likely behavioral tradition according to our criteria: (a) social context apparently contributed to the generation of the practice in new individuals; it was common in five groups, clearly absent in three groups (and one site), and appeared not to be universal at any site (Table 14.2); (b) it was possible to document an increase in the number of performers for two groups; and (c) the behavior was durable for six groups. Even at sites where it reached high frequencies, it did not remain a permanent part of the behavioral repertoire. For example, hand sniffing was common among female–female dyads at LB for a period of seven years and then disappeared from the repertoire when the most avid hand sniffer vanished from the group. In the Cerco de Piedra (Ce) group at SR, hand sniffing was common among male–male dyads in 1986 and vanished from the repertoire for a period of several years (approximately a decade) before reappearing primarily among male–female dyads. Hand sniffing was statistically associated with grooming in female–female dyads at LB (Perry, 1996a). At PV, dyads that hand sniffed spent more time in close proximity

Table 14.2. *Distribution of social conventions across study sites*

	Hand sniffing	Sucking	Finger game	Hair game	Toy game
Santa Rose					
Sendero	++	?	?	?	?
Cerco de Piedras	++	–	–	–	–
Los Valles	(+)	(+)	–	–	–
Nancite	–	–	–	–	–
Cafetal	?	++	?	?	?
Cuajiniquil	++	++	?	?	?
Bosque Humido	–	(+)	–	–	–
Lomas Barbudal					
Abby's group	++	++	++	++	++
Rambo's group	(+)	++	–	–	–
Palo Verde					
Station troop	++	(+)	–	–	–
Curú					
Bette's group	–	(+)	–	–	++

++, behavior common; (+), behavior seen extremely rarely; –, behavior never seen in over 250 hours of observation; ?, behavior not seen but data are inadequate to be confident of its absence.

than did dyads that never hand sniffed (Perry *et al.*, 2003). Although hand sniffing tended to be associated with particular age–sex classes within each social group, there was no consistency across social groups regarding which age–sex exhibited the behavior most predominantly. For example, hand sniffing was almost exclusively a female–female behavior in one group at LB, whereas it was seen primarily among male–male dyads at SR (with the exception of Ce group after 1996) and primarily among male–female dyads at PV and in SR's Ce group (post-1996). Further details about hand sniffing (intersite variation in form, temporal distribution, and distribution across social networks) are provided in Perry *et al.* (2003).

14.4.2.2 Sucking of body parts

In some groups, particular dyads sucked on one another's fingers, toes, ears, or tails for prolonged periods of time. Sucking, like hand sniffing, occurred during periods of relaxed socializing, such as grooming or resting in contact, when the pair was fairly isolated from other group members. Sucking was particularly common in one group (Rambo's group) at LB, in which monkeys often mutually sucked one another's body parts, sometimes for over an hour at a time. Over half of all observations at LB involved mutual sucking, and the behavior occurred in male–male and male–female dyads. Although 13 different individuals were seen to engage in sucking (11 of them taking an active sucking role), 88% of observations included a single young adult male. The behavior has virtually vanished since his disappearance. Some tail sucking was observed involving a male–male dyad in the neighboring group, but they performed this behavior infrequently. The only other site at which sucking of body parts was common was SR, where one male routinely sucked the fingers of his closest male associate; these two males migrated together from group to group. Sucking met our third criterion for being a social tradition (i.e., it was found in some sites and groups, but not others, and it was durable). However, it was difficult to document social contribution to acquisition except on the basis of its distribution, and it was difficult to document expansion in the number of performers (critierion 2) because we were not sure when the behavior entered the repertoire or how the acquisitions of the behavior by individuals coincided with the timing of field seasons. Most of the data on sucking at LB were collected during the first seven months for which behavioral data had been collected on Rambo's group, and we could not know whether we were seeing the first occurrences of sucking for any particular dyad.

14.4.2.3 Games

Three of the behaviors observed that were candidates for traditions were quite similar in their form and social context, and we termed them "games" because they were often initiated in a play context. Unlike rough-and-tumble play, they were of a quiet, relaxed nature and tended to occur when the two game partners were relatively isolated from the rest of the group. Grooming of the face or slow motion wrestling often preceded these games, and the two partners maintained a quiet focus on one another that is fairly unusual for capuchins. All three games involve two partners trying to extract something from one another's mouths. Another element they have in common is that there is frequently turn taking, with the partners switching roles repeatedly during a bout of game playing. Partner A will hold the prized object (partner B's finger or hair, or an inanimate object) tightly in his/her mouth, while the partner B uses hands, feet, and mouth to try to pry open A's mouth and retrieve the object. Once B has succeeded in prying the object from A's mouth, he either reinserts it to begin the game anew, or the monkeys switch roles. There were three basic variants of the game. In the "finger-in-mouth" (FIM) game, one partner

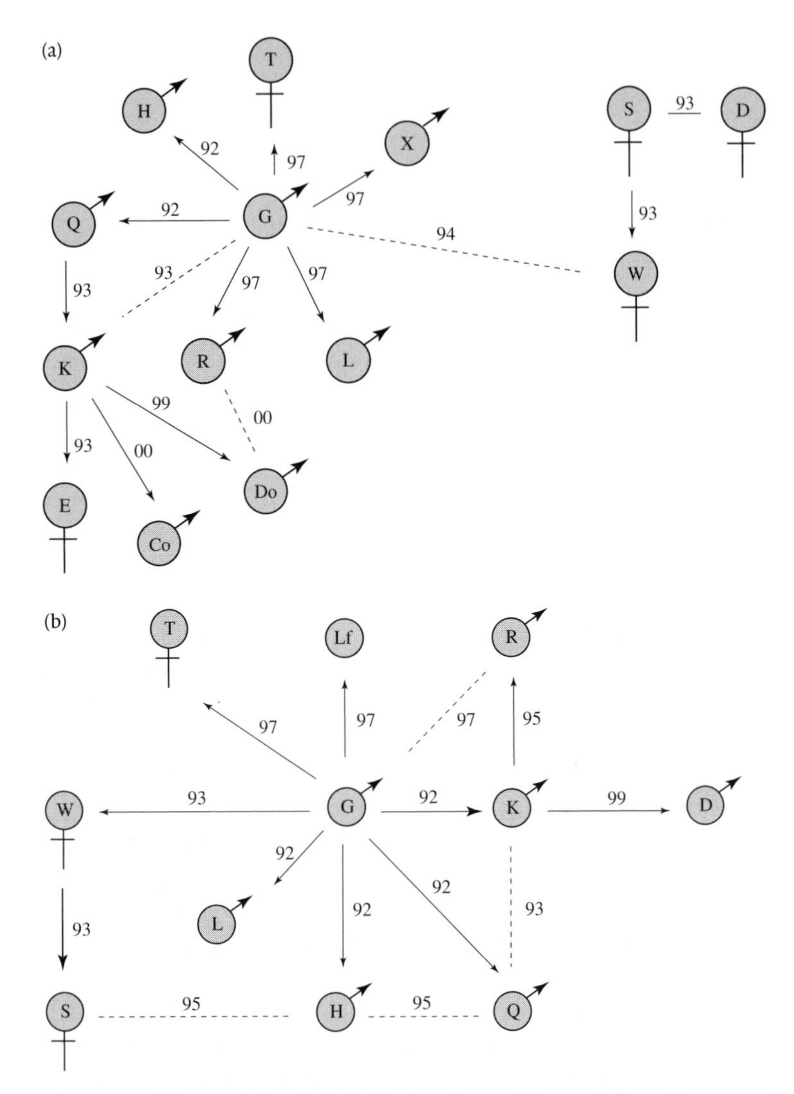

Fig. 14.2. Social transmission chain for the three "games". Arrows indicate the presumed direction of social transmission. The letters inside the male/female symbols indicate the names of the individuals. Solid arrows indicate those dyads in which one member has never previously been seen to play with other partners. Dotted lines connect dyads in which both members have previously played with other partners. Numbers indicate the year in which the game was first played by that dyad. (a) Finger-in-mouth game; (b) hair game; and (c) toy game. (Reprinted from Perry *et al.* (2003) with permission of the publisher, University of Chicago Press.)

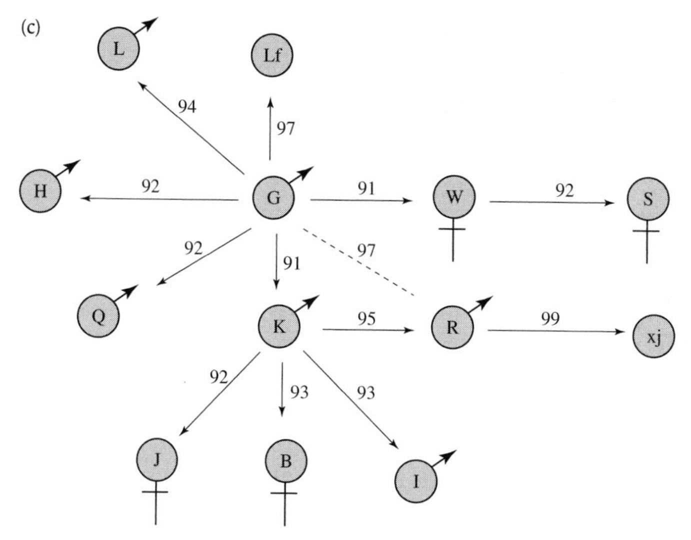

Fig. 14.2. (*cont.*)

bites down hard enough on the other's finger that it is quite difficult to remove but is not damaged. In the "hair" game, one monkey bites a large tuft of hair out of the face or shoulder of the other, and they forcibly pass the hair from mouth to mouth until it has all fallen out, at which time one of them bites another tuft of hair from the partner's shoulder. Although both the finger biting and hair biting look extremely uncomfortable, the animals apparently enjoy the activity enough to volunteer for another round of the game. In the "toy" game, the monkeys play with some object – a twig, green or otherwise inedible fruit, a piece of bark, or a leaf, for example – passing it back and forth until it is too mangled to use. No one eats the toy.

These games were observed exclusively in Abby's group at LB, with the exception of the toy game, which was also played in virtually identical form at Curú. Circumstantial evidence suggests that all three games at LB were invented by a single individual, Guapo, who was a subordinate young adult male at the time. Guapo was always the first player observed to play these games, and also the most frequent player, at least in the early years of observation. After becoming alpha male in 1999, he ceased to play games, but the games continued to be played by other monkeys (albeit at a lower rate) after Guapo ceased to play. It was possible to create social transmission chains for all three games at LB (Fig. 14.2). At Curú, however, the toy game was already widespread when Baker's observations began (i.e., approximately half of all group members played it), and so it was impossible to document the social transmission process. At Curú, only same-sexed adults played the game, though both sexes played with juveniles, who may have been responsible for transmitting the behavior from one sex to the other. The arrows in Fig. 14.3 show the probable transmission of the behavior from one individual to another at LB. Dotted lines connect those individuals who played the game with one another but who had previously played it with someone else. The males who acquired games from Guapo did so as juveniles, while most of the females learned the behavior as adults. The most avid players were dyads including one adult male (usually Guapo) and one juvenile male to our knowledge, most game players were not matrilineal kin. All three games qualified as traditions according to our criteria. (a) Social context contributed to their acquisition. (b) We were able to document social transmission in detail at LB: there were two documented links in the FIM game transmission chain, and three in the hair and toy games. (c) These games were highly durable, lasting for 10 years for the FIM game, 10 years for the hair game (which was still being played in 2001 by two male emigrants from Abby's group who are currently residing in an all-male group), and 9 years for the toy game.

14.4.3 Explaining the geographic and temporal patterning of social conventions

The geographic and temporal patterning of the observed traditions (Table 14.2) is puzzling in many ways. First, bizarre behaviors such as hand sniffing and the toy game spring up at multiple sites that are too far apart to permit migration between sites. The most likely explanation is that these behaviors were independently invented at multiple sites. As Huffman and Hirata point out (Ch. 10), each species has particular perceptual biases and a finite set of movement patterns that make them more likely to create certain types of innovation than others. They have a limited set of "building blocks" in their behavioral repertoire, which they can recombine to produce innovations. Many of the behavioral elements that are present in hand sniffing and in these games are elements that are borrowed from the foraging repertoire and are common motor actions in the monkeys. For example, capuchins frequently poke their fingers into small holes and crevices during foraging; therefore, it does not require much stretching of the imagination for them to insert fingers up their companions' nostrils or in their mouths. The task of removing fingers, toys, or hair from a partner's mouth also involves many of the skills that are typically practiced in the course of extractive foraging: prying, digging, probing, and pulling. Therefore, it is no surprise that the traditional social conventions found at different sites share many elements. The tricky part of creating new social conventions is not so much stringing together some "building blocks" from other parts of the behavioral repertoire in a novel combination, but rather persuading other group members to cooperate in the performance of these new "rituals". For example, whereas it may be perfectly comfortable for the innovator to stick his/her fingers up someone else's nostril, it may seem surprising and uncomfortable to the recipient of these actions initially to feel someone's long fingernails in his/her delicate nasal passages. And it may also be difficult to induce the partner to learn both roles in the interaction for those behaviors (such as the games) that involve role reversals.

14.4.4 What is the function of social conventions?

Although the social conventions described above show convergences with regard to their form, they may not share common function (as in the chimpanzee example in which leaf clipping is used for different purposes at different sites: Boesch, 1996b). However, all of the traditional social conventions described do have some noteworthy elements in common, which are suggestive of a common function: (a) these behaviors are performed in relaxed social contexts (grooming or slow play) by dyads that are fairly isolated from the rest of the group: (b) the monkeys slow down and concentrate on the activity, with trance-like expressions on their faces, for long periods of time, which is a striking deviation from the rest of the capuchin's daily routine; and (c) all of these behaviors involve a certain amount of risk or discomfort. Hand sniffers risk laceration of their nostrils from fingernails and also have their movement severely restricted. The hair game involves the presumably painful removal of large tufts of hair. The FIM game and the sucking conventions involve the insertion of body parts between the sharp teeth of another monkey. Since capuchins routinely lose digits and tail tips in bite injuries, it is safe to assume that a monkey would not voluntarily insert a finger or tail in another monkey's mouth unless she/he trusted that individual. The high level of risk involved in these conventions suggests that they may be ways of testing the bonds between individuals (Zahavi, 1977; see also Smuts and Watanabe, 1990).

Zahavi (1977) proposed that stressful stimuli are ideally suited for the testing of bonds. Such a signal would have a strong sensual component, such that it would be perceived as pleasurable if the relationship is on solid ground, but aversive if the relationship was not good. A French kiss is a good example of such a signal in humans. Unlike most signals, the important information is not contained in the signal itself but in the recipient's response to the signal. For example, if A sticks a finger up B's nose and B responds positively, this could indicate to A that B is positively disposed to A and likely to be supportive in the near future. If B responds apathetically or negatively to having A's fingers up her nose, this could tell A that B is not positively disposed to her and is unlikely to be a reliable source of aid in the near future.

Another body of theory relevant to explaining the function of traditional social conventions is Collins' theory regarding "interaction rituals", which comes from microsociology and was originally designed primarily to explain the role of human conversations in building up social structures (Collins, 1981, 1993). Collins assumes that, because humans cannot assess their exact positions in the power structure of the group, they use the emotional tones generated from conversations as cues to the way conversational partners value them relative to other people. The exact content of these conversations is virtually irrelevant compared with the affective displays of the participants. It is the enthusiasm, coordination or agreement, and engagement of the partners during the interaction that informs partners about their willingness to support one another in the future. In capuchins, therefore, it may not matter whether the task at hand is extracting a stick from someone else's mouth or inserting fingers in one another's noses. What is important is that both monkeys agree on what is to be done and who is to play which role, and that they focus deeply on this task for a long period of time, coordinating their movements.

14.4.5 Design features of social conventions

What design features in a signal would be optimal for providing information regarding emotional engagement and ability to cooperate? If the adaptive problem is the design of an interaction ritual that challenges the animals' abilities to coordinate their actions and to understand one another's behavioral goals (thus forcing them to devote their full attention to the social partner), then the following features seem desirable for a bond-testing behavior: (a) complex behavioral sequences, rather than simple ones; (b) turn taking and/or role reversals; and (c) flexibility and individually idiosyncratic forms, such that partners will need to familiarize themselves with one another's quirks and adjust their own behavior to produce a mutually satisfying interaction. It is important to note that such flexible bond-testing signals may be highly appropriate for eliciting information about the quality of relationships in dyads that already have a fairly comfortable relationship, because of the richness of the information the signals afford. They may not, however, always be superior to more stereotyped, species-universal signals. The use of such idiosyncratic signals would be too risky in the very earliest stages of relationship formation, when the two individuals have not yet sorted out their dominance ranks and do not know what to expect from one another or whether they can trust one another; in such cases, more stereotyped signals would be more appropriate for communicating about their relationship.

14.4.6 Stability of social conventions

Because the motor details of these traditional social conventions vary and are expected to "mutate" slightly as new practitioners are added, these social conventions are not expected to be highly stable in their form over long periods of time. Ontogenetic ritualization, the social learning process most likely to produce the social conventions described in this chapter, is a transmission process that affords low fidelity and, therefore, is unlikely to result in stable traditions spanning multiple generations (Boesch and Tomasello, 1998; Tomasello and Call, 1997, p. 309). Although the basic structure of the convention may stay the same for several links in a transmission chain, some of the fine details are expected to change, since part of the presumed adaptive value of this sort of signal is its malleability, which requires more focus on the part of the practitioners and hence provides more information about emotional engagement. Indeed, there is evidence in our data sets for interdyadic variation in the precise details of these rituals. Traditional social conventions are expected to dissolve when key members of social networks die or emigrate. If our hypothesis about the function of these social conventions is correct, then we might expect to find similar sorts of tradition in other species that have complex social relationships and signals that are apparently designed for communicating about their cooperative relationships.

14.5 Food processing

14.5.1 Methods and data

Because capuchins are so well known for their manipulative behavior and their skill at extractive foraging, we thought that we might find evidence for traditions among the wide range of techniques used for the processing of particularly hard-to-open fruits (see Ch. 13). Our first step in investigating food-processing techniques (begun in 1999) was to look for differences in the ways foods were prepared. We began by comparing the food lists for SR, LB, and PV to identify foods (both plant and animal) that were common to more than one site. Then we asked researchers to describe the techniques used by their monkeys to process each food on the list. Only one researcher (Panger) had collected detailed systematic data on food-processing techniques during focal follows. Perry and Rose did, however, collect *ad libitum* data on the typical patterns of food processing for each food type, even though they did not have data on each individual's processing style for each food. Gros-Louis, MacKinnon, and Baker also supplemented these data sets with additional observations. Once we had made a "short list" of foods for which we suspected extensive variation in processing technique, Perry (in 2001) and Gros-Louis (in 2000) went back to the field and collected more systematic observations on the range of techniques used by each individual for those foods we had identified as potentially interesting.

14.5.2 Intersite differences in processing

The results of this investigation of food processing are described in detail in Panger *et al.* (2002) and will be summarized briefly here. There were 49 plant and 12 animal foods that overlapped between at least two of the three main study sites. Of these, intersite variation in processing was noted for 20 foods, though in some cases only one individual at one site was observed to use the unusual technique. For 17 of the 20 foods that were processed differently at one or more sites, the difference consisted of the animals at one site pounding and/or rubbing the food, whereas animals at other sites declined to pound or rub those particular food species. Pounding (beating the food against a substrate with one or two hands), rubbing (sliding the food against a substrate while holding it in one or two hands), and tapping (rapid, rhythmic percussive contact of a finger tip against an object) are quite common elements in the standard behavioral repertoire of capuchins and, therefore, these do not represent particularly striking variants. However, six other processing variations involved less-common, and hence potentially more innovative, behavior patterns. The differences associated with two of the food species involved a behavior called "leaf wrap"; two others involved tapping with the fingertips, one involved fulcrum use, and one involved following army ant swarms.

The LB and SR capuchins sometimes wrap noxious *Automeris* caterpillars in leaves and then scrub them against a branch, removing urticating hairs before eating the caterpillar (i.e., "leaf wrap"). One SR monkey wrapped *Sloanea* fruits in leaves to scrub the irritating hairs off. The other striking intersite difference was army ant following. At Curú and SR, but not at PV or LB, monkeys actively followed columns of army ants as they foraged, capturing the insects flushed out by the ants. Army ants are quite common at LB and PV, and yet the ants are ignored by these monkeys.

Table 14.3 details the processing techniques that differed across sites, and the frequency of use of these patterns within groups in general terms, following Whiten *et al.* (1999) approximately. In Table 14.3, the category "Eat" means that the food was ingested, but no sophisticated processing technique was employed: the food was simply placed in the mouth, chewed, and swallowed. In all, there were a total of 40 different processing techniques reported for the three main study sites (since several of the 20 food species that showed food preparation differences across the sites were processed differently in more than one way). Of these 40 differences, five were exhibited by a single individual, 26 were exhibited by multiple individuals in one or more groups, and nine were exhibited by all members of at least one age–sex class in one or more groups.

14.5.3 The role of social learning in establishing individual differences in processing

It is premature to label particular food-processing behaviors as traditions in the absence of data indicating that social influence plays a role in the acquisition of particular foraging techniques. As argued by Fragaszy and Perry (Ch. 1), although it is impossible *directly* to establish a causal role for social learning in creating intersite variation and, in principle, to eliminate other potential factors such as the contribution of geographic variation in techniques, it should be possible to examine the role of social learning in establishing *within*-group differences in foraging techniques. This is the process model of traditions, as laid out in Ch. 1. Moreover, if social influence proves important for explaining within-group patterns of variation, then these behaviors can be identified as traditions. The inference is then stronger that social learning can be responsible at least in part for between-group variation in processing techniques, if one wants to explain variation at that level as well.

Table 14.3. *Processing techniques that vary across sites and their use patterns*

Food species	Lomas Barbudal		Palo Verde		Santa Rosa	
	Technique[a]	Use pattern	Technique[a]	Use pattern	Technique[a]	Use pattern
Acacia spp. (fruit)	Eat	C	NC		Rub	H
Acacia spp. (thorns)	Eat	C	Eat	C	Rub	H
Annona reticulata	Eat (rare)	H	Pound	H	Pound	H
			Rub	P	Rub	H
Apeiba tibouru	Rub	P	NC		Pound	H
					Rub	H
Bactris minor	Eat (rare)	C	Pound	H	NC	
Cecropia peltata	Eat	C	Eat	C	Pound	C
					Rub	C
Genipa americana	Eat	C	NC		Pound	C
					Rub	C
Mangifera indica	Pound	H	Rub	H	NC	
	Rub	H	Tap	H		
	Tap	H				
Manilkara chicle	Eat	C	Pound	P	Eat	H
Pithecellobium saman	Tap	P	Fulcrum	H	Rub	C
					Fulcrum	H
Quercus spp.	Eat	C	NC		Pound	H
Randia spp.	Pound	H	Pound	H	Eat	H
			Rub	P		
Sloanea terniflora	Rub	C	NC		Leaf wrap	P
					Rub	C

Stemmadenia donnell-smithii	Pound	P	Pound	H	Pound	C
					Rub	C
					Tap	C
Sterculia apetala	Rub (fruit inside of husk)	H	Rub (husk of fruit)	H	NC	
Tabebuia ochracea	Pound	H	Pound	H	Eat	H
	Rub	C	Rub	H		
Automeris spp. Caterpillar	Leaf wrap	H	Rub	H	Leaf wrap	H
	Rub	C			Rub	H
Insects in branches	Tap	C	Tap	C	Tap	H
			Pound	H		
Vertebrate prey (squirrels and coatis)	Pound	H	Eat (rare)	H	Pound	H
	Rub	H			Rub	H
Army ant following	No		No		Yes	H

C, customary (exhibited by all members of at least one age/sex class); H, habitual (not customary, but exhibited by multiple individuals); P, present (exhibited by one monkey); NC, food not consumed at this site.
[a] Eat is taken to mean that the food was ingested with no sophisticated processing technique.

In Panger's data set, for which there were detailed data on intragroup variation in processing techniques, we can examine the correspondence between proximity patterns and the distribution of processing techniques across group members. In many cases, dyads that shared a relatively rare foraging technique (i.e., the less common of two or more techniques used for the same food) also shared relatively high proximity scores (i.e., they spent a lot of time together). Statistical comparison was only possible for those foods for which more than one dyad shared a quirky processing technique. Four foods (*Annona reticulata, Mangifera indica, Randia* spp., and *Stemmadenia donnell-smithii*) met these criteria. In all four cases, the matched dyads (i.e., those sharing the odd technique) had higher proximity scores than the remaining dyads, and the difference was statistically significant in three cases.

The most difficult question to address in field studies is what role social learning plays in the production of shared practices. Clearly, many of the interindividual and intersite differences observed in this study could be attributable to individual experience independent of social influences. This is particularly true for variants such as pound or rub, in which the animal is merely utilizing a standard element of the behavioral repertoire in a slightly novel context. Slight differences in the physical properties of these fruits (e.g., harder rinds for fruits growing at some sites, as a result of differences in soil quality, or mature fruits being eaten rather than green ones, reflecting availability of alternative food sources influencing choice) could lead to intersite differences in the tendency to pound a particular species of fruit. However, the data from Panger *et al.* (2002) clearly showed that, even for relatively common processing techniques such as pound or rub, the distinctive technique is shared by those monkeys that spend more time together, which implicates social context as promoting shared usage.

14.6 Interspecific interactions

14.6.1 The datasets

Few studies have attempted to discern the role of social learning in interactions with allospecifics (members of other species), although it is certainly plausible that traditions could form in this behavioral domain. We examined datasets from three sites to assess the response of capuchins to vertebrates that could be classified as potential predators, potential prey, feeding competitors, or ecologically neutral. The data were drawn primarily from the following sources. Perry, Manson, and Gros-Louis systematically recorded responses to all allospecifics during 3703 hours of observation of one group at LB in 1991–1993; these observations were supplemented with anecdotes regarding rarely encountered animals from subsequent years and an additional monkey group from the dataset of Perry, Manson, and Gros-Louis. The majority of the SR observations came from Rose's observations (2682 hours of data from 1991 and 1995–1996) and Jack's observations (>1500 hours in 1997–1999) of five groups; K. MacKinnon contributed additional anecdotes. All of the PV data (>1200 observation hours of three groups) were contributed by M. Panger.

14.6.2 Potential prey

Capuchins were highly predatory at all three sites, though a wider range of prey was taken at SR and at LB than at PV (Rose *et al.*, 2003). There were some interesting differences between LB and SR in the ways that the monkeys interacted with adult squirrels, all of which are discussed in more detail in Rose *et al.* (2003). Adult squirrels were hunted at both sites, but hunting rates were higher at SR than at LB. LB monkeys encountered squirrels at lower rates than did SR monkeys, but they were more likely to hunt squirrels once they were encountered, and LB monkeys were more successful in their squirrel hunts than were SR monkeys. The most likely explanation for the difference in success rate between the two sites is that the LB monkeys have a different kill technique than do the SR monkeys. LB monkeys consistently kill squirrels with a rapid bite to the head or neck as soon as they catch them; in contrast, SR monkeys try to eat the squirrel before killing it, and squirrels are ultimately killed by a variety of inefficient techniques. SR monkeys also appear to engage in active search for squirrels (i.e., quiet, vigilant, stalking behavior without other forms of foraging in areas of high squirrel abundance), which is a hunting technique not observed at LB. Aside from these differences in squirrel hunting, there are no other striking differences between sites in predatory behavior that can plausibly be attributed to social learning. Although this is the only case in which a behavioral difference (e.g., squirrel neck biting) is homogeneous within a group, there are many other complex hunting techniques (e.g., drowning coati pups, baiting coati mothers off the nest, removing currassow eggs from under the mother via a hole created in the bottom of the nest) that are practiced by some but not all group members; these could conceivably be acquired at least in part via social learning. We currently lack the types of evidence necessary adequately to assess the role of social learning in establishing these variations in predatory behavior.

14.6.3 Potential predators

In general, there was strong intersite agreement about which animals were considered potential predators: boas, caiman, rattlesnakes, felids, canids, large raptors, and unfamiliar humans were consistently greeted with alarm calls and mobbing responses (Rose *et al.*, 2003). It is important to note that behavioral uniformity across individuals and sites does not necessarily imply that there is no social contribution to learning (see Chs. 5 and 6). If there is strong enough selective pressure for the monkeys to recognize a species as dangerous, they should all converge on a similar response to the species, for example alarm calling, fleeing, and/or mobbing from a safe distance. Given that it is very risky for young animals to discover the properties of potential predators via trial and error, it makes sense to rely heavily on social learning in this particular behavioral domain, at least in cases in which asocial cues are not necessarily reliable (Ch. 5). Even if fellow group mates are overly conservative in their assessment of the danger involved in interacting closely with predators, naïve animals will be better off using social cues than asocial cues as long as it is not too costly to exhibit an antipredator response or to give up the feeding opportunities entailed by avoiding the potential predator.

In many animals, juveniles' response to members of other species is influenced by adult responses. In vervets, for example (Cheney and Seyfarth, 1990), there are several different types of alarm call, including distinct calls for leopards, eagles, and pythons, and juveniles must learn the appropriate contexts for producing these calls. The appropriate response to hearing these calls is also learned: infants who had the opportunity to observe correct adult responses responded correctly to playbacks of alarm calls more frequently than did infants who did not look at adult models. There is some suggestive evidence that call usage is also influenced by the responses of adult models to calls (Cheney and Seyfarth, 1990). So far, the ontogeny of antipredator responses in capuchins has not been investigated in great detail for *C. capucinus*. Here we present some data on the demographic patterning of alarm call responses at LB, the only site for which we have detailed data on juveniles as well as adults (Table 14.4). One first step in demonstrating that social learning is taking place is to document a change in behavior developmentally; this we can accomplish in this chapter. The next step, which is more difficult, is to demonstrate that

Table 14.4. *Demographic patterning of alarm calls at Lomas Barbudal, 1991–1993.*

Alarm calls	Adult males ($n = 4$)	Adult females ($n = 5$)	Older juveniles[a] ($n = 7$)	Younger juveniles[b] ($n = 5$)
Calls/hour[c]	$\gg$0.70	0.44	0.28	0.30
Bouts/hour[d]	0.16	0.20	0.15	0.21
Calls/bout	$\gg$4.45	2.22	1.85	1.44
Percentage of bouts to harmless stimulii[e,f]	5.2	4.5	10.6	19.4
Percentage of bouts to dangerous referents[f,g]	31.7	20.2	29.5	14.7

[a]Juveniles at least 1.5 years of age and that have not yet reached reproductive age.
[b]Juveniles and infants < 18 months old.
[c]Number of calls per focal data hour, including multiple calls within a single alarm bout (i.e., response to a single stimulus). Note that adult male calls are underestimated because of frequent rapid calls within a single bout.
[d]Number of calling bouts (in response to a single stimulus) per focal data hour.
[e]Clearly directed toward harmless stimulus (e.g., nonpredatory birds, harmless snakes, primatologists, frogs, coatis).
[f]Some stimuli could not clearly be assigned to either the harmless or the dangerous category (e.g. medium-large birds).
[g]Clearly directed toward potentially dangerous referent (e.g., dogs, boa constrictors, felids, raptors, unfamiliar humans).

this change occurs as a consequence of social experience. We will propose some ways in which the role of social influence on the development of antipredator responses can be assessed in the field.

The data in Table 14.4 suggest that juveniles initially overgeneralize their alarm calls, frequently calling in response to harmless animals. Two findings stand out in particular. First, the proportion of alarm call bouts given in response to clearly harmless stimuli (e.g., nonpredatory birds, harmless snakes or even snakeskins, primatologists, frogs, coatis, and inanimate objects such as dolls) decreases with age, whereas the proportion of alarm calls given to dangerous animals (e.g., dogs, boa constrictors, rattlesnakes, felids, raptors, unfamiliar humans) increases, particularly between the younger juvenile (0–18 month) and older juvenile (18 month–puberty) stages. Rose *et al.* (2003) investigated the interactions of capuchins with mammals and found that only juveniles alarm call in response to agoutis and anteaters, and that 83% of alarm calls directed at coatis are by juveniles.

The second striking result emerging from Table 14.4 is that the rate of alarm calling increases with age, but the rate of alarm call bouts (i.e., responses to a particular stimulus) does not. This is the result of an age-related increase in the number of calls given per bout, which is more extreme for males than for females. One of the most striking aspects of antipredator behavior in capuchins is the amount of time and energy that the monkeys spend alarm calling and harassing predators, particularly snakes, when they could be devoting this time to foraging or other activities. Snake mobbings may prove to be a particularly productive context for the investigation of social learning. During a snake mobbing, the monkeys gather around the snake in response to hearing an initial alarm call by the discoverer. Each monkey looks for the snake, glancing back and forth between the alarm-calling monkey and the ground, as if following the gaze direction of the caller, until the snake is localized. Typically, the newcomers do not call until they have located the snake, but sometimes they give one or two tentative-sounding calls before locating the snake and begin calling with more insistence once they are looking in the right place to see the snake. Because boas and rattlesnakes do not move once they have been spotted, even when pelted with broken branches, they represent no danger to the monkeys once their location has been divulged to the rest of the group. Nonetheless, the monkeys remain at the site of a snake for extended periods of time, up to 45 minutes per session, calling repeatedly and menacing the snake from a safe distance. That the monkeys are willing to give up so much foraging time to stay near the snake is remarkable given the capuchins' highly constrained activity budgets (Terborgh, 1983, p. 49).

It is possible that lengthy snake-mobbing sessions are beneficial to young monkeys and their parents (whose reproductive success depends upon their offspring recognizing predators) because such sessions afford juveniles an opportunity to study the properties of snakes that elicit alarm or mobbing responses from adults and compare them with the properties of snakes that do not elicit concern from adults. Juvenile capuchins do occasionally mob nondangerous snakes. It might be worth considering the possibility that capuchins are engaging in teaching when they engage in prolonged snake-mobbing sessions. Caro and Hauser (1992, p. 153) define teaching as; "an individual actor A can be said to teach if it modifies its behavior only in the presence of a naive observer, B, at some cost or at least without obtaining an immediate benefit for itself. As a result, B acquires knowledge or learns a skill earlier in life or more rapidly or efficiently than it might otherwise do, or that it would not learn at all." It is possible that adult capuchins modify their behavior in the presence of juveniles and predators, by increasing call rate and lengthening mobbing duration in the presence of naïve animals, thus aiding naïve animals to learn how to identify predators and respond appropriately. Field experiments will help to resolve this issue.

Some naturalistic observations seem to suggest that adults do modify their behavior in such a way that they may aid the learner to learn (often termed "scaffolding", following Bruner, 1982; Wood, 1980). For example, L. M. Rose (unpublished data) reported the following intriguing anecdote from the Nancite group at SR, which is similar to instances that have been observed at LB. "In 1996, the group encountered and vigorously mobbed an unusually large boa (estimated at least 2 m long and thigh thickness) resting on the ground in a semi-clearing among mangrove roots. The mobbing monkeys had already begun to disperse when the beta male arrived on the scene. He alarm called briefly at the snake, scanned for a few moments, and then went to the edge of the clearing and picked up an infant on his back who had been hanging back from the activity. The male carried the infant to a branch well above the boa, stared at it, and again began to alarm call at it. He stayed in this position for about 5 minutes, during which time the infant also stared at the snake and alarm called at it." The male, in effect, aided the infant to inspect a dangerous predator and to practice the correct action from a safe position.

Another approach that can be taken to understanding the role of social learning in predator recognition is Dewar's cost–benefit model (Ch. 5). This approach could be useful in understanding, for example, the intersite differences in the reaction to indigo snakes that are reported in Rose *et al.* (2003). Indigo snakes (*Drymarchon corais*) are quite large (> 2 m) nonvenomous snakes that eat small vertebrates such as fish, turtles, lizards, frogs, rodents, and birds (Janzen, 1983). They have never been observed to prey on monkeys. At SR, monkeys of all age–sex classes routinely alarm call and harass indigo snakes. The LB monkeys do not alarm call in response to indigo snakes, but juveniles did harass them in six out of ten encounters. At PV, the monkeys do not exhibit any fear of indigo snakes. One adult female at PV was observed to drink from a waterhole while in physical contact with a large indigo snake that was also drinking (Rose *et al.*, 2003). These intersite differences are perplexing, but they might make more sense if we were able to clarify the relative values of social and asocial cues at the different sites. For example, the sites may differ with regard to variables such as the proportion of snakes encountered that are dangerous.

14.6.4 Potential feeding competitors

Dewar's approach (Ch. 5) might also illuminate intersite differences in capuchin decisions as to how to treat feeding competitors (such as coatis, howling monkeys, and spider monkeys) or species that are ecologically neutral. In the case of feeding competitors, the costs and benefits of ignoring the other species are different from the predator case. By ignoring the competitor when it is in a feeding tree, the capuchin may acquire less food, even if the capuchin is continuing to forage while ignoring the competitor. Most likely, asocial cues (such as whether the animal is currently feeding on a preferred food of capuchins) would be more useful than social cues for determining the optimal way to interact with allospecifics; but this is an empirical question. Rose *et al.* (2003) documented some fairly striking differences between SR and LB regarding interactions with howling monkeys. At LB, 80% of interactions between capuchins and howlers were aggressive, compared with 59% of interactions at SR. The LB capuchins were more vicious in their attacks on howlers, sometimes inflicting quite severe wounds on them. Almost all of the aggression was by capuchins against howlers, but howlers reciprocated in a few instances. To the extent that encounter rate between the two species can be used as a crude proxy for level of feeding competition, the higher levels of aggression at LB relative to SR cannot be attributed to greater feeding competition at LB (Rose *et al.*, 2003); however, encounter rate is a fairly crude means of measuring feeding competition. The SR monkeys exhibited affiliation (in the form of play) towards howlers in 10% of interactions, whereas the LB monkeys only exhibited affiliation towards howlers in 0.5% of their interactions with them. This result is more difficult to explain in terms of ecological costs and benefits to the monkeys. Even more striking is the tendency of some monkeys at SR to groom and otherwise affiliate with spider monkeys, a species not present at LB. With one exception, all 10 of the monkeys who groom spider monkeys had originally come from Los Valles group. It seems possible that the tendency to groom spider monkeys is a socially transmitted trait (Rose *et al.*, 2003).

14.6.5 Unfamiliar humans

Rose *et al.* (2003) also reported some interesting variation in the ways in which capuchins interact with unfamiliar humans. At PV, the monkeys rarely alarm called at humans, whereas the SR monkeys alarm called more frequently (25–75% of encounters), and the LB monkeys alarm called at humans the most frequently. Interestingly, however, the LB monkeys alarm called almost exclusively to local farmers, who were either traveling down the road bisecting the monkeys' home range or poaching in the reserve, and they virtually never alarm called at tourists. Although they often harassed tourists, they very rarely harassed farmers. Clearly the monkeys were using some asocial cues from the humans to decide whether they were dangerous or not. However, the monkeys may be using some social cues as well, either by observing how other monkeys interact with the unfamiliar humans, or by observing how the primatologists interact with them.

14.7 Longevity and biological significance of traditions

No doubt the longevity of traditions depends not only on the demographic characteristics of the population but also on the behavioral domain in question. For example, if the precise form of the behavioral trait is well suited to solving a particular and persistent problem, then it is likely that the tradition will be maintained with good fidelity for long periods of time. For example, antipredator responses to a predator or a particularly clever way of processing a desirable food are behaviors that would be expected to persist for generations. We do not yet know with absolute certainty that foraging or antipredator traditions exist in capuchins; nor do we know the durability of such putative traditions with any degree of certainty. We have better data on the longevity of traditions in the domain of social conventions. If our hypothesis about the function and design of these behaviors is correct (i.e., that the flexibility in form of such traditional social conventions is what makes them useful bond-testing signals), then it is no surprise that these traditions are short lived, lasting only a few years (approximately 7–10 years, though this may be a slightly low estimate because of censoring biases). Demographic considerations also affect the durability of traditional social conventions. Because capuchins live in small social groups, and because these conventions are practiced by only a subset of the group, the loss of one or two avid practitioners of a convention can cause the behavior to drop out of the repertoire entirely, even if many remaining group members know how to perform the behavior.

Another issue that is frequently raised is whether traditions lasting less than a generation time are biologically significant (e.g., Avital and

Jablonka, 2000; McGrew, 1992, p. 77, 1998; Whiten *et al.*, 1999). It is important to note that even short-lived traditions ("fads", as some would call them) can have fitness consequences for their practitioners. For example, let us assume for the moment that the traditional social conventions described in this chapter aid the monkeys enough in forming social bonds that they have positive fitness consequences. A monkey who has the capacity to develop these types of idiosyncratic bond-testing signal will have an advantage over monkeys who are less skilled at this form of social learning. Regardless of whether the particular traditions formed survive into the next generation, practicing this behavior during the monkey's own lifetime will have increased fitness.

The heuristic model of "traditions space" in Ch. 1 provides a useful framework for thinking about how the properties of traditions vary across behavioral domains in capuchins. Regarding social conventions, we can safely assume that social contribution is absolutely essential to the production of the behavior. Not all group members acquire the behavior: it is performed by members of cliques consisting of approximately 30–60% of all group members in most cases. We cannot accurately measure tradition duration in most cases, but they appear to last approximately 7–10 years (i.e., less than a generation time). For the behavioral domain of food processing, we have some limited evidence that social contribution is a factor affecting the distribution of foraging techniques within groups, at least for some food species; however, it may be the case that social contribution is fairly minimal. Our data on extractive foraging at LB indicate that there is heterogeneity within groups regarding processing techniques. For example, 29% of monkeys at LB scrub the *Luehea* fruits to extract the seeds, whereas the remaining 71% pound the fruits. Likewise, only about 77% of monkeys at LB tap branches when foraging for insects, and only 17% of LB monkeys rub the hairs off their panama fruits when foraging. At present, we have no data on the duration of food-processing traditions (in fact, we consider it premature to label these foraging variations traditions until we have more evidence regarding the social contribution to the acquisition of these behaviors). Likewise, we lack data on social context necessary to discern the social contribution to predator–prey interactions and thus label them as traditions, though these data can be collected in the future. There is extreme homogeneity both within and between groups in the reactions of capuchins to potential predators, and slightly less homogeneity in their responses to potential prey and feeding competitors. These patterns of behavior are stable over long periods of time (i.e., multiple generations), presumably because they are highly adaptive.

14.8 Conclusions

Many capuchin groups exhibit social conventions (e.g., hand sniffing, sucking of body parts, or games) that are specific to a large subset of a particular group. In some cases, it has been possible to document the social transmission process from the time of innovation until the "extinction" of the behavior. Some conventions sprang up in virtually identical form at multiple sites. Such social conventions are hypothesized to serve as tests of social bonds, and we speculate that the lability of these behaviors is a useful design feature for bond testing.

Numerous intersite and intragroup differences in food-processing techniques have been documented. In some cases, it is possible to demonstrate that those animals who spend the most time together also use the same foraging techniques, thus suggesting a role of social learning in the production of food-processing techniques.

There is considerable intersite homogeneity regarding the quality of the interactions of the monkeys with other species, particularly when they are interacting with potential predators. However, intergroup homogeneity does not necessarily imply lack of social contribution to the behavior pattern, and we propose some ways to assess the role of social learning in predator–prey interactions.

14.9 Acknowledgements

We are grateful to the Costa Rican National Park Service, the Area de Conservación Guanacaste (especially Roger Blanco Segura), the Area de Conservación Tempisque, the community of San Ramon de Bagaces, Hacienda Pelón, Rancho Jojoba/Brin D'Amor, and the Schutt family for permission to work in the areas occupied by these monkeys. Assistance in data collection was provided by Laura Sirot, Todd Bishop, Kathryn Atkins, Marvin Cedillos Amaya, Sarah Carnegie, Alisha Steele, Matthew Duffy, Maura Varley, Ryan Crocetto, Hannah Gilkenson, Jill Anderson, Craig Lamarsh, Sasha Gilmore, Dale Morris, Dusty Becker, three expeditions of Earthwatch volunteers, and seven groups of University Research Expedition Program volunteers. Susan Wofsy and Denise Alabart assisted in compiling the data. Barb Smuts, Joan Silk, Simon Reader, Mike Huffman,

Dorothy Fragaszy, Filippo Aureli, Eva Jablonka, and Andy Whiten commented on the manuscript, providing many helpful comments. Members of the University of California at Los Angeles (UCLA) Behavior, Evolution and Culture Group and the contributors to the *Traditions in Nonhuman Primates* conference at the University of Georgia at Athens provided useful comments as well. Joan Silk wrote the data collection program FOCOBS, used by many of the authors.

The UCLA Council on Research provided SP with funding for this project during the data analysis stage, and the Max Planck Institute for Evolutionary Anthropology provided funding during the write-up stage. Numerous other granting agencies inadvertently funded the fieldwork that gave rise to this project as well: SP thanks the National Science Foundation (NSF; for a graduate fellowship, an NSF–NATO postdoctoral fellowship, and a POWRE grant, SBR-9870429), the Leakey Foundation (three grants), the National Geographic Society, the University of Alberta (for an I.W. Killam Postdoctoral Fellowship), Sigma Xi, the University of Michigan Rackham Graduate School (four grants), the University of Michigan Alumnae Society, and UCLA (two Faculty Career Development grants). MB thanks the University of California–Riverside Graduate division (three grants), Earthwatch, UREP, and NIH-MIRT (grant to E. Rodriguez). LF's research is funded by an on-going grant (A7723) from the Natural Sciences and Engineering Research Council of Canada (NSERCC). JG thanks the University of Pennsylvania and NSF for a graduate fellowship and a dissertation improvement grant. KJ's research was supported by the National Geographic Society, NSERCC (postgraduate scholarship), Alberta Heritage Scholarship fund (Ralph Steinhauer Award), Ruggles–Gates fund for Biological Anthropology (Royal Anthropological Institute), Sigma-Xi, Faculty of Graduate Studies and Research/Department of Anthropology at the University of Alberta, and the above-mentioned NSERCC grant to LF. KM's fieldwork was supported by an NSF dissertation improvement grant (SBR-9732926) and a University of California–Berkeley Social Sciences and Humanities Research Grant. JM was supported by a UCLA Faculty Career Development Award. MP thanks the Costa Rican National Park System (park fee exemption grant), the Organization for Tropical Studies (especially M. Quesada and K. Stoner), and the University of California–Berkeley's Anthropology Department (three grants) for helping to make her research possible. LR was supported by an NSERCC postgraduate scholarship, NSF, National

Geographic Society, the L.S.B. Leakey Foundation, Sigma Xi, and Ammonite Ltd.

References

Avital, E., and Jablonka, E. 2000. *Animal Traditions: Behavioural Inheritance in Evolution.* Cambridge: Cambridge University Press.

Boesch, C. 1996a. The emergence of cultures among wild chimpanzees. In *Proceedings of The British Academy*, Vol. 88, *Evolution of Social Behaviour Patterns in Primates and Man*: ed. W. G. Runciman, J Maynard Smith, and R.I.M. Dunbar, pp. 251–268. Oxford: Oxford University Press.

Boesch, C. 1996b. Three approaches for assessing chimpanzee culture. In *Reaching into Thought: The Minds of the Great Apes*, ed. A.E. Russon, K. Bard, and S.T. Parker, pp. 404–429. Cambridge: Cambridge University Press.

Boesch, C., and Boesch-Achermann, H. 2000. *The Chimpanzees of the Taï Forest: Behavioural Ecology and Evolution.* Oxford: Oxford University Press.

Boesch, C., and Tomasello, M. 1998. Chimpanzee and human cultures. *Current Anthropology*, **39**, 591–614.

Bruner, J. 1982. The organization of action and the nature of the adult–infant transaction. In *Social Interchange in Infancy*, ed. E.Z. Tronick, pp. 23–35. Baltimore, MD: University Park Press.

Caro, T.M. and Hauser, M.D. 1992. Is there teaching in nonhuman animals? *Quarterly Review of Biology*, **67**, 151–174.

Chapman, C. and Fedigan, L. 1990. Dietary differences between neighboring *Cebus capucinus* groups: local traditions, food availability, or responses to food profitability? *Folia Primatologica*, **54**, 177–186.

Cheney, D.L. and Seyfarth, R.M. 1990. *How Monkeys See the World.* Chicago, IL: University of Chicago Press.

Collins, R. 1981. On the microfoundations of macrosociology. *American Journal of Sociology*, **86**, 984–1014.

Collins, R. 1993. Emotional energy as the common denominator of rational action. *Rationality and Society*, **5**, 203–230.

de Waal, F.B.M. and Seres, M. 1997. Propagation of handclasp grooming among captive chimpanzees. *Americal Journal of Primatology*, **43**, 339–346.

Emmons, L. 1997. *Neotropical Rainforest Mammals: A Field Guide*, 2nd edn. Chicago, IL: University of Chicago Press.

Fedigan, L.M., Rose, L.M., and Avila, R.M. 1996. See how they grow: tracking capuchin monkey populations in a regenerating Costa Rican dry forest. In *Adaptive Radiations of Neotropical Primates*, ed. M. Norconck, A. Rosenberger, and P. Garber, pp. 289–307. New York: Plenum Press.

Fragaszy, D., Visalbergh, E., and Fedigan, L. 2003. *The Complete Capuchin: The Behavioral Biology of the Genus* Cebus. Cambridge: Cambridge University Press, in press.

Frankie, G.W., Vinston, S.B., Newstrom, L.E., and Barthell, J.F. 1988. Nest site and habitat preferences of *Centris* bees in the Costa Rican dry forest. *Biotropica*, **20**, 301–310.

Freese, C.H. 1976. Food habits of white-faced capuchins *Cebus capucinus* (Primates: Cebidae) in Santa Rosa National Park, Costa Rica. *Brenesia*, **10/11**, 43–56.

Hartshorn, G.S. 1983. Plants. In *Costa Rican Natural History*, ed. D. Janzen, pp. 118–157. Chicago, IL: University of Chicago Press.

Itani, J. 1958. On the acquisition and propagation of a new food habit in the troop of Japanese monkeys at Takasakiyama. In *Japanese Monkeys: A Collection of Translations*, ed. K. Imanishi and S. Altmann, pp. 52–65. Edmonton: University of Alberta Press.

Janzen, D.H. (ed.) 1983. *Costa Rican Natural History.* Chicago, IL: Chicago University Press.

McGrew, W.C. 1992. *Chimpanzee Material Culture: Implications for Human Evolution.* Cambridge: Cambridge University Press.

McGrew, W.C. 1998. Culture in nonhuman primates? *Annual Review of Anthropology*, **27**, 301–328.

McGrew, W. C. and Tutin, C. 1978. Evidence for a social custom in wild chimpanzees? *Man*, **12**, 234–251.

Nakamura, M., McGrew, W.C., Marchant, L.F., and Nishida, T. 2000. Social scratch: another custom in wild chimpanzees? *Primates*, **41**, 237–248.

Panger, M. 1997. Hand preference and object-use in free-ranging white-faced capuchins (*Cebus capucinus*) in Costa Rica. PhD Thesis, University of California at Berkeley.

Panger, M.A. 1998. Object-use in free-ranging white-faced capuchins (*Cebus capucinus*) in Costa Rica. *American Journal of Physical Anthropology*, **106**, 311–321.

Panger, M., Perry, S., Rose, L., Gros-Louis, J., Vogel, E., MacKinnon, K.C., and Baker, M. 2002. Cross-site differences in the foraging behavior of white-faced capuchins (*Cebus capucinus*). *American Journal of Physical Anthropology*, **119**, 52–66.

Perry, S. 2003. Coalitionary aggression in white-faced capuchins. In *Animal Social Complexity: Intelligence, Culture, and Individualized Societies*, ed. F.B.M. de Waal and P. Tyack. Boston, MA: Harvard University Press, in press.

Perry, S. 1996a. Female–female social relationships in wild white-faced capuchin monkeys, *Cebus capucinus. American Journal of Primatology*, **40**, 167–182.

Perry, S. 1996b. Intergroup encounters in wild white-faced capuchins, *Cebus capucinus. International Journal of Primatology*, **17**, 309–330.

Perry, S. 1997. Male–female social relationships in wild white-faced capuchin monkeys, *Cebus capucinus. Behaviour*, **134**, 477–510.

Perry, S. 1998a. Male–male social relationships in wild white-faced capuchins, *Cebus capucinus. Behaviour*, **135**, 1–34.

Perry, S.E. 1998b. A case report of a male rank reversal in a group of wild white-faced capuchins (*Cebus capucinus*). *Primates*, **39**, 51–69.

Perry, S. and Rose, L.M. 1994. Begging and transfer of coati meat by white-faced capuchin monkeys, *Cebus capucinus. Primates*, **35**, 409–415.

Perry, S., Baker, M., Fedigan, L., Gros-Louis, J., Jack, K., MacKinnon, K.C., Manson, J.H., Panger, M., Pyle, K., and Rose, L. 2003. Social conventions in wild capuchin monkeys: evidence for behavioral traditions in a neotropical primate. *Current Anthropology*, in press.

Rose, L. 1994. Benefits and costs of resident males to females in white-faced capuchins, *Cebus capucinus. American Journal of Primatology*, **32**, 235–248.

Rose, L.M. 1998. Behavioral ecology of white-faced capuchins (*Cebus capucinus*) in Costa Rica. PhD Thesis, Washington University-St. Louis.

Rose, L.M., Perry, S., Panger, M., Jack, K., Manson, J., Gros-Louis, J., MacKinnon, K.C., and Vogel, E. 2003. Interspecific interactions between *Cebus capucinus* and other species at three Costa Rican sites. *International Journal of Primatology*, in press.

Seyfarth, R.M. and Cheney, D.L. 1997. Some general features of vocal development in nonhuman primates. In *Social Influences on Vocal Development,* ed. C.T. Snowdon and M. Hausberger, pp. 249–273. Cambridge: Cambridge University Press.

Smuts, B. B. and Watanabe, J.M. 1990. Social relationships and ritualized greetings in adult male baboons (*Papio cynocephalus anubis*). *International Journal of Primatology,* **11**, 147–172.

Tanaka, I. 1995. Matrilineal distribution of louse egg-handling techniques during grooming in free-ranging Japanese macaques. *American Journal of Physical Anthropology,* **98**, 197–201.

Tanaka, I. 1998. Social diffusion of modified louse egg-handling techniques during grooming in free-ranging Japanese macaques. *Animal Behaviour,* **56**, 1229–1236.

Terborgh, J. 1983. *Five New World Primates: A Study in Comparative Ecology.* Princeton, NJ: Princeton University Press.

Tomasello, M. and Call, J. 1997. *Primate Cognition.* Oxford: Oxford University Press.

van Schaik, C.P., Deaner, R.O., and Merrill, M.Y. 1999. The conditions for tool use in primates: implications for the evolution of material culture. *Journal of Human Evolution,* **36**, 719–741.

Westergaard, G. C. 1994. The subsistence technology of capuchins. *International Journal of Primatology,* **15**, 899–906.

Westergaard, G. C. and Suomi, S.J. 1995. The stone tools of capuchins (*Cebus apella*). *International Journal of Primatology,* **16**, 1017–1024.

Whiten, A., Goodall, J. , McGrew, W.C. , Nishida, T. , Reynolds, V. , Sugiyama, T., Tutin, C.E.G. , Wrangham, R.W.W., and Boesch, C. 1999. Cultures in chimpanzees. *Nature,* **399**, 682–685.

Wood, D.J. 1980. Teaching the young child: some relationships between social interaction, language and thought. In *The Social Foundations of Language and Thought,* ed. D.R. Olson, pp. 280–298. New York: Norton.

Zahavi, A. 1977. The testing of a bond. *Animal Behaviour,* **25**, 246–247.

SUSAN PERRY

15

Conclusions and research agendas

15.1 Current state of knowledge regarding the biology of traditions

15.1.1 What is the biological importance of social learning and traditions to the animals?

Recently, biologists have become aware that social learning may play a pivotal role in the behavioral biology and evolution of many animal species. Animals may alter their environments in such a way that they create new selective pressures for the next generation; in other words, they take an active role in shaping the environments that determine the course of their species' evolution (Avital and Jablonka, 2000; Laland, Odling-Smee and Feldman, 2000; Pulliam, 2000; see Ch. 12 for examples of ways in which nonhuman primates may construct their niches). Nevertheless, there are astonishingly few data, particularly from the field, regarding the prevalence of traditions in nature and the fitness consequences of engaging in social learning or practicing particular traditions. Consequently, modelers continue to rely heavily on thought experiments and hypothetical examples to convince readers of the logic of their arguments (e.g., Avital and Jablonka, 2000). Currently, there are very few species and behavioral domains for which the topic of traditions has been thoroughly addressed (i.e., with adequate methodology to assess the role of social learning) in the wild, and we know little about the biological importance of social learning in nature. There are, no doubt, taxonomic biases regarding which species and topics have been targeted for study (e.g., biologists regularly look for tool use in primates and vocal traditions in birds). In this volume, we do not attempt the same taxonomic breadth covered by Box and Gibson (1999) or by Avital and Jablonka (2000), but rather we focus primarily on a few taxa in which researchers have tailored their methodologies specifically to examine social learning and traditions.

The chapters in this volume provide evidence of the utility of social learning in a variety of behavioral domains. Reader's comparative analysis of brain size and the frequency of reported incidence of social learning (Ch. 3) suggests that the ability to make use of social context during learning may have been instrumental in the evolution of intelligence in primates. But under what circumstances is social learning most adaptive? In which behavioral domains do traditions most frequently occur? Several of the authors in this volume suggest (with varying degrees of certainty regarding the amount of social influence necessary to produce the trait) that social learning is useful in aiding animals to exploit foods that are difficult to process (Chs. 9, 12–14). In some species, social influence may also be important in making decisions about which items to incorporate into the diet (rats: Ch. 6; capuchin monkeys: Ch. 7) or which plants have medicinal value (chimpanzees: Ch. 10). Similarly, conspecifics may guide one another's choices about where to look for food. Knowledge regarding the use of space and fixed resources (e.g., travel and migration routes, nesting sites, foraging sites, or food storage sites) is known to be transmitted socially in some species: for example, in two species of fishes (guppies and French grunts), individuals learn travel routes from one another as a consequence of shoaling (Helfman and Schultz, 1984; Laland and Williams, 1997).

Social cues may also be particularly important in aiding animals to learn things that are too dangerous to learn via individual experimentation, such as predator recognition (Avital and Jablonka, 2000; Chs. 2, 5, and 14). For example, fathead minnows and brook stickleback learn to fear northern pike when they are paired with minnows who have prior experience with pike (Mathis, Chivers and Smith, 1996), and naïve fish may acquire this antipredator response quickly in the presence of experienced minnows (see also research on blackbirds by Vieth, Curio and Ernst (1980)).

Vocal traditions can help to define groups (e.g., Boughmann, 1998), thus identifying who are appropriate social or mating partners (Ch. 8). Call-matching is useful for assessment of rivals and signaling of aggressive intent in some songbirds (Bradbury and Vehrencamp, 1999; Greenfield, 1994; Vehrencamp, 2001); use of call-matching in territorial encounters results in local dialects. Matching or coordination of particular signals may also be useful in eliciting information about the quality of a social bond. The performance of dyadic gestural social conventions in capuchins occurs among dyads that have relaxed social relationships, and these rituals seem to provide information about the degree of commitment and tolerance between partners (Ch. 14).

Although much progress has been made in attributing function to particular traditional behaviors, there are, of course, some traditional behaviors for which researchers cannot discern a function (e.g., stone handling in Japanese macaques: Ch. 10), and for which there may, in fact, be no particular adaptive value to the animals. Even when the function is known for a particular behavior, it is difficult if not impossible in most cases to know the degree to which performance of a traditional behavior enhances or detracts from an organism's fitness. An important challenge for the future is to figure out how to demonstrate fitness differences between socially acquired variants and individually acquired variants of behaviors. Such information would be useful when trying to assess whether traditions are contributing to a niche-construction process (Laland *et al.*, 2000).

15.1.2 How does variation promote traditions?

Variation across time and across individuals is an intrinsic property of behavior, reflecting constitutional differences (e.g., size, strength, dexterity) and psychological differences (e.g., motivation). Behavioral variation across time and individuals is the wellspring of adaptive behavior within a species, as biologists have recognized since Darwin's time. Behavioral variations can support the exploitation of new resources and adaptation to altered conditions (an idea going at least as far back as Baldwin (1896)). In addition to its direct consequences for the individuals involved, the propensity to produce behavioral variation is hypothesized to associate positively with rates of speciation (Wilson, 1985). There is some evidence linking the propensity to vary behaviorally with brain size (see Chs. 3 and 4).

As Huffman and Hirata point out (Ch. 10), the types of behavioral variation possible are limited to some extent by the number and types of basic element (e.g., possible motor patterns) in the species-typical behavioral repertoire. These basic elements can be combined in novel ways to create "innovations" (novel behavioral variants). The more basic elements there are in the species-typical repertoire, the more behavioral variants will be possible. As we have argued in Ch. 1, and contributors to this volume illustrate repeatedly, the uneven distribution of variations across individuals, and their uneven acquisition by new practitioners in a group, underlie the creation of traditions.

Evolutionary biologists are profoundly interested in the relation between variation in behavior and the occurrence of adaptations to altered conditions. The term "innovation" is widely used to emphasize that a behavioral variant seems to solve a problem in a way that was not recognized before (see Chs. 3 and 4). As observers looking at mere slices of time in a population's history, we cannot know for certain that any behavior we see is truly new (although we are most sure when the behavior concerns a newly produced human artefact, for example foil caps on milk bottles: Fisher and Hinde, 1949). Nevertheless, in principle, behavioral innovations are sure to arise, and those behaviors that we notice for the first time may be true innovations. Thus, we return to the core issues in the origins of traditions: what individual and social processes generate new combinations of basic behavioral elements, transmit them across social networks, and maintain them through time? Understanding the origins of traditions, whether they are based on unusual and conspicuous behaviors (those likely to be termed "innovations" by researchers) or on more common behavioral variants, requires an appreciation for the origins of behavioral variation. This appreciation can be garnered through careful study of the species involved (as illustrated in Ch. 10). With this background knowledge, we can focus on those kinds of variant deemed most unusual (e.g., "innovations").

15.1.3 What factors promote the spread of variations via social learning?

Two primary factors promoting social learning are gregariousness and tolerance (Coussi-Korbel and Fragaszy, 1995; van Schaik, Deaner, and Merrill, 1999; Ch. 11). Certainly for some behaviors, opportunities for learners to be near experienced individuals increase the chances that social context can support learning. This is obviously as true for common variants as for uncommon ones (those recognized as innovations). Tolerance is probably more important for some behavioral domains than others. For example, transmission of one of the social conventions described for capuchins (Ch. 14) requires physical contact between individuals, as well as a great deal of trust on the part of the participants. In contrast, one can learn to recognize predators through detecting alarm calls, mobbing, and flight responses, all conspicuous behaviors that can be perceived at a distance. Tolerance may contribute minimally to socially supported learning in these latter circumstances.

Boyd and Richerson (1985) and several other modelers (reviewed in Ch. 2) have demonstrated that social learning is expected to be most prevalent in environments with intermediate rates of environmental change. It is also likely to be more important in behavioral domains for which there is a great degree of similarity between model and learner regarding the way in which they are affected by the relevant environmental variables (Ch. 2). The amount of risk involved in individual experimentation could also affect the tendency to rely on social as opposed to asocial cues (Giraldeau and Caraco, 2000; Ch. 5).

15.1.4 What factors affect the fidelity and longevity of a tradition?

There is expected to be greater selection for transmission fidelity in some behavioral domains than others. For example, in the domain of predator recognition, there could be quite high costs to "drift" if learners fail to focus on the precise characteristics that are necessary to identify an animal correctly as a predator, such that they fail to exhibit antipredator behavior. In contrast, consider the social conventions described for capuchins (assuming for the moment that the authors' adaptive explanation is correct). In this case, a certain amount of flexibility in form is actually desirable, to a point, because the time and focus required by both interactants to detect new modifications of the ritual is what gives the interaction its signaling value.

Note that the fidelity of a tradition appears to be independent of the learning mechanisms that support its maintenance. For example, Galef (Ch. 6) notes that feeding traditions in rats have quite impressive durations and fidelity in the absence of the more complex transmission mechanisms such as imitation. Research on the types of transmission mechanism present in various species and also on the qualities of social traditions possible given particular transmission mechanisms would be useful to evolutionary biologists and anthropologists interested in explaining the origins of human culture (e.g., Boyd and Richerson, 1996) and the patterning of traditions across the animal kingdom generally.

Obviously, demographic factors will also have an impact on the duration, geographic spread, and rates of spread of traditions (Chs. 10 and 12). It seems probable that traditions will last longer if they form in large groups and are practiced by a large subset of the group, such that there is less chance of the behavior being extinguished owing to the disappearance of one or two individuals (Ch. 14). However, it may be the case that group size and number of practitioners is a factor in longevity only below a certain (fairly low) threshold (Ch. 10). As Russon points out (Ch. 12), it is important to consider not only the total numbers of models and learners but also the developmental stages of the potential models and learners, because group members have differential tolerance of different age–sex classes, which affects the opportunity for social transmission. Consequently, a more thorough understanding of how social interaction patterns shift over the course of the lifespan will be useful in predicting and interpreting patterns of social transmission across age–sex classes (see Chs. 10 and 12).

15.2 Methodological challenges and research agendas

15.2.1 Contributions of modelers, field researchers, and laboratory researchers to the study of traditions

As Laland and Kendal (Ch. 2) make clear, very few quantitative models pertaining to social learning are easily transported to the field, and hence few studies have attempted to test such models of social learning empirically. Dewar's model (Ch. 5) is a notable exception, and it is our hope that more modelers will make serious attempts to guide empirical researchers by proposing testable predictions and making suggestions as to how variables could be operationalized.

As noted in Ch. 1, it is important to realize that laboratory and field data provide different types of information, and different methodologies are feasible in the two research settings. Following are some examples of research tasks that are appropriate to particular research settings.

1. *Population dynamics of traditions.* Although longevity/durability can be examined in both field and captive settings, field settings give a better idea of the longevity of traditions in natural demographic conditions: those most likely to have been present during the evolutionary history of the organism. Field studies can also inform us about the likelihood of diffusion between social groups.
2. *Range of behavioral variation.* In order to understand the origins of traditions, it is necessary to understand not only how behavioral variants spread but also where they come from. It is critical to understand what types of circumstance or what characteristics of individuals make them more prone to innovation (i.e., producing novel combinations of behaviors existing in their repertoires, or applying old behaviors to novel situations). In order to enable a "socioecology of innovation", field and laboratory researchers need to document systematically the range of variation in a variety of behavioral domains.
3. *Role of demography in establishing and maintaining traditions.* The wide range of demographic parameters possible in captive colonies affords,

on the one hand, an opportunity to examine the role of demography in social transmission that is lacking in the field. On the other hand, captive settings force close proximity on animals that might decline to interact closely with one another in the wild, thus permitting more social learning opportunities than are typical in nature. By studying the natural population dynamics and the social dynamics as they change over the lifespan, we can gain a better understanding of the pathways by which social transmission is likely to occur.

4. *Fitness consequences of traditions.* By observing the direct consequences of adopting socially learned (apparently) variants of behavior versus individually learned variants, field researchers can formulate some hypotheses about fitness consequences of social learning, or, at least, of adopting particular traditional behaviors. Field researchers are best equipped to provide information about the niche construction process, which is an essential part of evolutionary models of social learning.
5. *Opportunities for social learning in nature.* Field data are useful for providing insights into the contexts in which social learning might occur in nature by identifying the range of adaptive challenges faced by the animals, and documenting the social contexts in which they occur. It is important to realize that wild animals typically have a wider range of problems to solve than do captive animals. For example, they interact with a wide range of animal species (as prey, predators, and feeding competitors), and they have to solve far more foraging challenges than captive animals confront. Information generated by field researchers can help laboratory researchers to design ecologically relevant experiments that will motivate the animals and better display their abilities to learn. While we cannot be certain that current environments are identical with the most common environments in the evolutionary history of the organisms (i.e., the environments that shaped their current perceptual and cognitive machinery), we can at least be confident that current field conditions are a better match to their evolutionarily relevant environments than are laboratory conditions.
6. *Determination of social influence on learning.* In general, tighter experimental control of what stimuli the animals are exposed to, and of accessibility of the data to models, makes some methodologies possible in the laboratory that are not possible in the field (see, for example, the experiments performed by Visalberghi and Addessi (Ch. 7) in which novel foods were introduced). Fieldworkers are continually hampered by incomplete information regarding animals' prior experience with tasks, and their prior social influences.

7. *Perceptual biases*. Laboratory research on the perceptual biases of animals could be useful to researchers such as Dewar (Ch. 5), who need to know this information in order to assess the reliability and salience of cues of various sorts.
8. *Mechanisms of social learning*. It is essentially impossible to do controlled experiments designed to distinguish between mechanisms of social learning under wild conditions, but numerous methodologies are available for investigating such issues in the laboratory (e.g., Whiten and Ham, 1992; Zentall, Sutton, and Sherbourne, 1996; Ch. 1).
9. *Socio-ecology of social learning and traditions*. Perhaps the most exciting discoveries about the role of individual variation, social learning, and traditions will come from comparative studies because these will address the core questions of evolutionary contributions of social learning. Presently, comparative studies of this scope are represented by literature-based retrospective comparative analyses, such as those conducted by Reader (Ch. 3) and Lefebvre and colleagues (Lefevbre *et al.*, 1997; Ch. 4). It is only by conducting such comparative work that we can hope to develop a socio-ecological theory of social learning. Of course, the quality of the results obtained from comparative studies is affected by the quality of data used in the analyses, and it is extraordinarily difficult to judge the quality of data derived from the literature, especially when using papers that were written on topics other than social learning. Another way to generate hypotheses is to use survey questionnaires, as Whiten and Byrne (1988) did for their research on tactical deception. Once promising trends are isolated in the literature, perhaps field researchers will be more inclined to coordinate their research efforts to conduct prospective studies with shared orientation and shared variables. By coordinating methods among field researchers, we can begin to tackle these comparative questions with quantitative data, having greater confidence that results are not skewed by methodological biases.

15.2.2 Determining the extent of social influence in the field

Although field researchers cannot hope to assess the role of social learning with the accuracy that laboratory researchers can, they can at least investigate whether patterns of association conform to the distribution of the trait in a manner consistent with the assumption that the behavior is shared between individuals as a result of social learning. This method will work, of course, only in social groups for which there is heterogeneity in the form of a particular trait within the group. Several variants of this general approach are employed in this book. At the most general level, van

Schaik (Ch. 11) conducts intersite comparisons for two species in which he correlates "tolerance" (operationalized in many different ways but considered a proxy for the amount of relaxed time spent with conspecifics, and hence social learning opportunity) with a number of documented traditions for the population. Other papers focus on within-site comparisons. Perry *et al.* (Ch. 14) look at the distribution of two alternative foraging techniques (one typically far less common than the other) and test whether individuals who spend more time together are more likely to exhibit the same (rarer) variant. Russon (Ch. 12) employs detailed case studies of orangutans in which she documents the association between individuals and the timing of their skill acquisition. Laland and Kendal (Ch. 2) propose a test by which kinship (or time spent in association) and time to acquire a particular trait are correlated; the stronger the association between kinship and time of learning, the greater the likelihood that the spread of the trait is a result of social learning. Their method, they propose, can help to distinguish genetic transmission, social transmission, and asocial learning. By examining the level of concordance between parents and offspring, their technique can also be used to examine whether vertical or horizontal learning is more prevalent.

Although the chapters in this book have made some advances towards assessing the probable contribution of social influence to behavioral variation in field situations, it is possible to "fine tune" methodologies more than we have so far. Because association patterns vary according to activity and also according to life-history stage, it will be important for future field researchers to document the activities both of focal animals and of animals near the focal animal, to enable a more accurate assessment of social learning opportunities during particular activities. For example, a monkey mother and her juvenile son may spend all of their rest periods in close association yet tend to forage far from one another, while the son tags along after an adult male and observes him as he forages on foods that are rarely eaten by adult females. In this case, the overall degree of association between the juvenile male and others would not accurately reflect the opportunities for social influence to affect his learning of foraging skills. Researchers who work with large social groups may find it useful to employ group scans (interindividual distances, gaze orientations, and activities) when a large portion of the group is engaged in an activity (e.g., a particularly challenging foraging situation) for which social influence is suspected. This will enable researchers to obtain larger sample sizes of proximity data for any particular task and will also reduce the chance of missing instances of social learning opportunities. Focal sampling can be supplemented by regular sampling of the proximities and activities of animals that are close enough to plausibly provide social influence to the focal animal. It may, in some circumstances, be productive to note the gaze and postural orientations of the focal animal relative to nearby animals as well.

Methodological and analytical challenges will vary according to the social structure of the species under study. Those researchers working with species that live in fairly stable, cohesive social groups (e.g., capuchins, macaques), rather than fission–fusion groups (e.g., dolphins, chimpanzees, orangutans), have the advantage of being able to track more reliably what all individuals in the group are doing and whether group members have been exposed to basically the same challenges (e.g., whether they have encountered a particular conspicuous predator or been exposed to a large, novel fruit tree). However, researchers of fission–fusion species who choose to focus on case studies (as Russon has done in Ch. 12) have the advantage of being able to narrow down potential social influences to a very small number of group members.

15.2.3 Ideas for investigation of social learning processes in particular behavioral domains

15.2.3.1 Predator–prey interactions

Placing real or model predators such as snakes in the path of wild social groups might enable investigators to assess the role of social learning in the acquisition and expression of antipredator behavior. Such experiments could use both harmful and harmless snakes and have treatments in which recordings of alarm calls are played in order to draw attention to the snake. Laboratory studies such as those performed by Mineka and Cook (Cook and Mineka, 1989; Mineka and Cook, 1986) have demonstrated a role of social learning in the acquisition of fear of snakes (even harmless toy snakes) in rhesus monkeys. But there are several issues regarding social learning about predators that have not yet been explored in depth for many species, for example (a) the characteristics of models that make them more likely to be attended to as reliable informants about predators, and (b) whether the discoverers of predators alter their antipredator response as a function of which group members are nearby and thereby capable of benefiting from their actions (e.g., prairie dogs: Hoogland, 1996).

Dewar's model could be used to determine whether social cues are likely to be more reliable than asocial cues in any given situation for deciding which animals are dangerous (see Chs. 5 and 14 for details).

15.2.3.2 Foraging techniques

There are various lines of evidence that can be employed in the field for examining social contribution to food-processing techniques. One is simply to note the correlation between association patterns and distribution of techniques, as discussed in Section 15. 2.2. Assume that it takes multiple exposures to learn a complex technique by a combination of observation and practice, and the following conditions are met: (a) there is a dense sampling schedule, especially during the early phases of life; and (b) the researcher makes an effort to record virtually every observation of foraging on particular resources that have short periods of availability, such that little information is missed. Under such conditions, it can be assumed that the proximity data collected when the animal is doing this particular task fairly accurately represent opportunities for social learning, particularly if the proximity dataset is fairly homogeneous over time.

Another (far more time-consuming) type of analysis that could be done is to test whether social learning opportunity affects the rate at which animals narrow their range of foraging techniques so as to specialize on one or two effective strategies. This can be accomplished by measuring social learning opportunity (by proximity data during this particular type of task, as described above) over the juvenile period until the number of techniques used plateaus for any given task (as Russon illustrates in Ch. 12). If social influence is important in shaping the foraging techniques used, then those animals that have frequent associations with others should more rapidly converge on the techniques used by their associates, thus arriving at a preferred technique(s) earlier than animals having less opportunity for observation. It is hypothesized that the process of social convergence will eliminate some of the variants in the individual's repertoire, thus causing the repertoire size to plateau earlier than it would otherwise.

15.2.3.3 Food choice

Dewar's cost–benefit model (Ch. 5) was explicitly designed for answering questions about the likely role of social influence in food choice. We will not discuss this method in depth here, as it is detailed in her chapter. Her methods would be relevant not only to food choice but also to medicinal plant use. Instead, we will focus on suggestions for exporting experimental methods from the laboratory to the field. Specifically, we wish to draw attention to the potential of experimental designs in which food-stressed wild animals are provided with large quantities of nutritious foods in circumscribed areas such as platforms (as Janson and DiBitetti (1997) have done in their investigation of brown capuchin spatial cognition). Using this technique, researchers could (a) try to induce behavioral traditions involving food choice, documenting the transmission chain; and (b) perform experiments analogous to those performed by Visalberghi, Addessi, and Fragaszy (Fragaszy and Visalberghi, 1996; Ch. 7). Naturally occurring variation in the number of other animals feeding at the platform at any one time could be used to investigate the influence of bystanders on food intake.

15.2.3.4 Vocal communication

Much excellent work has been done on vocal learning and population differences in birds and in cetaceans (see Ch. 8 for a review). The literature is too vast and the methodologies too numerous to describe in detail here. There is remarkably little research on the role of social learning in vocal development in other taxa so far, and we would like to encourage researchers of other taxa (e.g., bats, social carnivores, elephants, primates) to pay more attention to social influences in this domain. For example, despite the widespread interest among primatologists in communication (and in vocal learning in particular: see Seyfarth and Cheney, 1999), little attempt has been made to see whether nonhuman primates exhibit intragroup or intersite differences in communication in species other than Japanese macaques and chimpanzees (Arcadi, 1996; Arcadi, Robert, and Boesch, 1998; Green, 1975; Mitani, Hunley, and Murdoch, 1999). Although primates tend to be rather inflexible in the domain of vocal production, they are more flexible regarding vocal usage (Seyfarth and Cheney, 1999), and it seems likely that intersite examinations would yield interesting differences in the usage of different calls.

15.2.3.5 Social conventions

Social conventions have been neglected in the laboratory, but this need not be the case. Often, captive animals spend more time socializing than do wild animals (Kummer, 1995). They might arguably have greater need to negotiate about social relationships, since they cannot escape from their group mates so easily (either by migrating to a new group or by moving away from unpleasant social situations) and, therefore, more frequently need mechanisms of forestalling disputes or reconciling them (Judge, 2000). Coalitionary aid might be more important in such living situations as well. Given these circumstances, it seems that captive animals might have opportunities to develop novel ways to communicate about social relationships. Perhaps the lack of published reports of group-idiosyncratic social conventions in captivity is because of a reticence on the part of laboratory researchers to report quirky new communicative behaviors because they fear that these behaviors will be viewed as aberrations produced by captive conditions. If so, this is unfortunate. Although the social situations of captive animals are not, most likely, typical of the conditions that the ancestors of these animals experienced in their evolutionary history, they may represent one subset of a distribution of conditions to which the organism was exposed over the course of evolution.

Tests of hypotheses about the functions of social conventions (e.g., bond testing: Ch. 14) require data about the initiation, termination, duration, recipient's response, initiator's and recipient's baseline social relationship, and initiator's and recipient's interactions during the hours preceding and following the conventional interactions of interest. These variables can be measured in the wild (S. Perry and J. Manson, unpublished data), but they are easier to measure in captivity because of the possibility of uninterrupted observation over periods of several hours.

15.3 Conclusions

The biology of social learning and traditions is a field still in its infancy. Now that modelers have succeeded in persuading zoologists, psychologists, and biological anthropologists that social learning is an exciting *evolutionary* phenomenon, they need to follow up with models that can be more readily tested with empirical data. Solid empirical evidence for traditions in nature is still fairly scant, and heavily biased towards particular taxa and behavioral domains. While psychologists have made much progress toward understanding the cognitive processes involved in social learning, their studies are also heavily concentrated on a few taxa (e.g., rats, humans, chimpanzees, and capuchins). In order to employ the comparative method successfully to answer the questions posed in this chapter, many more researchers looking at diverse taxa will have to be recruited in a cooperative effort to collect the necessary data. It is the hope of the editors that this book will encourage other researchers to investigate the biology of traditions.

15.4 Acknowledgements

I thank all of the participants of the *Traditions in Nonhuman Primates* conference and the contributors to this volume for stimulating discussion and for their willingness to provide useful commentary on one another's chapters. The Max Planck Institute for Evolutionary Anthropology and the University of California at Los Angeles Academic Senate funded SP during the writing of this chapter. Dorothy Fragaszy, Joe Manson, and Mike Huffman provided many helpful comments.

References

Arcadi, A. C. 1996. Phrase structure of wild chimpanzee pant hoots: patterns of production and interpopulation variability. *American Journal of Primatology*, **39**, 159–178.

Arcadi, A. C., Robert, D., and Boesch, C. 1998. Buttress drumming by wild chimpanzees: temporal patterning, phrase integration into loud calls, and preliminary evidence for individual distinctiveness. *Primates*, **39**, 505–518.

Avital, E. and Jablonka, E. 2000. *Animal Traditions: Behavioural Inheritance in Evolution*. Cambridge: Cambridge University Press.

Baldwin, J. M. 1896. A new factor in evolution. *American Naturalist*, **30**, 441–451, 536–553.

Boughman, J. W. 1998. Vocal learning by greater spear-nosed bats. *Proceedings of the Royal Society of London, Series B*, **265**, 227–233.

Box, H. and Gibson, K. R. 1999. *Mammalian Social Learning: Comparative and Ecological Perspectives*. Cambridge: Cambridge University Press.

Boyd, R. and Richerson, P. J. 1985. *Culture and the Evolutionary Process*. Chicago, IL: University of Chicago Press.

Boyd, R. and Richerson, P. J. 1996. Why culture is common, but cultural evolution is rare. In *Proceedings of the British Academy*, Vol. 88, *Evolution of Social Behaviour Patterns in Primates and Man*, ed. W. G. Runciman, J. M. Smith, and R. I. M. Dunbar. pp. 77–93. Oxford: Oxford University Press.

Bradbury, J. W. and Vehrencamp, S. L. 1998. *Principles of Animal Communication*. Sunderland, MA: Sinauer.

Cook, M. and Mineka, S. 1989. Observational conditioning of fear to fear-relevant versus fear-irrelevant stimuli in rhesus monkeys. *Journal of Abnormal Psychology*, **98**, 448–459.

Coussi-Korbel, S. and Fragaszy, D. M. 1995. On the relation between social dynamics and social learning. *Animal Behaviour*, **50**, 1441–1453.

Fisher, J. and Hinde, R. A. 1949. The opening of milk bottles by birds. *British Birds*, **42**, 347–357.

Fragaszy, D. and Visalberghi, E. 1996. Social learning in monkeys: primate "primacy" reconsidered. In *Social Learning in Animals: The Roots of Culture*, ed. C. M. Heyes and B. G. Galef Jr., pp. 65–84. San Diego, CA: Academic Press.

Giraldeau, L.-A. and Caraco, T. 2000. *Social Foraging Theory*. Princeton, NJ: Princeton University Press.

Green, S. 1975. Dialects in japanese monkeys: vocal learning and cultural transmission of locale-specific vocal behavior? *Zeitschrift für Tierpsychologie*, **38**, 304–314.

Greenfield, M. D. 1994. Cooperation and conflict in the evolution of signal interactions. *Annual Review of Evolution and Systematics*, **25**, 97–126.

Helfman, G. S. and Schultz, E. T. 1984. Social transmission of behavioural traditions in a coral reef fish. *Animal Behaviour*, **32**, 379–384.

Hoogland, J. L. 1996. Why do Gunnison's prairie dogs give anti-predator calls? *Animal Behaviour*, **51**, 871–880.

Janson, C. H. and DiBitetti, M. S. 1997. Experimental analysis of food detection in capuchin monkeys: effects of distance, travel speed, and resource size. *Behavioral Ecology and Sociobiology*, **41**, 17–24.

Judge, P. 2000. Coping with crowded conditions. In *Natural Conflict Resolution*, ed. F. Aureli and F. B. M. de Waal, pp. 129–154, Berkeley, CA: University of California Press.

Kummer, H. 1995. *In Quest of the Sacred Baboon: A Scientist's Journey*. Princeton, NJ: Princeton University Press.

Laland, K. N. and Williams, K. 1997. Shoaling generates social learning of foraging information in guppies. *Animal Behaviour*, **53**, 1161–1169.

Laland, K. N, Odling-Smee, J., and Feldman, M. W. 2000. Niche construction, biological evolution, and cultural change. *Behavioral and Brain Sciences*, **23**, 131–175.

Lefebvre, L., Whittle, P., Lascaris, E., and Finkelstein, A. 1997. Feeding innovations and forebrain size in birds. *Animal Behaviour*, **53**, 549–560.

Mathis, A., Chivers, D. P, and Smith, R. J. F. 1996. Cultural transmission of predator recognition in fishes: intraspecific and interspecific learning. *Animal Behaviour*, **51**, 185–201.

Mineka, S. and Cook, M. 1986. Immunization against the observational conditioning of snake fear in rhesus monkeys. *Journal of Abnormal Psychology*, **95**, 307–318.

Mitani, J. C., Hunley, K. L., and Murdoch, M. E. 1999. Geographic variation in the calls of wild chimpanzees: a reassessment. *American Journal of Primatology*, **47**, 133–151.

Pulliam, H. R. 2000. On the relationship between niche and distribution. *Ecology Letters*, **3**, 349–361.

Seyfarth, R. M. and Cheney, D. L. 1999. Production, usage, and response in nonhuman primate vocal development. In *The Design of Animal Communication*, ed. M. D. Hauser and M. Konishi, pp. 392–417, Cambridge, MA: MIT Press.

van Schaik, C. P., Deaner, R. O., and Merrill, M. Y. 1999. The conditions for tool use in primates: implications for the evolution of material culture. *Journal of Human Evolution*, **36**, 719–741.

Vehrencamp, S. L. 2001. Is song-type matching a conventional signal of aggressive intentions? *Proceedings of the Royal Society of London, Series B*, **268**, 1637.

Vieth, W., Curio, E. and Ernst, U. 1980. The adaptive significance of avian mobbing: III. Cultural transmission of enemy recognition in blackbirds: cross-species tutoring and properties of learning. *Animal Behaviour*, **28**, 1217–1229.

Whiten, A. and Byrne, R. W. 1988. Tactical deception in primates. *Behavioral and Brain Sciences*, **11**, 233–273.

Whiten, A. and Ham, R. 1992. On the nature and evolution of imitation in the animal kingdom: reappraisal of a century of research. *Advances in the Study of Behavior*, **21**, 239–283.

Wilson, A. C. 1985. The molecular basis of evolution. *Scientific American*, **253**, 148–157.

Zentall, T. R., Sutton, J. E., and Sherburne, L. M. 1996. True imitation in pigeons. *Psychological Science*, **7**, 343–346.

Further reading

The following reading list is arranged according to topic. Following each reading is a list of those chapters whose authors specifically recommend this reading as relevant to their chapter.

GENERAL BACKGROUND ON THE BIOLOGICAL SIGNIFICANCE OF SOCIAL LEARNING

Aunger, R. 2000. *Darwinizing Culture: The Status of Memetics as a Science*. New York: Oxford University Press. (*This volume is not only of interest to meme enthusiasts, but also to those interested in the evolution of culture and social learning: Chs. 1, 2, 3, and 15.*)

Avital, E. and Jablonka, E. 2000. *Animal Traditions: Behavioural Inheritance in Evolution*. New York: Cambridge University Press. (*This volume is written for an undergraduate audience new to animal behavior, with many examples (conceptual and empirical) of social contributions to learning in natural settings in species less represented in the empirical literature. It emphasizes a process conception of social learning and the potential for social learning to contribute to niche construction. Chs. 1, 10, 14, and 15.*)

Box, H. and Gibson, K. 2000. *Mammalian Social Learning*. New York: Cambridge University Press. (*This edited collection of chapters emphasizes social learning as a possible component in the natural history and behavioral ecology of a broad spectrum of mammalian orders: Chs. 1, 4, 5, and 15.*)

Dugatkin, L. A. 2000. *The Imitation Factor: Evolution Beyond the Gene*. New York: Free Press. (*This book is written for the nonscientist interested in the role of imitation and culture in animals and its link to evolution and survival. Some of the research of contributors to the present volume is introduced and reviewed here. Chs. 10 and 14.*)

Giraldeau, L.-A. 1997. The ecology of information use. In *Behavioral Ecology: An Evolutionary Approach*, edn, ed. J. R. Krebs and N. B. Davies, pp.42–68. Oxford: Blackwell Scientific. (*This is a comprehensive review of learning and information transfer from a behavioral ecology perspective: Ch. 7.*)

Heyes, C. and Galef, B. G., Jr. 1996. *Social Learning in Animals: The Roots of Culture*. San Diego, CA: Academic Press. (*This is an updated collection of empirical chapters by both *psychologists and biologists spanning the issues in animal social learning in the 1990s: Chs. 1, 2, 5, and 7.*)

Heyes, C., and Huber, L. (ed.) 2000. *The Evolution of Cognition*. Cambridge, MA: MIT Press. (*This collection of essays deals with the evolutionary processes that have shaped cognition, as well as various aspects of cognition such as causal reasoning, culture, consciousness, and categorization: Chs. 3 and 7.*)

Shettleworth, S. J. 1998. *Cognition, Evolution, and Behavior*. New York: Oxford University Press. (*This is the best comprehensive overview of learning and cognition from both psychological and biological perspectives: Chs. 7 and 9.*)

Zentall, T. and Galef, B. G., Jr. 1988. *Social Learning: Psychological and Biological Perspectives*. Hillsdale, NJ: Erlbaum. (*This seminal collection of conceptual and empirical chapters on social learning in nonhuman species influenced the research agenda in this field for several years to follow. This volume is a good starting point for the person new to this area: Chs. 1, 5, and 15.*)

SOCIAL LEARNING IN THE DOMAIN OF COMMUNICATION

Catchpole, C. K. and Slater, P. J. B. 1995. *Bird Song: Biological Themes and Variations*. Cambridge: Cambridge University Press. (*This book gives an overview of what is known about bird song. Several chapters refer to traditions in bird song and their possible causes and consequences: Ch. 8.*)

Janik, V. M. and Slater, P. J. B. 1997. Vocal learning in mammals. *Advances in the Study of Behavior*, **26**, 59–99. (*This is the most recent review of vocal learning in mammals. It also reports all known cases of geographic variation in mammalian calls and explores their possible causes: Ch. 8.*)

Mann, J., Connor, R., Tyack, P., and Whitehead, H. 2000. *Cetacean Societies: Field Studies of Dolphins and Whales*. Chicago, IL: University of Chicago Press. (*This book provides an overview of behavioral diversity in cetaceans and includes a chapter on vocal communication: Ch. 9.*)

Payne, R. B. 1996. Song traditions in indigo buntings: origin, improvisation, dispersal, and extinction in cultural evolution. In *Ecology and Evolution of Acoustic Communication in Birds*, ed. D. E. Kroodsma and E. H. Miller, pp. 198–220. Ithaca: Comstock Publishing. (*This review of Payne's work summarizes the results of the most exhaustive study on vocal traditions in animals. It describes the stability and transmission processes of traditions and discusses cultural mutation rates and cultural evolution in bird song: Ch. 8.*)

Rendell, L. and Whitehead, H. 2001. Culture in whales and dolphins. *Behavior and Brain Sciences*, **24**, 309–382. (*This paper argues that there is good evidence for traditions in cetaceans in a variety of behavioral domains, particularly in the domain of vocal communication: Ch. 9.*)

Zahavi, A. and Zahavi, A. 1999. *The Handicap Principle: A Missing Piece of Darwin's Puzzle*. Oxford: Oxford University Press. (*This book, written for the layperson, details the applications of Zahavi's main theoretical contributions to biology, including his "testing of a bond" idea: Ch. 14.*)

SOCIAL LEARNING IN THE DOMAIN OF PREDATOR RECOGNITION

Mineka, S. and Cook, M. 1988. Social learning and the acquisition of snake fear in monkeys. In *Social Learning: Psychological and Biological Perspectives,* ed. T. R. Zentall and B. G. Galef Jr., pp. 51–73.Hillsdale, NJ: Erlbaum. (*This chapter summarizes a research program designed to detect social influences on the development of antipredator behavior in monkeys: Chs. 4 and 14.*)

SOCIAL LEARNING IN THE DOMAIN OF FORAGING

Galef, B. G., Jr. and Giraldeau, L.-A. 2001. Social influences on foraging in vertebrates. causal mechanisms and adaptive functions. *Animal Behaviour,* **61**, 3–15. (*This is a very recent brief review of social influences on foraging behavior: Chs. 4 and 5.*)

Giraldeau, L.-A. and Caraco, T. 2000. *Social Foraging Theory.* Princeton, NJ: Princeton University Press. (*This volume concerns how animals search and compete for food in groups. The book reviews the large literature in this area and presents a new theory of social foraging as an economic interaction among individuals: Ch. 1.*)

EVOLUTION OF THE BRAIN AND OF INTELLIGENCE

Byrne, R. W. and Whiten, A. (ed.) 1988. *Machiavellian Intelligence: Social Expertise and the Evolution of Intellect in Monkeys, Apes and Humans.* Oxford: Oxford University Press. (*This is an excellent overview of the various hypotheses put forward to explain the evolution of intelligence in primates: Ch. 3.*)

Lee, P. C. (ed.) 1999. *Comparative Primate Socioecology.* Cambridge: Cambridge University Press. (*Particularly relevant contributions include those from Barton on primate brain evolution and Purvis and Webster on comparative methods: Ch. 3.*)

Whiten, A. and Byrne, R. W. (ed.) 1997. *Machiavellian Intelligence II: Extensions and Evaluations.* Cambridge: Cambridge University Press. (*This book details the progress made on the social intelligence hypothesis since the publication of the first edition: Ch. 3.*)

TRADITIONS AND BEHAVIORAL VARATION IN PRIMATES

The following publications concern the possibility of traditions in nonhuman primates and their potential significance for the evolution of culture in humans.

de Waal, F. 2001. *The Ape and the Sushi Master: Cultural Reflections of a Primatologist.* New York: Basic Books. (*A very accessible book for the interested reader new to the fields of animal behavior and primatology. The second section of this book includes a popular account of the development of the field of "cultural primatology", including descriptions of the pioneering research on traditions in Japanese macaques conducted in the 1950s and 1960s: Ch. 14.*)

Matsuzawa, T. 2001. *Primate Origins of Human Cognition and Behavior.* New York: Springer. (*This book has several contributions pertinent to social learning and traditions in nonhuman primates: Chs. 1 and 14.*)

McGrew, W. C. 1992. *Chimpanzee Material Culture: Implications for Human Evolution*. Cambridge: Cambridge University Press. (*This book presents a detailed analysis of between-site population variation in material culture in chimpanzees: Chs. 11–14.*)

Quiatt, D. and Itani, J. 1994. *Hominid Culture in Primate Perspective*. Denver, CO: University of Colorado Press (*This edited volume contains papers on various aspects of the evolution of culture in humans and non-human primates: Ch. 10.*)

Whiten, A. and C. Boesch. 2001. The cultures of chimpanzees. *Scientific American*, **284**, 61–67. (*A good introduction to the group contrast approach to the documentation of traditions, and some of the best-known examples used to support the notion of traditions from this perspective: Chs. 1 and 14.*)

Whiten, A., Goodall, J., McGrew, W. C., Nishida, T., Reynolds, V., Sugiyama, T., Tutin, C. E. G., Wrangham, R. W., and Boesch, C. 1999. Cultures in chimpanzees. *Nature*, **399**, 682–685. (*This article summarizes variations in behavior observed across sites in Africa where chimpanzees have been studied. See also Whiten* et al. *(2001): Chs. 11 and 14.*)

Whiten A., Goodall J., McGrew, W. C., Nishida, T., Reynolds, V., Sugiyama, Y., Tutin, C. E. G., Wrangham, R. W., and Boesch, C. 2001. Charting cultural variations in chimpanzees. *Behaviour*, **138**, 1481–1516. (*A more detailed version of the data in Whiten* et al. *(1999): Chs. 11 and 14.*)

The following set of readings concerns behavioral diversity in two primate taxa that are the focus of several chapters in this volume: chimpanzees (*Pan troglodytes*) and capuchin monkeys (genus *Cebus*).

Boesch, C. and Boesch-Achermann, H. 2000. *The Chimpanzees of the Tai Forest*. Oxford: Oxford University Press. (*This book describes the behavioural ecology of chimpanzees at one long-term study site and provides detailed comparisons with findings from studies of chimpanzees at other long-term study sites: Chs. 10 and 14.*)

Fragaszy, D. M., Visalberghi, E., and Fedigan, L. 2003. *The Complete Capuchin. The Biology of a Genus*. Cambridge: Cambridge University Press. (*This book summarizes the behavioral biology of capuchin monkeys* (Cebus) *in the laboratory and in the field, including a discussion of social learning: Chs. 6, 13, and 14.*)

McGrew, W. C., Marchant, L. F., and Nishida, T. 1996. *Great Ape Societies*. Cambridge: Cambridge University Press. (*This edited volume is a valuable compendium of research about the ecology, behavior, and psychology of the great apes: Chs. 10 and 12.*)

Wrangham, R. W. W., McGrew, W. C., de Waal, F. B. M., and Heltne, P. G. 1994. *Chimpanzee Cultures*. Cambridge, MA: Harvard University Press (*This edited volume presents the behavior and ecology of chimpanzees from a comparative perspective: Ch. 10.*)

Index